T0185931

Nanocosmetics

Jean Cornier · Cornelia M. Keck ·
Marcel Van de Voorde

Editors

Nanocosmetics

From Ideas to Products

 Springer

Cornier
onsult
desheim, Germany

Marcel Van de Voorde
Faculty of Natural Sciences
University of Technology DELFT
Crans-Montana, Switzerland

Cornelia M. Keck
Institut für Pharmazeutische
Technologie und Biopharmazie
Philipps-Universität Marburg
Marburg, Germany

ISBN 978-3-030-16575-8   ISBN 978-3-030-16573-4   (eBook)
https://doi.org/10.1007/978-3-030-16573-4

© Springer Nature Switzerland AG 2019
Chapter 6 is licensed under the terms of the Creative Commons Attribution 4.0 International License
(http://creativecommons.org/licenses/by/4.0/). For further details see license information in the chapter.
This work is subject to copyright. All rights are reserved by the Publisher, whether the whole or part
of the material is concerned, specifically the rights of translation, reprinting, reuse of illustrations,
recitation, broadcasting, reproduction on microfilms or in any other physical way, and transmission
or information storage and retrieval, electronic adaptation, computer software, or by similar or dissimilar
methodology now known or hereafter developed.
The use of general descriptive names, registered names, trademarks, service marks, etc. in this publication does not imply, even in the absence of a specific statement, that such names are exempt from the
relevant protective laws and regulations and therefore free for general use.
The publisher, the authors and the editors are safe to assume that the advice and information in this
book are believed to be true and accurate at the date of publication. Neither the publisher nor the
authors or the editors give a warranty, expressed or implied, with respect to the material contained
herein or for any errors or omissions that may have been made. The publisher remains neutral with regard
to jurisdictional claims in published maps and institutional affiliations.

This Springer imprint is published by the registered company Springer Nature Switzerland AG
The registered company address is: Gewerbestrasse 11, 6330 Cham, Switzerland

Preface

The cosmetics industry has long been at the forefront of applying nanotechnology into its products. As the nanotechnology revolution progresses, the possibilities offered by new technologies for modification of particles at the nanoscale to provide new properties and features are growing dramatically. This gives great opportunities to the cosmetics industry, but also challenges in assuring consumers that products are safe and in being able to demonstrate safety to regulatory bodies.

Nanoscience in Cosmetics aims at applying nano-enabled technologies to the development and production of innovative cosmetic products. More precisely, it involves the preparation and delivery of substances in the molecular and nanometer size range to the skin tissues, providing maximum efficacy and benefit while minimizing potential negative side effects. The importance of this emerging field of research and development relies on the fact that nanoparticles exhibit unique characteristics for protecting and repairing the skin.

This book deals with a broad range of topics, from the description of nano-sized materials, their formulation, production, characterization, delivery to the skin, to toxicity and safety. Research in this field requires a multidisciplinary approach, involving materials and chemical engineers, cellular biologists, pharmacists, and dermatologists.

The goal of this unique book is to present an overall picture of the use of nanotechnology in cosmetics. It is designed to be a reference textbook on the application of nanotechnology in the development of nanostructures for use in the development of innovative cosmetics products. Focus is placed on the description, research and manufacturing of candidate nanostructures, as well as their translation and use into marketable products by industry. We also review the most interesting and promising developments in this emerging but fast-developing field.

Part I gives an *introduction into cosmetics and into the nanocosmetics revolution*. They address the science behind cosmetics, provide an overview about the history, potential, challenges, and most recent developments in cosmetics and address the potential of the emerging nanotechnological applications in cosmetics science.

In Part II, *a systematic review of the various nanoparticles used in cosmetics* is provided with a description of currently used nanostructures (like inorganic nanoparticles, micelles and nanoemulsions, polymeric nanostructures, liposomes, lipid nanoparticles and nanocrystals), and their applications.

In Part III, *the characterization methods* of isolated particles as well as of particles in dermal formulations and of nanocosmetics on the skin are presented.

It also gives *a detailed overview of the preparation and manufacturing methods and associated issues.* Established processes as well as new exploratory methods are reviewed and concrete actual examples of utilization given. All relevant aspects are addressed including scale-up from laboratory to factory, and the requirements for occupational health in cosmetics production.

Part IV addresses *the governance of nanocosmetics.* Emphasis is laid on nanotoxicology and nanosafety aspects as well as regulatory issues for translation to the market of the most promising nanostructures.

Finally, the last chapter lays emphasis on *advances and potentials for use of nanomaterials in cosmetics.* Market prospects and commercialization aspects are also addressed, with special focus on the commercial translation and its bottlenecks, including the protection of intellectual property. Actual information about current commercialized products and market figures is also provided.

The chapters are written by leading researchers in cosmetics, chemistry, pharmacy, biology, chemistry, physics, engineering, and medicine, as well as law and social science. The authors come from a range of backgrounds including academia, industry, and national and international laboratories, from Europe, Israel, the USA, Brazil and India.

It is expected that this book will become a standard work for cosmetics scientists, pharmacists, dermatologists, and the cosmetics industry, but also a reference work for scientists, researchers, and students, as well as for agencies, government, and regulatory authorities.

This book aims to bring inspiration for scientists, new ideas for cosmetics developers, innovation in industry, and guidelines for toxicologists and finally will result in the development of guidelines for agencies and government authorities to establish safety rules in using this new promising technology. The book will also stimulate breakthroughs in the cosmetic sciences, leading to improved skin products for healthcare, skin protection, and beauty.

Hildesheim, Germany   Jean Cornier
Marburg, Germany   Cornelia M. Keck
Crans-Montana, Switzerland   Marcel Van de Voorde

Contents

About the Editors

Jean Cornier is presently Consultant to several companies and R&D organizations in the areas of life science, new technologies and business development. Based in Munich, Germany, he obtained his state diploma "Doctor of Pharmacy"—Pharm. D.—from the University of Caen, France, and an MSc degree in Pharmaceutical Medicine from the University of Duisburg-Essen, Germany. Since 1986, he has worked in the space industry as Expert on Materials and Life Science research and projects and participated in commercialization initiatives supported by the European and German space agencies, as well as several EU-funded projects on biotechnology and civil security research. He was involved in the first skin research project in space and is Co-editor of a book on nanopharmacy (Pharmaceutical Nanotechnology: Innovation and Production (2017)).

Cornelia M. Keck is Pharmacist and obtained her PhD from the Freie Universität (FU) in Berlin in 2006. In 2009, she was appointed as Adjunct Professor for Pharmaceutical and Nutritional Nanotechnology at the University Putra Malaysia (UPM), and in 2011, she completed her postdoctoral degree (Habilitation) at the Freie Universität Berlin and was appointed as Professor for Pharmacology and Pharmaceutics at the University of Applied Sciences Kaiserslautern. Since 2016, she has been Professor of Pharmaceutics and Biopharmaceutics at the Philipps-University Marburg. Her chief field of research is the development and characterization of innovative nanocarriers for improved delivery of poorly soluble actives for healthcare and cosmetics. She is Vice-Chair of the "Dermocosmetics" unit at the German Society of Dermopharmacy, Active Member of many pharmaceutical societies, and Member of the Committee for Cosmetics at the Federal Institute for Risk Assessment (BfR).

Marcel Van de Voorde has 40 years experience at European research organizations, including CERN (Geneva), European Commission Research (Brussels), and 10 years working at the Max Planck Institute in Stuttgart, Germany. For many years, he was involved in research and research strategies, policy and management, especially in European research institutions. He is currently Professor at the University of Technology in Delft, the Netherlands, and holds a doctor *honoris causa* and various honorary professorships. He is Senator of the European

Academy for Sciences and Arts in Salzburg, Fellow of the World Academy for Sciences, and Member of the Science Council of the French Senate and National Assembly. He is Fellow of various scientific societies and has been decorated for European merits by the King of Belgium. He has authored numerous scientific and technical publications and co-edited several books in the fields of nanoscience and nanotechnology.

Part I
Introduction to Nanocosmetics

Science Behind Cosmetics and Skin Care

1

Becky S. Li, John H. Cary and Howard I. Maibach

Abstract

Women and men worldwide use an abundance of skin care and cosmetic products, in pursuit of cleanliness with soaps and shampoos, or everlasting youth with creams and serums. No matter what age, location or socioeconomic background, we are all exposed to the widespread cosmetics industry. The global cosmetic market was worth $460 billion in 2014 and is estimated to reach $675 billion by 2020. While marketing and advertising efforts have forged a path for the industry's continued growth, this momentum has not translated to equally refreshed regulatory practices in some parts of the world. This chapter reviews the regulatory status of cosmetics, adverse event reporting and studies attempting to link cosmetics with contact dermatitis. It aims to set the foundation for the emergence of nanocosmetics.

Keywords

Cosmetics · Cosmetic regulation · Cosmetic research · FDCA · VCRP · CIR · RIFM · PCPSA · Contact dermatitis · Adverse effects · Nanocosmetics

B. S. Li
Howard University College of Medicine, 520 W Street NW, Washington, DC 20059, USA
e-mail: Becky.Li@ucsf.edu; beckysiyunli@gmail.com

J. H. Cary
Louisiana State University School of Medicine, 433 Bolivar Street, New Orleans, LA 70112, USA
e-mail: John.Cary@ucsf.edu; havenscary@gmail.com

H. I. Maibach (✉)
School of Medicine, Department of Dermatology, University of California, San Francisco, 90 Medical Center Way, Surge Building, Room 110, Box 0989, San Francisco, CA 94143-0989, USA
e-mail: Howard.Maibach@ucsf.edu

© Springer Nature Switzerland AG 2019
J. Cornier et al. (eds.), *Nanocosmetics*,
https://doi.org/10.1007/978-3-030-16573-4_1

1.1   Introduction

Women and men worldwide use an abundance of skin care and cosmetic products
—whether in pursuit of cleanliness with soaps and shampoos or everlasting youth
with creams and serums. No matter age, location, or socioeconomic background,
we are all exposed. President of the Cosmetic, Toiletry, and Fragrance Association
(CTFA) notes the wide-encompassing nature of the industry as well as its potential
for seemingly limitless expansion: "In an era of globalization, we are truly one of
the world's most global industries. Our products and our innovation know no
boundaries. Whether it's Bangkok or Beijing, Baton Rouge or Bagdad, the products
that we make are the products that women and families use every single day." The
global cosmetic market was $460 billion in 2014 and is estimated to reach $675
billion by 2020 [1]. Marketing and advertising efforts have forged a path of continued growth for the industry. However, industrial growth failed to bring about
new cosmetic research and regulations.

1.2   Science of Safety of Cosmetics: Laws Enacted and Review Boards Established

Due to growing public concern regarding safety of cosmetics in the twentieth
century, congress enacted the Food, Drug, and Cosmetic Act of 1938 (FDCA).
The FDCA brought the cosmetic industry under the regulatory jurisdiction of the
FDA; however, with limited budget and other constraints, the FDA has relied on
the cosmetic industry to regulate itself in order to ensure consumer safety. In fact,
the FDA does not review or approve ingredients or products before they are sold
to the public, instead it functions like a highway patrol: "its inspectors look out for
products that are dangerous to health, about which it can, like a highway patrolman,
do something..." [2]. It relies on the public and physicians to alert the agency about
problem products.

While select state legislatures have opted for tighter control over cosmetics, the
federal government has not enacted significant new regulatory cosmetic legislation
or substantially amended the FDCA [3]. The FDCA defined a cosmetic and outlined
procedural aspects of cosmetic regulation. Under the FDCA, cosmetics are defined
as, "articles intended to be rubbed, poured, sprinkled or sprayed on, introduced into,
or otherwise applied to the human body or any part thereof for cleansing beauti-
fying, promoting attractiveness or altering the appearance and articles intended for
use as a component of any such articles; except that such term shall not include
soap." In addition, the FDCA states that a cosmetic should not be adulterated or
misbranded. The Fair Packaging and Labeling Act of 1973 established further
labeling provisions, requiring listing of the cosmetic name and quantity, the manufacturer name and place of business, the ingredients listed in order of predomi-
nance, and a warning label for untested ingredients. However, no established
pre-approval process for cosmetics currently exists. The FDA has established the
Voluntary Cosmetic Regulation Program (VCRP), which is available to

manufacturers, packagers, and distributors of cosmetics wishing to participate. The FDA may use the VCRP to direct the Cosmetic Ingredient Review (CIR) program in prioritizing specific ingredient testing [3]. Essentially, current legislation leaves the responsibility of determining cosmetic ingredient safety to the manufacturer of the product.

In 1976, the cosmetics industry established a self-regulatory panel, the Cosmetic Ingredient Review (CIR) . While the CIR directs independent review of current literature and research available on cosmetic ingredients, they neither conduct nor fund any of the laboratory research or toxicology tests [4]. Ingredients that receive priority testing include those most frequently used, new ingredients or those with novel use, those with significant new information, and those that have not been reviewed in over 15 years [4]. The CIR selects ingredients from the International Cosmetic Ingredient Dictionary and Handbook, which lists over 21,000 ingredients that were used in the past, are in current use, or are planned for future use [4].

The US FDA and the independent fragrance review board known as Research Institute for Fragrance Materials (RIFM) have reviewed approximately 30% of ingredients listed; however, only 7000 of the 21,000 ingredients are in current use [4]. As of March 2017, the CIR has completed safety assessments of 4740 individual cosmetic ingredients [4]. Of the 4740 reviewed cosmetic ingredients, 4611 were determined safe, 12 determined unsafe, and 117 with insufficient information [4]. While their annual review rate has increased from 100 to 400 safety assessments, thousands of ingredients are in current use without formal evaluation [4]. Should an ingredient be determined dangerous, neither the CIR nor the FDA possesses the power to remove cosmetic products from the market; however, most manufacturers tend to recall products to due to liability concerns [3].

RIFM was founded in 1966 in order to help evaluate the safety of ingredients in fragrances. RIFM begins the process of evaluation with an exhaustive literature search for relevant research on the ingredient and may also perform in silico and in vitro testing for fragrances when necessary [5]). Like the CIR, RIFM publishes its ingredient evaluation, but has little authority in ensuring that dangerous products are removed from the marketplace. In addition, there is no outside review of the primary documents upon which the reports are based.

Lack of scientific evidence and low incidence of successful lawsuits have limited movements for further regulation (Termini 2008). However, there are major shortcomings to the current cosmetic regulation in place. Neither the FDA nor the CIR have the authority to recall products from the marketplace, while cosmetic manufacturers are also not obligated to register their facilities or adhere to Good Manufacturing Practices (GMPs). According to Linda Katz (Director for the Office of Cosmetics and Colors in FDA's Center of Foods and Applied Nutrition), the FDA is "…hampered in tracking down tainted products in situations like these by the lack of facility registration requirement," and while there is voluntary facility registration, "…we estimate that it only covers a fraction of what is marketed."

Recently, California Senator Diane Feinstein introduced the Personal Care Products Safety Act (PCPSA), which seeks to give the FDA authority to recall unsafe cosmetics, impose mandatory adverse event reporting for manufacturers, and propose an annual safety review of five ingredients [6].

1.3 Surveillance Data

In 2003, cosmetics were responsible for 9.3% of calls to poison centers, with 2484 or 1% of those cases having moderate, major, or deadly effects [3]. Aside from an outlier of 9% in 2014, cosmetics are responsible for a decreasing amount of calls to poison control centers with the latest released data estimating 7.4% of poison control center calls due to cosmetics [7]. However, percent calls to poison center are not necessarily reflective of cosmetic adverse events, as the majority of adverse reactions to cosmetics do not result in calls to poison control centers and the long-term effects of cosmetic and skin care ingredients remain unknown [3].

In order to encourage additional adverse event reporting, the FDA opened its reporting system, Center for Food Safety and Applied Nutrition's Adverse Event Reporting System (CFSAN), to the public [6]. There has since been a spike in adverse events filed with CFSAN with 436 events reported in 2014, 706 in 2015, and 1591 in 2016 [6]. The most commonly implicated cosmetic categories in adverse events include hair care followed by skin care and tattoos [6]. Linda Katz attributes the increase in adverse event reports to a few high profile cases as well as the increased FDA outreach that has raised awareness to consumers and healthcare professionals to report such adverse events.

Our personal experience at UCSF is that most documented such adverse reactions—mainly irritant and allergic contact dermatitis—are not reported.

1.4 Cosmetics and Contact Dermatitis

As previously mentioned, cosmetic and skin care products may sometimes trigger an adverse reaction, commonly inflammation of skin. Dermatitis is the general term used for inflammation of skin causing pruritus and erythema. When dermatitis is caused by exogenous material (e.g., cosmetics) coming into direct contact with skin, it is termed contact dermatitis.

Contact dermatitis consists of both irritant contact dermatitis (ICD) and allergic contact dermatitis (ACD), with ICD being more common and accounting for a majority of cases [8–12]. Whereas ACD is a type IV hypersensitivity reaction and requires prior sensitization, ICD predominantly depends upon the [13]). Most acute presentations are characterized by erythema and papules, coalescent vesicles, bullae, and, in severe cases, oozing [13]. ACD more frequently presents with an ill-defined lesion with extension beyond the site of allergen exposure, whereas ICD is often

Table 1.1 Time line of laws enacted and review boards established

Year	Legislation or review board establishment	Description
1938	Food, Drug, and Cosmetic Act (FDCA)	Brought the cosmetic industry under the regulatory jurisdiction of the FDA
1966	Research institute fragrance material	Help evaluate the safety of ingredients in fragrances
1973	Fair Packaging and Labeling Act	Provide labeling provisions, including the cosmetic name, quantity, name and place of business of the manufacturer, ingredients listed in order of predominance, and a warning label for untested ingredients
1976	Cosmetic ingredient review	Self-regulation panel of the cosmetic industry
2017	Personal Care Products Safety Act (PCPSA)	Gives the FDA authority to recall unsafe cosmetics, impose mandatory adverse event reporting for manufacturers, and propose an annual safety review of five ingredients

sharply demarcated and restricted to the site of the irritant [13]. Considerable overlap exists between the two in clinical, histological, and molecular presentation [13]. However, distinction between the two conditions is critical, as allergens should be generally avoided, while irritants can often be tolerated in small amounts.

It is difficult to estimate the prevalence of ACD induced by cosmetics in the general population due to low rates of patients seeking medical advice, as many simply discontinue use of the culprit product. Menkart [14] suggested that an adverse reaction to cosmetics occurs approximately once every 13.3 years per person. Another review found that the pooled prevalence rate of ACD to cosmetics in 7 different studies to be 9.8% [15]. The North American Contact Dermatitis Group (NACDG) patch tested 10,061 patients with suspected ACD over 7 years, 23.8% of female patients and 17.8% of male patients had at least one allergic patch-test reaction associated with a cosmetic source [16] (Table 1.1).

Park and Zippin [17] outline the common sites of cosmetically induced ACD and their most common allergens (Table 1.2). For patients with ACD, the cornerstone of management is avoidance of the triggering allergen. This has been made more feasible with the Fair Packaging and Labeling Act of 1973, requiring cosmetic ingredients to be listed on packaging. However, there remain challenges in fragrances, a leading cause of contact allergy to cosmetics. Ingredients are not individually listed in fragrances as they are considered trade secrets.

Formaldehyde avoidance also appears to be difficult. In 2000, Rastogi found that formaldehyde content was incorrectly labeled in 23–33% of products tested.

An invaluable tool for physicians and patients to manage cosmetically induced ACD is the Contact Allergen Management Program, developed and managed by American Contact Dermatitis Society. It is a computerized database of thousands of cosmetics and personal care products. After a patient has been patch tested and the allergen pinpointed, the offending agent may be entered into the database and a list will be generated of all the safe products the patient can use. This makes avoidance of the triggering agent much more manageable for patients.

Table 1.2 Common sites of ACD due to cosmetics and their most common allergens

Body region	Allergen	References
Face	Tosylamide formaldehyde resin in nail lacquer	Lazzarini et al. [46]
Eyelid	Nail lacquer	Rietschel et al. [47]
	p-Phenylenediamine (PPD), ammonium persulfate	Rietschel and Fowler [48]
	Mascara, eyeliner, eye shadow, false eyelashes, and eyelash curler metal	Brandrup [49], Guin [50]
	Shellac	Le Coz et al. [51], Shaw et al. [52]
	Gold	Nedorost and Wagman [53]
Neck	Tosylamide formaldehyde resin in nail lacquer	Lazzarini et al. [46]
	Fragrance	Jacob and Castanedo-Tardan [54]
Scalp	PPD in hair dyes	Rietschel and Fowler [48]
Anogenital	Fragrances, preservatives, corticosteroids	Warshaw et al. [55]
	Balsam of Peru, nickel	Bhate et al. [56], Warshaw et al. [55], Kugler et al. [57]
	Spices and flavorings (i.e., nutmeg, peppermint oil, coriander, curry mix, peppermint oil, and onion powder)	Vermaat et al. [58], Vermaat et al. [59]

1.5 Hazardous Health Outcomes

Regarding some cosmetic ingredients, studies have shown adverse health outcomes, including malignancy, reproductive and developmental toxicity, and neurotoxicity. Besides consumers, cosmetologists are particularly at risk.

More than one million cosmetologists have licenses in the USA and several million worldwide. Most are female and of reproductive age. Common specialties in cosmetology include hairdressers and nail technicians. Hairdressers use a variety of products, including shampoos, conditioners, hair sprays, straighteners, bleaches, and of particular harm, hair dyes. Nitrosamine is the primary ingredient in hair dye and requires bioactivation, thereby allowing for adverse effects at locations other than the initial exposure site [18]. Nitrosamine has shown mutagenicity in vitro and carcinogenic properties in vivo. In 1975, Ames et al. designed a simple and sensitive bacterial test for detecting chemical mutagen, in which nitrosamine tested positive. 4-Nitroquinoline-N-oxide was plated on petri dishes with specially constructed mutants of *Salmonella typhimurium*, and with histidine, reverted the mutant strains back to wild type. This suggests the mutagenicity of nitrosamine in vitro. Sontag [19] later showed the carcinogenic properties of nitrosamine in vivo. Eleven substituted-benzenediamines were tested in rats and mice, with four inducing significant incidences of tumors in both rats and mice. However, while nitrosamine may have been carcinogenic in animal studies, there remains "inadequate evidence" to determine carcinogenicity in humans [20].

For nail technicians, nail polish is the primary source of chemical exposure. They include toluene, plasticizers (i.e., dibutyl phthalate), and formaldehyde. Hairdressers and nail technicians typically share workspace and sometimes perform some of the same tasks. The space and role overlap increase occupational chemical exposures for both groups.

While some studies have indicated an association between hairdressing and health risks, others have not. Early on, Nohynek et al. [21] conducted a review, which showed no correlation between occupational exposure to hair dyes and carcinogenic or other health outcomes. This review, however, focused on acute toxicity and health effects of hair dyes and not on potential long-term outcomes of exposure. Similarly, Hougaard et al. [22] found that hairdressers and the general public exhibited comparable rates of infertility. On the other hand, Ronda et al. [23] found an increased risk for subfertility and menstrual disorders among female hairdressers in comparison with female office worker and shop assistant controls. Harling et al. [24] performed a meta-analysis of 42 studies, which showed an increased and statistically significant risk for bladder cancer among hairdressers, especially those who held the job for more than 10 years [24].

Formaldehyde, another chemical found in skin care and cosmetics, provides more definitive data. According to the International Agency for Research on Cancer (IARC), the U.S. National Toxicology Program (NTP), the Environmental Protection Agency (EPA), and the Occupational Safety and Health Administration (OSHA), formaldehyde is considered a human carcinogen. Duong et al. [25] conducted a meta-analysis of 18 human studies to understand the relationship between maternal formaldehyde exposure and adverse reproductive and development effects. Results revealed an increased risk of spontaneous abortion in formaldehyde-exposed women. Evaluation of animal studies presented similar findings: a positive association between formaldehyde exposure and reproductive toxicity, particularly in males. Although a positive relationship was found, mechanisms underlying formaldehyde-induced reproductive and development toxicities are unclear.

The effects of toluene, a commonly used organic solvent, have been studied extensively. Toluene inhalation during pregnancy has been shown to lead to growth restriction, congenital malformations, development delay, and premature delivery [26]. In one study, where male rats inhaled 4000–6000 ppm toluene two hours a day for one month, rats with higher toluene concentration exposure showed significantly reduced sperm count, motility, quality, and penetration, as well as increased rates of abnormal sperm morphology [27]. When pregnant rats were exposed to toluene once per day from GD6 through GD19, results revealed a 15% reduction in cell number of the offspring forebrain. There were also reductions in neocortical myelination [28]. These findings are comparable to the developmental delays (mental retardation, language impairment, hyperactivity) found in children with early toluene exposure. Moreover, decreased birth weight and small for gestational age at birth are the most commonly reported outcomes of occupational solvent exposure during pregnancy [29]. Toluene not only appears to affect individuals exposed but also offsprings of women who have inhaled excessive concentrations, and these toluene-exposed women appear to be more likely to have negative pregnancy outcomes, i.e., intrauterine growth restriction, decreased birth weight, developmental delays.

Other ingredients in cosmetic products, mainly those that have stabilizing and preserving properties, also have health impacts. Parabens, for example, have been shown to have endocrine-disrupting activity in vitro and in vivo [30–32]. There also seems to be a connection between parabens and male infertility in animals. Furthermore, parabens appear to have the capability of transforming human breast cancer cells in vitro [33]. Wrobel and Gregoraszczuk [34] furthered the association between parabens and breast carcinogenesis by showing the ability of parabens to stimulate proliferation of human breast cancer cells via increased estradiol secretion and aromatase activity.

Continued research into the effects of cosmetic and skin care ingredients is crucial as the findings above have led to increasing discontinuation of harmful ingredients in cosmetics [35].

1.6   Cosmetic and Skin Care Ingredients for Thought

The current state of the cosmetic industry bears many questions to ponder upon.

Should there be more regulation?

The FDA heavily relies on the cosmetic industry to self regulate in order to assure consumer safety with neither review nor approval necessary before ingredients and products are placed on the market. Although the FDA continues to make strides toward bettering consumer safety (i.e., establishing labeling provisions, reviewing cosmetic ingredients), the FDA, CIR, and RIFM do not have the jurisdiction to remove dangerous products from the market. However, Senator Diane Feinstein appears to be working to resolve this problem with the establishment of PCPSA, which seeks to give the FDA authority to recall unsafe cosmetics, impose mandatory adverse event reporting for manufacturers, and propose an annual safety review of five ingredients [6]. Regardless, current regulation and legislation in place are guidelines at best.

Should there be systems in place to report problematic products and adverse effects?

Currently, the cosmetic industry relies on the public and physicians to alert the agency about problematic products through its reporting system, CFSAN Adverse Event Reporting System (CAERS). There has been a gradual increase in adverse events filed in the recent years, to which Linda Katz, director of the FDA's Office of Cosmetics and Colors, attributes to increased FDA outreach and raised awareness to consumers and healthcare professionals to report such adverse events.

The American Contact Dermatitis Society has the Contact Allergen Management Program, a computerized database of thousands of cosmetics and personal care products. After a patient has been patch tested and the allergen pinpointed, the offending agent may be entered into the database and a list will be generated of all the safe products the patient can use. This makes avoidance of the triggering agent much more manageable for patients. *Is CAERS and Contact Allergen Management Program enough? Are outreach and raised awareness sustainable methods to increase reporting of cosmetic-induced adverse events? How do we capture the patients who simply discontinue usage of culprit cosmetic products without seeking medical attention?*

Should there be an independent review board?

In 1976, the cosmetic industry established a self-regulation panel, the CIR, while the FDA continues to monitor the cosmetic industry from a more indirect approach, conducting and directing review of ingredients and advising companies with potentially dangerous ingredients in cosmetics [2]. The CIR, as well as other review boards like RIFM, continues to provide guidelines, lacking jurisdiction over dangerous ingredients in the market. All the while, cosmetically induced ACD continues to prevail, and the long-term effects of cosmetics remain unknown. [14] suggested that an adverse reaction to cosmetics occurs approximately once every 13.3 years per person. Another review found that the pooled prevalence rate of ACD to cosmetics in 7 different studies to be 9.8% [15].

Would more ingredients under review decrease incidence and prevalence? This also calls pretesting and efficacy into question. *Should we require pretesting before market? Are there measures of efficacy for cosmetic products? Should products be rated on efficacy as sunscreens are classified by sun protection factor (SPF)?*

1.7 Efficacy of Skin Care and Cosmetics

The science behind efficacy claims is substantial and offers the consumer some confidence when properly planned and executed. Baran, Sivamani, Barel, and the UCSF Cutaneous Skin Bioengineering Series summarize some of this science [36–45].

References

1. Wood L. Research and Markets: Global Cosmetics Market 2015–2020: Market was $460 Billion in 2014 and is Estimated to Reach $675 Billion by 2020. [online] Businesswire.com; 2015. Available at: https://www.businesswire.com/news/home/20150727005524/en/Research-Markets-Global-Cosmetics-Market-2015-2020-Market. Accessed 11 Apr. 2018.
2. Merrill R. FDA regulatory requirements as tort standards. J. L. & POL'Y. 2004;12(1):552–3.
3. Termini RB, Tressler LB. American beauty: an analytical view of the past and current effectiveness of cosmetic safety regulations and future direction. Food Drug Law J. 2008;63 (1):257–74.
4. Boyer IJ, Bergfeld WF, Heldreth B, Fiume MM, Gill LJ. The cosmetic ingredient review program-expert safety assessments of cosmetic ingredients in an open forum. Int J Toxicol. 2017; 36 (5_suppl2):5S–13S. https://doi.org/10.1177/1091581817717646.
5. Api AM, Belsito D, Bruze M, Cadby P, Calow P, Dagli ML, Dekant W, Ellis G, Fryer AD, Fukayama M, Griem P, Hickey C, Kromidas L, Lalko JF, Liebler DC, Miyachi Y, Politano VT, Renskers K, Ritacco G, Salvito D, Schultz TW, Sipes IG, Smith B, Vitale D,

Wilcox DK. Criteria for the research institute for fragrance materials, Inc. (RIFM) safety evaluation process for fragrance ingredients. Food Chem Toxicol, 82 (Supplement):S1–S19. https://doi.org/10.1016/j.fct.2014.11.014.

6. Kwa M, Welty LJ, Xu S. Adverse events reported to the us food and drug administration for cosmetics and personal care products. JAMA Intern Med. 2017;177(8):1202–4. https://doi.org/10.1001/jamainternmed.2017.2762.

7. Mowry JB, Spyker DA, Brooks DE, Zimmerman A, Schauben JL. 2015 Annual report of the American Association of Poison Control Centers' National Poison Data System (NPDS): 33rd annual report. Clin Toxicol (Phila). 2016;54(10):924–1109. https://doi.org/10.1080/15563650.2016.1245421.

8. Adams RM, Maibach HI. A five-year study of cosmetic reactions. J Am Acad Dermatol. 1985;13(6):1062–9. https://doi.org/10.1016/S0190-9622(85)70258-7.

9. de Groot AC. Contact allergy to cosmetics: causative ingredients. Contact Dermatitis. 1987;17(1):26–34. https://doi.org/10.1111/j.1600-0536.1987.tb02640.x.

10. Eiermann HJ, Larsen W, Maibach HI, Taylor JS. Prospective study of cosmetic reactions: 1977–1980. North American Contact Dermatitis Group. J Am Acad Dermatol. 1982;6 (5):909–17. https://doi.org/10.1016/S0190-9622(82)70080-5.

11. Nielsen NH, Menne T. Allergic contact sensitization in an unselected Danish population. The Glostrup allergy study, Denmark. Acta Derm Venereol. 1992;72(6):456–60.

12. Romaguera C, Camarasa JM, Alomar A, Grimalt F. Patch tests with allergens related to cosmetics. Contact Dermatitis. 1983;9(2):167–8. https://doi.org/10.1111/j.1600-0536.1983.tb04346.x.

13. Lachapelle J-M, Maibach HI, editors. Patch testing and prick testing: a practical guide official publication of the ICDRG. Berlin: Springer Science & Business Media; 2012.

14. Menkart J. An analysis of adverse reactions to cosmetics. Cutis. 1979;24(6):599–662.

15. Biebl KA, Warshaw EM. Allergic contact dermatitis to cosmetics. Dermatol Clin. 2006;24 (2):215–32. https://doi.org/10.1016/j.det.2006.01.006.

16. Warshaw EM, Buchholz HJ, Belsito DV, Maibach HI, Fowler JF Jr, Rietschel RL, Zug KA, Mathias CG, Pratt MD, Sasseville D, Storrs FJ, Taylor JS, Deleo VA, Marks JG Jr. Allergic patch test reactions associated with cosmetics: retrospective analysis of cross-sectional data from the North American Contact Dermatitis Group, 2001-2004. J Am Acad Dermatol. 2009;60(1):23–38. https://doi.org/10.1016/j.jaad.2008.07.056.

17. Park ME, Zippin JH. Allergic contact dermatitis to cosmetics. Dermatol Clin. 2014;32(1):1–11. https://doi.org/10.1016/j.det.2013.09.006.

18. Holly EA, Bracci PM, Hong MK, Mueller BA, Preston-Martin S. West Coast study of childhood brain tumours and maternal use of hair-colouring products. Paediatr Perinat Epidemiol. 2002;16(3):226–35. https://doi.org/10.1046/j.1365-3016.2002.00420.x.

19. Sontag JM. Carcinogenicity of substituted-benzenediamines (phenylenediamines) in rats and mice. J Natl Cancer Inst. 1981;66(3):591–602. https://doi.org/10.1093/jnci/66.3.591.

20. La Vecchia C, Tavani A. Epidemiological evidence on hair dyes and the risk of cancer in humans. Eur J Cancer Prev. 1995;4(1):31–43.

21. Nohynek GJ, Fautz R, Benech-Kieffer F, Toutain H. Toxicity and human health risk of hair dyes. Food Chem Toxicol. 2004;42(4):517–43. https://doi.org/10.1016/j.fct.2003.11.003.

22. Hougaard KS, Hannerz H, Bonde JP, Feveile H, Burr H. The risk of infertility among hairdressers. Five-year follow-up of female hairdressers in a Danish national registry. Hum Reprod. 2006;21(12):3122–6. https://doi.org/10.1093/humrep/del160.

23. Ronda E, Garcia AM, Sanchez-Paya J, Moen BE. Menstrual disorders and subfertility in Spanish hairdressers. Eur J Obstet Gynecol Reprod Biol. 2009;147(1):61–4. https://doi.org/10.1016/j.ejogrb.2009.07.020.

24. Harling M, Schablon A, Schedlbauer G, Dulon M, Nienhaus A. Bladder cancer among hairdressers: a meta-analysis. Occup Environ Med. 2010;67(5):351–8. https://doi.org/10.1136/oem.2009.050195.

25. Duong A, Steinmaus C, McHale CM, Vaughan CP, Zhang L. Reproductive and developmental toxicity of formaldehyde: a systematic review. Mutat Res. 2011;728(3):118–38. https://doi.org/10.1016/j.mrrev.2011.07.003.
26. Hannigan JH, Bowen SE. Reproductive toxicology and teratology of abused toluene. Syst Biol Reprod Med. 2010;56(2):184–200. https://doi.org/10.3109/19396360903377195.
27. Ono A, Kawashima K, Sekita K, Hirose A, Ogawa Y, Saito M, Naito K, Yasuhara K, Kaneko T, Furuya T, Inoue T, Kurokawa Y. Toluene inhalation induced epididymal sperm dysfunction in rats. Toxicology. 1999;139(3):193–205. https://doi.org/10.1016/S0300-483X(99)00120-1.
28. Gospe SM Jr, Zhou SS. Toluene abuse embryopathy: longitudinal neurodevelopmental effects of prenatal exposure to toluene in rats. Reprod Toxicol. 1998;12(2):119–26. https://doi.org/10.1016/S0890-6238(97)00128-7.
29. Ahmed P, Jaakkola JJ. Exposure to organic solvents and adverse pregnancy outcomes. Hum Reprod. 2007;22(10):2751–7. https://doi.org/10.1093/humrep/dem200.
30. Boberg J, Taxvig C, Christiansen S, Hass U. Possible endocrine disrupting effects of parabens and their metabolites. Reprod Toxicol. 2010;30(2):301–12. https://doi.org/10.1016/j.reprotox.2010.03.011.
31. Crinnion WJ. Toxic effects of the easily avoidable phthalates and parabens. Altern Med Rev. 2010;15(3):190–6.
32. Park CJ, Nah WH, Lee JE, Oh YS, Gye MC. Butyl paraben-induced changes in DNA methylation in rat epididymal spermatozoa. Andrologia. 2012;44(Suppl 1):187–93. https://doi.org/10.1111/j.1439-0272.2011.01162.x.
33. Khanna S, Darbre PD. Parabens enable suspension growth of MCF-10A immortalized, non-transformed human breast epithelial cells. J Appl Toxicol. 2013;33(5):378–82. https://doi.org/10.1002/jat.2753.
34. Wrobel A, Gregoraszczuk EL. Effects of single and repeated in vitro exposure of three forms of parabens, methyl-, butyl- and propylparabens on the proliferation and estradiol secretion in MCF-7 and MCF-10A cells. Pharmacol Rep. 2013;65(2):484–93. https://doi.org/10.1016/S1734-1140(13)71024-7.
35. Konduracka E, Krzemienieccki K, Gajos G. Relationship between everyday use cosmetics and female breast cancer. Pol Arch Med Wewn. 2014;124(5):264–9.
36. Baran R, Maibach HI, editors. Textbook of cosmetic dermatology. 5th ed. Boca Raton: CRC Press; 2017.
37. Barel AO, Paye M, Maibach HI, editors. Handbook of cosmetic science and technology. 4th ed. Boca Raton: CRC Press; 2014.
38. Berardesca E, Elsner P, Wilhelm K-P, Maibach HI, editors. Bioengineering of the skin: methods and instrumentation. Boca Raton: CRC Press; 1995.
39. Elsner P, Berardesca E, Maibach HI, editors. Bioengineering of the skin: water and the stratum corneum. Boca Raton: CRC; 1994.
40. Elsner P, Berardesca E, Wilhelm K-P, Maibach HI, editors. Bioengineering of the skin: skin biomechanics. Boca Raton: CRC Press; 2002.
41. Fluhr J, Elsner P, Berardesca E, Maibach HI, editors. Bioengineering of the skin. Water and the stratum corneum. 2nd ed. Boca Raton: CRC Press; 2004.
42. Shai A, Maibach HI, Baran R, editors. Handbook of cosmetic skin care. 2nd ed. Boca Raton: CRC Press; 2009.
43. Sivamani RK, Jagdeo JR, Elsner P, Maibach HI, editors. Cosmeceuticals and active cosmetics. 3rd ed. Boca Raton: CRC Press; 2016.
44. Wilhelm K-P, Elsner P, Berardesca E, Maibach HI, editors. Bioengineering of the skin: skin imaging & analysis. 2nd ed. New York: Informa; 2007.
45. Wilhelm K-P, Elsner P, Berardesca E, Maibach HI, editors. Bioengineering of the skin: skin surface imaging and analysis. Boca Raton: CRC Press; 1996.

46. Lazzarini R, Duarte I, de Farias DC, Santos CA, Tsai AI. Frequency and main sites of allergic contact dermatitis caused by nail varnish. Dermatitis. 2008;19(6):319–22. https://doi.org/10.2310/6620.2008.08009.
47. Rietschel RL, Warshaw EM, Sasseville D, Fowler JF, DeLeo VA, Belsito DV, Taylor JS, Storrs FJ, Mathias CG, Maibach HI, Marks JG, Zug KA, Pratt M, North American Contact Dermatitis Group. Common contact allergens associated with eyelid dermatitis: data from the North American Contact Dermatitis Group 2003–2004 study period. Dermatitis. 2007;18 (2):78–81. https://doi.org/10.2310/6620.2007.06041.
48. Rietschel RL, Fowler JF, editors. Fisher's contact dermatitis. 6th ed. Hamilton: BC Decker; 2008.
49. Brandrup F. Nickel eyelid dermatitis from an eyelash curler. Contact Dermatitis. 1991;25 (1):77. https://doi.org/10.1111/j.1600-0536.1991.tb01788.x.
50. Guin JD. Eyelid dermatitis: experience in 203 cases. J Am Acad Dermatol. 2002;47(5):755–65. https://doi.org/10.1067/mjd.2002.122736.
51. Le Coz CJ, Leclere JM, Arnoult E, Raison-Peyron N, Pons-Guiraud A, Vigan M, Members of R-G. Allergic contact dermatitis from shellac in mascara. Contact Dermatitis. 2002;46 (3):149–52. https://doi.org/10.1034/j.1600-0536.2002.460304.x.
52. Shaw T, Oostman H, Rainey D, Storrs F. A rare eyelid dermatitis allergen: shellac in a popular mascara. Dermatitis. 2009;20(6):341–5. https://doi.org/10.2310/6620.2009.09009.
53. Occupational Safety and Health Administration. Stay healthy and safe while giving manicures and pedicures: A guide for nail salon workers. (n.d.b). Retrieved from: www.osha.gov/Publications/3542nail-salon-workers-guide.pdf.
54. Jacob SE, Castanedo-Tardan MP. A diagnostic pearl in allergic contact dermatitis to fragrances: the atomizer sign. Cutis. 2008;82(5):317–8.
55. Warshaw EM, Furda LM, Maibach HI, Rietschel RL, Fowler JF Jr, Belsito DV, Zug KA, DeLeo VA, Marks JG Jr, Mathias CG, Pratt MD, Sasseville D, Storrs FJ, Taylor JS. Anogenital dermatitis in patients referred for patch testing: retrospective analysis of cross-sectional data from the North American Contact Dermatitis Group, 1994-2004. Arch Dermatol. 2008;144(6):749–55. https://doi.org/10.1001/archderm.144.6.749.
56. Bhate K, Landeck L, Gonzalez E, Neumann K, Schalock PC. Genital contact dermatitis: a retrospective analysis. Dermatitis. 2010;21(6):317–20. https://doi.org/10.2310/6620.2010.10048.
57. Kugler K, Brinkmeier T, Frosch PJ, Uter W. Anogenital dermatoses—allergic and irritative causative factors. Analysis of IVDK data and review of the literature. J Dtsch Dermatol Ges. 2005;3(12):979–86. https://doi.org/10.1111/j.1610-0387.2005.05763.x.
58. Vermaat H, Smienk F, Rustemeyer T, Bruynzeel DP, Kirtschig G. Anogenital allergic contact dermatitis, the role of spices and flavour allergy. Contact Dermatitis. 2008;59(4):233–7. https://doi.org/10.1111/j.1600-0536.2008.01417.x.
59. Vermaat H, van Meurs T, Rustemeyer T, Bruynzeel DP, Kirtschig G. Vulval allergic contact dermatitis due to peppermint oil in herbal tea. Contact Dermatitis. 2008;58(6):364–5. https://doi.org/10.1111/j.1600-0536.2007.01270.x.
60. Ames BN, McCann J, Yamasaki E. Methods for detecting carcinogens and mutagens with the Salmonella/mammalian-microsome mutagenicity test. Mutat Res. 1975;31(6):347–64. https://doi.org/10.1016/0165-1161(75)90046-1.
61. National Toxicology Program. Final report on carcinogens: Background document for formaldehyde. 2010 Retrieved from: http://ntp.niehs.nih.gov/ntp/roc/twelfth/2009/November/Formaldehyde_BD_Final.pdf.
62. Nicolopoulou-Stamati P, Hens L, Sasco AJ. Cosmetics as endocrine disruptors: are they a health risk? Rev Endocr Metab Disord. 2015;16(4):373–83. https://doi.org/10.1007/s11154-016-9329-4.
63. Occupational Safety and Health Administration. Health hazards in nail salons. (n.d.a). Retrieved from: www.osha.gov/SLTC/nailsalons/chemicalhazards.html.

64. Occupational Safety and Health Administration. Health hazards in nail salons. 2013 Retrieved from: www.osha.gov/SLTC/nailsalons/chemicalhazards.html.
65. Nedorost S, Wagman A. Positive patch-test reactions to gold: patients' perception of relevance and the role of titanium dioxide in cosmetics. Dermatitis. 2005;16(2):67–70 (quiz 55–66).
66. Pak VM, McCauley LA, Pinto-Martin J. Phthalate exposures and human health concerns: a review and implications for practice. AAOHN J. 2011;59(5):228–33. https://doi.org/10.3928/08910162-20110426-01 (quiz 234–225).
67. Pak VM, Powers M, Liu J. Occupational chemical exposures among cosmetologists: risk of reproductive disorders. Workplace Health Saf. 2013;61(12):522–8. https://doi.org/10.3928/21650799-20131121-01 (quiz 529).
68. Rastogi SC. Analytical control of preservative labelling on skin creams. Contact Dermatitis. 2000;43(6):339–43. https://doi.org/10.1034/j.1600-0536.2000.043006339.x.
69. Research and Markets: Global Cosmetics Market 2015–2020: Market was $460 Billion in 2014 and is Estimated to Reach $675 Billion by 2020 [database on the Internet] 2015. Available from: https://www.businesswire.com/news/home/20150727005524/en/Research-Markets-Global-Cosmetics-Market-2015-2020-Market. Accessed Nov 2017.
70. Roelofs C, Azaroff LS, Holcroft C, Nguyen H, Doan T. Results from a community-based occupational health survey of Vietnamese-American nail salon workers. J Immigr Minor Health. 2008;10(4):353–61. https://doi.org/10.1007/s10903-007-9084-4.
71. World Health Organization, International Agency for Research on Cancer. IARC monographs on the evaluation of carcinogenic risks to humans: volume 57. Occupational exposures of hairdressers and barbers and personal use of hair colourants; some hair dyes, cosmetic colourants, industrial dyestuffs and aromatic amines. 1993 Retrieved from http://monographs.iarc.fr/ENG/Monographs/vol57/volume57.pdf.
72. World Health Organization, International Agency for Research on Cancer. IARC monographs on the evaluation of carcinogenic risks to humans: volume 88. Formaldehyde, 2-butoxyethanol and 1-tert-butoxypropan-2-ol. 2006 Retrieved from: http://monographs.iarc.fr/ENG/Monographs/vol88/mono88.pdf.

Nanotechnology in Cosmetics

Birgit Huber and Jens Burfeindt

Abstract

The manufacturing and application of nanomaterials and nanostructures—nanotechnologies—are the subject matter of research activities which are increasingly gaining in significance all over the world. At present, nanomaterials can be found in many everyday products, including in cosmetics. In sunscreens, for instance, pigments with a nanodimension serve as UV filters: titanium dioxide and zinc oxide and several other substances reflect and absorb the invisible UV radiation of the sunlight and hence protect the skin from its damaging effects. These substances are used as nanomaterial since they present decisive advantages over the same substance with larger dimensions.

Keywords

Carbon black · Cosmetic ingredient · Cosmetic products regulation · Definition · Emulsion · INCI · Labelling · Liposome · MBBT · Microemulsion · Multiple emulsion · Nanoemulsion · Nanopigment · Notification · Silica · Silicon dioxide · TBPT · Titanium dioxide · Zinc oxide

2.1 Introduction

The most important area for nanotechnological applications in cosmetics is currently skincare. The effect of skincare products depends on the composition of their ingredients and essentially on whether they reach the areas where they are needed. The product base has, as a rule, to ensure that this actually works. The task of a

B. Huber (✉) · J. Burfeindt
IKW, Mainzer Landstr. 55, 60329 Frankfurt am Main, Germany
e-mail: bhuber@ikw.org

© Springer Nature Switzerland AG 2019
J. Cornier et al. (eds.), *Nanocosmetics*,
https://doi.org/10.1007/978-3-030-16573-4_2

"base", i.e. the main ingredients of a cream, lotion, gel, etc., in which the active substances are integrated, is, on the one hand, the efficient transport of the essential ingredients to the target site. On the other hand, it has to be ensured that the active ingredients remain there in order to achieve an effect with the longest possible persistence.

In order to meet the many different skin and application requirements, there are different types of formulations. These include nanotechnological applications which are used at the manufacturing of cosmetic products. They are essentially **nanoemulsions** and **nanopigments**.

2.2 Emulsions

2.2.1 Types of Emulsions

The simplest known emulsion is milk. The term itself is derived from milk: emulsion: the milked fat in water. A classical basis of skincare is the emulsion. The emulsion is in practice referred to as cream or lotion, depending on its consistency. Emulsions are made up of two main components: oil and water phase which mix permanently with the support of a surface-active substance (emulsifier). Depending on the mixing ratio, these emulsions have different properties; the type of emulsion is decisive for the effect. A distinction is made between oil-in-water (O/W) and water-in-oil emulsions (W/O emulsions). In the oil-in-water emulsion (O/W emulsion), oil and water droplets are finely distributed; the outer phase consists of water. This type of emulsion delivers a lot of moisture, can be easily distributed and is readily absorbed by the skin. It is particularly suited for normal and rather oily skin. Most day creams and light moisturising creams are part of this category.

The water-in-oil emulsion (W/O emulsion) works according to the reverse principle: here, too, oil and water droplets are finely distributed; however, the coherent or external phase consists of oil. Because of the high release of oil onto the skin, the lipid-replenishing properties and the reduction in water loss, this composition is, for instance, recommended in the form of rich creams or rich oil-based creams in particular for dry and sensitive skin.

2.2.2 Multiple Emulsions

Emulsions are currently increasingly refined. Apart from the classical emulsions, there are today more differentiated systems with which the effect is to be improved and extended: the multiple emulsions. These are complex oil-plus-water systems in which further small droplets of the outer phase are in the inner phase of two intermixed liquids. A distinction is made between water-in-oil-in-water systems (W/O/W) for a lot of moisture and oil-in-water-in-oil systems (O/W/O) for rich lipid content. With multiple emulsions, oil- and water-soluble ingredients can be introduced into the skin next to one another and can be released one after the other. The effect can be prolonged through the delayed release of the ingredients. The goal is to moisturise the skin more lastingly and support the build-up of a lipophilic (oil-loving) protective film. The multiple emulsions used most frequently consist of different oil phases in combination with water.

2.2.3 Microemulsions or Nanoemulsions

So-called microemulsions or nanoemulsions are particularly finely distributed. They are widely spread in nature, for instance in milk. In cosmetic products, they are macroscopic preparations which contain oil and water droplets, which are reduced to nanometric dimensions in order to increase the content of caring oils and maintain at the same time the transparency and lightness of the formulation. As opposed to traditional emulsions, they are characterised by a smaller droplet diameter. Since the superfine resolution is not discernible for the eye, microemulsions appear externally as a transparent, monophase system which is not at all perceived as a "turbid" mixture. The advantage of this type of emulsion is that it can penetrate the skin more easily. Experts also refer to this as the "increased ability to penetrate" the skin. However, nanoemulsions are not penetrating the skin barrier.

2.2.4 Liposomes

Liposomes are amongst the most important cosmetic novelties of the late twentieth century. They serve as a transport system, primarily in dispersions, for the care of dry, dehydrated skin. Liposomes are tiny spheres whose shell consists of hydrophilic and lipophilic substances. They are mostly filled with water and water-soluble active ingredients such as vitamins. Due to their tiny size (less than 1/10,000 mm) and their cell-like external structure, they can penetrate the skin. In this way, liposomes are to channel active ingredients, such as vitamins, into the epidermis and fix them there. Depending on the manufacturer, different liposomes are used. They differ in terms of the composition of the fat globules and active ingredients they transport. In some variants, it is not only the inside of the sphere which is filled with water or water-soluble active ingredients, but liposoluble active ingredients are integrated in addition to the outer skin. In this way, their effect is to be additionally improved. Liposomes, which also include the so-called nanospheres, can be integrated into any type of emulsion. Due to their size, liposomes can be considered as nanoparticles. (Some manufacturers, therefore, also use the term "nanosomes".) As opposed to inorganic pigments, which consist of chemically stable solid particles, liposomes are, however, structures of physiological materials which are, by comparison, loosely bound, which dissolve again into their components as a rule already in the upper skin layer and release the transported active ingredients. Liposomes are particularly suited in order to transport sensitive active ingredients such as for instance vitamins in small bubbles (vesicles) with a size in a nanometre range, protected from air, into the skin, because they are only released at the time of application in contact with the skin.

2.3 Nanomaterials in Cosmetics

According to the general definition, "nanomaterial" is a designation for particles with dimensions of less than 100 nm (1 nm = 1 billionth of a metre). The term "nanomaterial" hence only describes the size range of the substances concerned, but not their further properties. Some substances occur naturally in the form of nano-materials, whilst others are especially manufactured as nanomaterials, since these substances present with their low size other, special and possibly novel properties.

The manufacturing and application of nanomaterials and nanostructures—nanotechnologies—are the subject matter of research activities which are increasingly gaining in significance all over the world. At present, nanomaterials can be found in many everyday products, including in cosmetics. In sunscreens, for instance, pigments with a nanodimension serve as UV filters: titanium dioxide and zinc oxide and several other substances reflect and absorb the invisible UV radiation of the sunlight and hence protect the skin from its damaging effects. These substances are used as nanomaterial since they present decisive advantages over the same substance with larger dimensions.

2.3.1 The Legal Framework

In the uniform EU cosmetics legislation in force since 1976 (Cosmetics Directive 76/768/EEC), nanomaterials were initially not explicitly regulated. Titanium dioxide, for instance, was already approved as a UV filter pigment within the EU since 2002—also in a nanoscale form—without expressly emphasising this. During the past years, EU cosmetics law was intensely revised and newly published as *Cosmetic Products Regulation* (EC) No. 1223/2009 at the end of 2009. On that occasion, the term "nanomaterial" was anchored in cosmetics law for the first time [1].

Consequently, cosmetics have been the first product area in which the size of substances, when it is in the nanometre range, is considered separately, and substances produced accordingly are subject to own rules. In the Cosmetics Regulation, the substances are referred to as nanomaterials if they have been produced on purpose, are indissoluble or biologically resistant and have at least in one dimension (length, width or height) a size between 1 and 100 nm.

Moreover, the Cosmetics Regulation requires for all ingredients contained in the form of nanomaterials to be explicitly labelled as such. This means that cosmetic products play a pioneer role by informing consumers on their product packages actively about the use of these technologies.

Apart from labelling, there are other conditions imposed on cosmetic products containing nanomaterials. Cosmetics with nanomaterials which are not yet expressly approved in the Cosmetics Regulation must be notified to the European Commission no later than six months before they are placed on the market. In this way, the European Commission obtains a European overview of the nanomaterials used. In June 2017, the Commission published a catalogue of all nanomaterials in cosmetic products which have been placed on the market within the EU [2]. With a certain delay, it meets its obligation under Article 16.10a of the EC Cosmetics Regulation. The first catalogue submitted contains formally 43 entries. Some substances (titanium dioxide, copper, silver, gold, platinum) are, however, listed several times depending on their function or with different INCI designations. The published catalogue of all nanomaterials in cosmetic products is to be considered as a first draft which has a mere information character and does not at all constitute a directory of approved nanomaterials. The substances covered must be reviewed in a more detailed manner, and the catalogue must possibly be revised.

The notification also requires submission of detailed information about the properties and safe use of the nanomaterials so that the European Commission receives sufficient information in order to request, if necessary, an opinion on the safety of the respective substances from the competent Scientific Committee on Consumer Safety (SCCS).

Certain substance groups may only be used in cosmetic products if the individual substances are expressly approved: this concerns colourants, preservatives and UV filters. If such substances are used as nanomaterials in cosmetic products, they must be expressly assessed and approved as nanomaterials. The UV filter pigments, titanium dioxide and zinc dioxide, which have already been approved and used for many years, have recently been intensely examined once more in view of their properties as nanomaterial and were assessed by the competent Scientific Committee on Consumer Safety (SCCS).

Nano or not nano—the principle that applies is that all ingredients must equally meet the applicable high safety and quality standards for cosmetic products. The basis for the production of safe cosmetic products is the safety assessment which must be carried out for every cosmetic product by an expert. With a view to safety, it is necessary already at that stage to explicitly also take into account the particle size, including that of nanomaterials.

2.3.2   Labelling and Its Significance

Nanomaterials are labelled within the framework of the already established full declaration of ingredients in accordance with INCI (*International Nomenclature Cosmetic Ingredients*) in line with the provisions of the Cosmetics Regulation with the addition "(nano)". On many sunscreens, the ingredient is stated as *titanium dioxide (nano)*.

The mentioning of "(nano)" is a contribution to market transparency: in this way, the consumers are informed about part of the manufacturing practice; they learn whether and where nanoparticles have been used and may consider the information for their buying decision. At the same time, all other market players receive an overview of the products in which nanomaterials are used and this benefits ongoing safety research and risk assessment.

2.3.3 The Substances

The catalogue of nanomaterials published by the EU Commission in June 2017 includes formally 43 entries. Some substances (titanium dioxide, copper, silver, gold, platinum) are, however, listed several times depending on their function and different INCI designations. Apart from the known, comparatively frequently used nanomaterials, the catalogue includes further entries: some of the mentioned substances may occur possibly on a nanoscale and are used in this form. However, they are usually used in non-nanoform, i.e. in typical pigment particle size in a micrometre range. Since many different methods are used for the identification of nanomaterials and the term "nanomaterial" is not clearly defined in EU cosmetics law and leaves, therefore, a certain room for interpreting, there may also be different interpretations and implementations of the provisions in individual cases.

For some of the mentioned substances (retinol, tocopheryl acetate), the classification as nanomaterial is doubtful. These substances themselves cannot occur as nanomaterials. There is a need for further clarification concerning their classification as nanomaterials. These substances are possibly included in other—biologically unstable—nanoscale transport systems such as liposomes. The latter are, however, not to be classified as nanomaterials within the meaning of the EC Cosmetics Regulation.

2.3.3.1 Titanium Dioxide and Zinc Oxide

Titanium dioxide and zinc oxide are minerals which are widespread in nature and have been used already for many years as UV filters since they can particularly well scatter and reflect UV light. With the originally used particle sizes in a micrometre range (1 micrometre—1 thousandth of a millimetre), thick pastes were, however, developed, which were not accepted by the consumer. As a result of the reduction in the particle size, titanium dioxide and zinc oxide became transparent for the human eye so that the formation of an undesired white film on the skin can be avoided.

Moreover, the lower dimensions are able to significantly improve the protective effect of the two UV filters to UV radiation. In most products, the use of nanopigments actually allows for the achievement of the desired, very high sun protection factors. In sunscreens, they are used in the form of nanoparticles because they reflect and scatter the UV light. Consequently, they contribute towards protecting human skin from the negative effects of UV radiation, including from skin cancer. In sunscreen lotions, nanoscale titanium dioxide is present in large clusters (aggregates and agglomerates) whose size is above 100 nm in order to safeguard optimum protection for the skin. Numerous studies, including those which were carried out within the framework of the NANODERM research programme of the European Union, reached the conclusion that nanopigments do not pass through the skin barrier even in cases in which the skin is damaged, for instance in the event of psoriasis or sunburn [3–7]. Furthermore, studies carried out in Europe as well as by the Food and Drug Administration (FDA) in the USA have shown that no adverse effects are observed even when titanium dioxide nanopigments are injected into the bloodstream [8]. Titanium dioxide as a highly inert substance is, for instance, also used as a colourant in foods and toothpaste. Its toxicological safety has been comprehensively documented. Titanium dioxide and zinc oxide are the most important nanopigments in cosmetics. The Scientific Committee on Consumer Safety of the EU Commission (SCCS) has confirmed the safety of the two substances for the application in cosmetic products [9–13].

2.3.3.2 Other Important Nanomaterials

Apart from the two most important substances, titanium dioxide and zinc oxide, there are several other important nanoparticles which are used as UV filters in skincare products.

UV filters MBBT and TBPT

The organic nanoscale UV filter pigments MBBT (2,2'-methylene-bis-(6-(2H-benzotriazol-2-yl)-4-(1,1,3,3-tetramethylbutyl)phenol) and tris-biphenyl-triazine (TBPT, 2,4,6-tris([1,1'-biphenyl]-4-yl)-1,3,5-triazine) place themselves as a protective film on the upper skin layer and scatter, reflect and absorb the UV radiation of the sun. In this way, the skin is protected from UV radiation and its health-damaging effects (sunburn, DNA damage, skin ageing, etc.). The two substances are so-called broadband UV filters with a distinct filter effect not only in the UV-B but also in the longer wave UV-A area. The SCCS has recently confirmed the safety of the two substances for the application in cosmetic products [14–16].

Carbon black

Carbon black is used in cosmetic products as a black pigment. It is compatible and mixable with almost all other ingredients and solvents and is highly resistant to temperature and light. It has a particularly high colour strength and depth. Carbon black is practically approved all over the world as a cosmetic colourant. Its safety for use in cosmetic products has also been recently confirmed by the SCCS [17].

Silicon dioxide or silica

Synthetic silicon dioxide is not manufactured as a nanomaterial in a targeted manner. Whether it is to be considered as nanomaterial or not depends essentially on the underlying definition. Due to the manufacturing process and its build-up from smaller primary particles with dimensions of less than 100 nm, it can be classified— depending on the individual specification of the material at stake—possibly as nanomaterial according to the currently applicable definition recommendation of the European Commission. The deviating definition in the EC Cosmetics Regulation includes several additional criteria on top, which must all be met (e.g. insolubility and resistance in biological media; deliberate production as a nanomaterial). These criteria are not all met in many cases for the silicon dioxide variants used in cosmetics. Silica is used in many different cosmetics, e.g. in toothpaste, decorative cosmetics, make-up, powder, hair colourations, hair styling products, etc.

2.4 Outlook

The most important task of skincare products continues to be supporting the natural functions of the skin such as moisture regulation and protective shield formation. Highly effective care, mitigation of existing wrinkles and intensive skin regeneration are other objectives. Against this background, an important trend emerges in skincare: the industry makes increasing efforts to develop new technologies which permit claims in respect of specific product performances. In particular in the field of anti-ageing and products for the mature skin, this has resulted in novelties and permitted the development of skincare products with multiple benefits. Nanotechnology is an important component of research in this connection. These new technologies allow the research departments of the manufacturers to investigate the effect of ingredients in constantly new combinations. In this way, they try to exhaust all possible effects. With the liposomes, they try to imitate the natural biochemical structures and activities of the epidermis in a biological way. In this connection, the care products are no longer used exclusively as the carriers of active ingredients, but also as an active principle of their own. New developments and findings impact the existing formulations. Heavy oils and waxes are no longer used to an increasing extent. Their task was to form a water-resistant film on the skin and hence avoid the evaporation of moisture. Now ingredients are used instead which are compatible with the natural moisture barrier of the skin. It consists of several lipids and cell layers. In this connection, liposomes are used, for instance, which are similar to the skin's own lipids, as well as fatty acids which occur in skin membranes and many new ingredients. Milk lipids, for instance, are a technologically relatively novel combination of lipid components (fats) from the milk. All these new ingredients can enormously reinforce the moisture barrier and hence avoid the transepidermal water loss or dehydration of the skin.

References

1. Regulation (EC) No. 1223/2009 of the European Parliament and of the Council of 30 November 2009 on cosmetic products: http://eur-lex.europa.eu/en/index.htm.
2. Catalogue of nanomaterials used in cosmetic products placed on the EU market (document date: 15/06/2017—Publication date: 12/07/2017): http://ec.europa.eu/docsroom/documents/24521.
3. Nohynek GJ, Lademann J, Ribaud C, Roberts MS. Grey goo on the skin? Nanotechnology, cosmetic and sunscreen safety. Crit Rev Toxicol. 2007;37:251–77.
4. Schilling K, Bradford B, Castelli D, Dufour E, Nash JF, Pape W, Schulte S, Tooley I, van den Bosch J, Schellauf F. Human safety review of "nano" titanium dioxide and zinc oxide. Photochem Photobiol Sci. 2010;9:495–509. https://doi.org/10.1039/b9pp00180h.
5. Filipe P, Silva JN, Silva R, Cirne de Castro JL, Marques Gomes M, Alves LC, Santus R, Pinheiro T. Stratum corneum is an effective barrier to TiO2 and ZnO nanoparticle percutaneous absorption. Skin Pharmacol Physiol. 2009;22:266–75. https://doi.org/10.1159/000235554.
6. NANODERM—Quality of skin as a barrier to ultra-fine particles, Final Report, 2007.
7. Monteiro-Riviere NA, Wiench K, Landsiedel R, Schulte S, Inman AO, Riviere JE. Safety evaluation of sunscreen formulations containing titanium dioxide and zinc oxide nanoparticles

in UVB sunburned skin: an in vitro and in vivo study. Toxicol Sci. 2011;123:264–80. https://doi.org/10.1093/toxsci/kfr148.

8. Fabian E, Landsiedel R, Ma-Hock L, Wiench K, Wohlleben W, van Ravenzwaay B. Tissue distribution and toxicity of intravenously administered titanium dioxide nanoparticles in rats. Arch Toxicol. 2008;82:151–7.

9. Scientific Committee on Cosmetic Products and Non-Food Products Intended for Consumers (SCCNFP): Opinion on Titanium Dioxide; adopted by the SCCNFP during the 14th plenary meeting of 24 October 2000 (SCCNFP/0005/98).

10. Scientific Committee on Consumer Safety (SCCS): Opinion on Titanium Dioxide (nano form); adopted by written procedure on 22 July 2013, Revision of 22 April 2014 (SCCS/1516/13).

11. Scientific Committee on Consumer Safety (SCCS): Opinion on Titanium Dioxide (nano form) coated with Cetyl Phosphate, Manganese Dioxide or Triethoxycaprylylsilane as UV-filter in dermally applied cosmetic; adopted by written procedure on 7 November 2016 (SCCS/1580/16).

12. Scientific Committee on Consumer Safety (SCCS): Opinion on Zinc Oxide (nano form); adopted by the SCCS at its 16th plenary meeting of 18 September 2012, Revision of 11 December 2012 (SCCS/1489/12).

13. Scientific Committee on Consumer Safety (SCCS): Addendum to the Opinion SCCS/1489/12 on Zinc oxide (nano form); adopted by written procedure on 23 July 2013, Revision of 22 April 2014 (SCCS/1518/13).

14. Scientific Committee on Consumer Safety (SCCS): Opinion on 2,2'-Methylene-bis-(6-(2H-benzotriazol-2-yl)-4-(1,1,3,3-tetramethylbutyl)phenol) (nano form); adopted by the SCCS on its 9th plenary meeting on 25 March 2015, Revision of 25 June 2015 (SCCS/1546/15).

15. Scientific Committee on Cosmetic Products and Non-Food Products Intended for Consumer (SCCNFP): Opinion on 2,2'-Methylene-bis-(6-(2H-benzotriazol-2-yl)-4-(1,1,3,3-tetramethylbutyl)phenol); adopted by the plenary session of the SCCNFP of 17 February 1999.

16. Scientific Committee on Consumer Safety (SCCS): Opinion on 2,4,6-tris[1,1'-biphenyl]-4-yl-1,3,5-Triazine, adopted by the SCCS at its 12th Plenary Meeting of 20 September 2011, Revision of 13/14 December 2011 (SCCS/1429/11).

17. Scientific Committee on Consumer Safety (SCCS): Opinion on Carbon Black (nano form), adopted by the SCCS at its 4th plenary meeting of 12 December 2013, 2nd Revision of 15 December 2015 (SCCS/1515/13).

Part II
Nanoparticles in Cosmetics

Inorganic Nanoparticles in Cosmetics

3

T. P. Vinod and Raz Jelinek

Abstract

Inorganic nanomaterials of different chemical compositions and morphologies have been applied in cosmetic products due to their size- and shape-dependent properties which can improve the performance of the products. This chapter discusses the application of inorganic nanoparticles in cosmetic products with an emphasis on the characteristic features of nanoparticles suitable for cosmetic applications. In particular, applications of inorganic nanoparticles as UV filters and antimicrobial materials are discussed in detail with a basic overview of the fundamental scientific basis related to these applications. Types of nanoparticles used in commercial cosmetic products are enlisted, reflecting the range of applications and property modifications. Applications of inorganic nanoparticles in cosmetic formulations as active components and nanocarriers are also discussed along with relevant examples.

Keywords

Nanoparticles · Inorganic nanoparticles · Cosmetics · UV filters · Sunscreens · Antimicrobial nanoparticles

T. P. Vinod
Department of Chemistry, CHRIST (Deemed to be University),
Hosur Road, Bangalore, Karnataka 560029, India

R. Jelinek (✉)
Department of Chemistry, Ben-Gurion University of the Negev,
84105 Beer Sheva, Israel
e-mail: razj@bgu.ac.il

Ilse Katz Institute for Nanotechnology, Ben-Gurion University of the Negev,
84105 Beer Sheva, Israel

© Springer Nature Switzerland AG 2019
J. Cornier et al. (eds.), *Nanocosmetics*,
https://doi.org/10.1007/978-3-030-16573-4_3

29

3.1 Introduction

Nanoparticles (NPs) are viewed as valuable additions to the repertoire of industrially relevant materials due to the variety, tunability, and integrability of their properties pertinent to a wide range of applications [1–3]. Interest in the functionality and utility of nanoparticles emanate from their unique characteristics such as high surface-to-volume ratio, hardness, quantum confinement effects, electronic and magnetic properties, catalytic properties, and biological activity which are not manifested in either the bulk state or molecular level. Inorganic nanoparticles are among the most extensively studied class of materials due to the potential applications they can realize as functional and structural constituents in materials and devices. The variety of inorganic materials which can form particles of nanoscale dimensions has expanded with the proliferation of physical and chemical methods applicable to the *top-down* and *bottom-up* synthesis of nanomaterials.

In particular, some features of NPs are suitable for performance enhancements in cosmetic products used for dermatological, hair care, and dental care applications [4]. Accordingly, commercial cosmetic products which utilize the potentials of nanomaterials have already reached the market [5, 6]. NPs are generally used in cosmetic products to harness the advantages of size- and shape-related properties [7, 8]. Nanomaterials are mainly used as active components, nanocarriers, and as formulation aids in various cosmetic products. Two types of nanoscale systems have been generally used in cosmetic applications: (i) systems which disintegrate to their molecular form upon application to the skin and (ii) insoluble particles which retain their structural integrity even after being applied to the skin. While the first category constitutes mainly of nanocarrier systems such as liposomes and niosomes, most of the inorganic NPs used in cosmetic formulations fall under the second category.

Even though the terminology and concepts of nanotechnology became prevalent only a few decades ago, materials in nanoscale dimensions were used in ancient times for cosmetic applications. A recent study has revealed that a 2000-year-old recipe of hair dye used in Greco-Roman period utilized PbS NPs to modify the color of human hair [9]. In modern times, publications related to the study of potential application of nanomaterials in cosmetic products started to emerge in the early 1980s [10]. Cosmetic materials containing titanium dioxide were presented in a patent application in 1983, and a patent application for the use of ZnO nanopowder in ultraviolet (UV) protection of the skin was published in 1985 [10]. As apparent in Fig. 3.1, there is a rapid increase in the number of scientific publications related to nanoscale ingredients in cosmetic materials. A significant portion of research in this field is dedicated to study the health/biological impact of NPs. Guidelines and restrictions for the usage of nanomaterials in consumer products are being established, monitored, and re-evaluated by various bodies and agencies in different parts of the world.

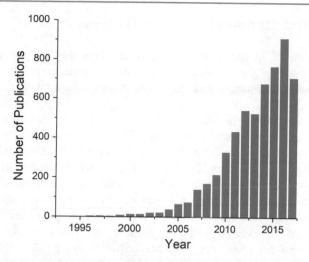

Fig. 3.1 Publications related to nanoparticles in cosmetics. Data obtained from https://www. scopus.com. Search was carried out for documents containing the terms "nanoparticles" and "cosmetic" in their title, abstract, and key words. Data was acquired in November 2017

3.2 Synthesis of Inorganic Nanoparticles

Methods for the synthesis of inorganic nanoparticles pertinent to cosmetic applications are generally classified into two as *top-down* and *bottom-up* methods. Top-down methods convert material in the bulk state to nanoscale mostly through physical or mechanical means. Ball milling, high-pressure extrusion, and ultra-sonication are some of the common *top-down* methods used for the synthesis of nanoparticles. While the *top-down* methods are suitable for large-scale production of nanomaterials, they have disadvantages such as the formation of defects and contaminations. *Bottom-up* methods are based on the self-assembly of molecules and ions in suitable reaction medium resulting in the formation of nanocrystals. Sol-gel chemistry, hydrothermal reactions, electrochemical synthesis, ion exchange reactions, coprecipitation reactions, etc. are some of the commonly used methods for the *bottom-up* synthesis of inorganic nanoparticles. *Bottom-up* processes generally result in nanoparticles with less defects and more homogeneity. Limitations in scalability and the additional purification steps required are considered as the drawbacks of these methods. In practice, a combination of *bottom-up* and *top-down* approaches is used for preparing nanomaterials for application in cosmetics. Nanoparticles prepared through *bottom-up* methods tend to aggregate and agglomerate in solution, thereby changing the particle size and properties. These agglomerates and aggregates are therefore subjected to *top-down* methods which will disassemble them to the primary particle size.

3.3   Characterization of Inorganic Nanoparticles

Characterization of inorganic nanomaterials involves the analysis of size, size distribution, shape, physical properties, crystallinity, surface charges, and composition. Various spectroscopic and microscopic methods are being used for this analysis (Fig. 3.2).

Electron microscopy techniques such as scanning electron microscopy (SEM) and transmission electron microscopy (TEM) can provide information about the size and morphology of the particles. Visual representation of the particles is obtained as digital images through these techniques, which can be analyzed further with image analysis software to quantify the particle sizes and particle size distributions. SEM utilizes the secondary electrons emitted by atoms upon excitation with a high-energy electron beam while TEM uses electron beams transmitted through a sample for creating an image. Sample for analysis in SEM is normally made by drying a dilute solution of the nanoparticles on a solid surface. In case of materials which are not electrically conductive, a metal coating is usually done by sputter coating technique prior to doing the SEM analysis. TEM samples are usually prepared by drying a dilute solution of the particles on a metallic grid. Atomic force microscopy (AFM) is a scanning probe technique which works by measuring the forces between an atomic size probe tip and a surface as a function of separation between them. In contact mode AFM, the topographical graph is generated by taping the probe on to the surface while in noncontact mode AFM, the probe is moved over the surface without physical contact. AFM images can provide information about the 3D morphology of the nanoparticles being analyzed. AFM measurements are usually done on samples deposited on a smooth solid surface. Dynamic light scattering (DLS) analysis is an efficient tool to get particle size distribution in a sample. This measurement is based on the Brownian motion of nanoparticles in solution. When a beam of monochromatic laser is incident on a solution or dispersion of nanoparticles, the random motion of the particles causes changes in the intensity of light scattered. Hydrodynamic diameter of the particle

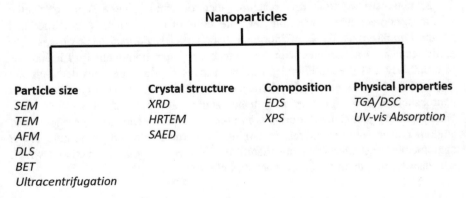

Fig. 3.2 Analytical techniques used for the characterization of inorganic nanoparticles

can be estimated from the measurement of translational diffusion coefficient through DLS at a given temperature and viscosity of the medium. Software associated with the DLS instruments performs these calculations, and a distribution of particle size is delivered as read out. As it collects data from a large number of particles simultaneously, DLS can give a comprehensive picture of the particle size distribution in a single experiment and in shorter durations, unlike the electron microscopy techniques (SEM, TEM) and AFM. Scattering of radiation (laser) can be used to quantify the surface charge of the particle (zeta potential) also. When an electric field is applied to a solution of nanoparticles, they move in response to the applied field. The velocity of particle movement is related to the surface charge of the particle, and this can be quantified by analyzing the scattering from the particles and standard equations. There are DLS instruments with combined capabilities of particle size measurement and zeta potential of nanoparticles. Brunauer–Emmett–Teller (BET) surface area analysis is a technique to get specific surface area of the samples through the study of adsorption of the gases. It gives information about the pore size of solid samples, and it can be used to calculate the average particle size of a nanoparticle sample. Analytical ultracentrifugation techniques such as differential sedimentation analysis and integral sedimentation analysis also can be used to obtain the measure of average particles. The principle behind these techniques is that the sedimentation velocity of a particle depends on its mass and size. Ultracentrifugation of a solution of nanoparticle causes fractionation of the sample as particles of different sizes sediment at different rates. The sedimentation velocities of different fractions can be measured through optical methods, and these velocities can be directly correlated to the particles consisting of each fraction.

Crystal structure of the nanoparticles can be analyzed with powder X-ray diffraction (XRD). In this, a dry sample of the nanoparticle is analyzed to get peaks corresponding to specific crystallographic planes. The XRD pattern thus obtained is then compared with patterns in the standard XRD database to identify the crystal structure of the nanoparticles. High-resolution transmission electron microscopy (HRTEM) helps the direct visualization of the crystal planes in a crystalline sample of nanoparticles. The inter-planar spacing deduced from these images will indicate the crystal structure of the nanoparticles. Selected area electron diffraction (SAED) is another tool to investigate the crystalline nature of nanoparticle samples. SAED measurement is usually done in TEM, and the corresponding diffraction pattern of electrons obtained is analyzed with suitable analytical software and database.

Elemental composition of the nanoparticles can be quantitatively analyzed using energy-dispersive X-ray spectroscopy (EDS). In this technique, a high-energy electron beam is used to stimulate the characteristic X-ray emissions of elements present in a sample. Analysis of the resultant X-ray emission gives information about the types of elements present in a specimen. Relative percentages of different elements are obtained by comparing the intensities of characteristic peaks of each element. EDS analytical setup is usually coupled with SEM or TEM. Site-specific analysis on individual nanostructures also is possible by focusing the electron beam on selected area of the structure. X-ray photoelectron spectroscopy (XPS) can be used to identify the oxidation states of different ions present in an inorganic nanoparticle. XPS analysis is obtained by irradiating a material with a beam of X-rays and measuring the kinetic energy and number of electrons that escape from the surface of the material. The measure of kinetic energy of the electrons will give a measure of the binding energy of them, which is characteristic of an element in a particular electronic state or oxidation state. Comparison of relative percentages of atomic species present in a sample also can be obtained from XPS.

Thermal stability and thermochemistry of a nanoparticle sample can be analyzed with thermogravimetric analysis (TGA) and differential scanning calorimetry (DSC), which is usually performed simultaneously with a single instrument. TGA analyzes the weight change in a sample upon increasing the temperature. This gives information such as decomposition temperature of the particles, temperature at which organic ligands are removed from the system, etc. DSC records the heat changes happening in a system upon increasing the temperature. This allows to monitor exothermic and endothermic processes happening in the sample upon raising the temperature.

Absorption or transmittance properties of inorganic nanoparticles can be obtained from UV–Vis absorption spectroscopy. A solution of the sample or a film of the sample made on suitable substrate can be used for recording the absorbance or transmittance. From the absorption spectra, band gap of the inorganic nanocrystals can also be estimated.

3.4 Functional Roles of Inorganic Nanoparticles in Cosmetic Materials

Inorganic nanoparticles have characteristics that are suitable for fulfilling a variety of functions in cosmetic products. These roles fall mainly into three categories: (i) active substances, (ii) nanocarriers, and (iii) modifiers of the appearance and/or rheology of the end product. Various classes of inorganic nanomaterials such as noble metals, metal oxides, nanoclays, and mesoporous nanostructures are being used in commercial cosmetic products in this regard. Figure 3.3 indicates the types of nanomaterials prominently used in cosmetic products and their relative abundance. Uses of these materials for respective functionalities is envisaged to be advantageous due to factors like the propensity of the nanomaterials to remain on the skin without being absorbed, relatively smaller concentrations required in the formulation, and their effect in facilitating desired changes in texture and solubility [11]. Table 3.1 summarizes the types of inorganic nanomaterials currently used in cosmetic products. As seen in the table, inorganic nanoparticles are mainly used for UV protection and antimicrobial activity. Among the main classes of inorganic materials, metal and metal oxide nanoparticles are most commonly used.

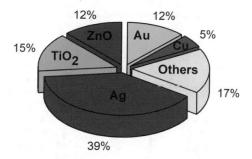

Fig. 3.3 Types of inorganic nanomaterials prominently used in cosmetic formulations and the relative percentage of number of products containing them. Data obtained from catalog of nanomaterials used in cosmetic products placed on the market [5] and the Project on Emerging Nanotechnologies [6]. Data was acquired in November 2017

3.4.1 Inorganic Nanomaterials as Active Substances in Cosmetic Products

Inorganic nanomaterials are used as active ingredients in cosmetic formulations based on their capacity to improve skin appearance, protect the skin from external factors, and for cleaning roles. Several products rely upon the characteristic physicochemical properties of inorganic nanoparticles to achieve desired functionalities.

3.4.1.1 Inorganic Nanoparticles as Sunscreens and UV Filters

About 10% of daylight incident on earth is ultraviolet (UV) light. Overexposure to sunlight and UV light can cause adverse effects to human skin such as mutations, photocarcinogenesis, photoaging, inflammation, pigmentation, hyperplasia, immunosuppression, and vitamin D synthesis [12]. Both the UVA (320–400 nm) and UVB (290–320 nm) regions of the ultraviolet spectrum can cause damage to skin, and they need to be blocked for the protection of the skin [12]. UVC radiation (wavelength less than 290 nm) is also harmful, but it is blocked by the ozone layer of the earth's atmosphere. Cosmetic sunscreen products are used to prevent harmful exposure to UVA and UVB radiations. These products protect the skin by blocking the exposure to UV radiation either by absorbing the UV light or by scattering and/or reflecting it. UV filters which function based on the absorption of UV radiation are referred as chemical UV filters and corresponding filters which operate through scattering and/or reflection of UV light are called physical UV filters. Absorption of UV light is generally a chemical property associated with molecular structure and functional groups in chemical compounds. Organic compounds such as salicylates, benzophenones, cinnamates, and camphor derivatives are used as chemical UV filters in sunscreens to absorb UV light. Upon absorption of UV radiation, these molecules are excited to a higher energy state and the energy difference is dissipated through subsequent events such as heat transfer to the surroundings, fluorescence, and phosphorescence.

Table 3.1 Inorganic nanomaterials used in cosmetic products and their applications

Nanoparticle	Applications	Nanoparticle	Applications
Ag	Antimicrobial protection Coating of hair Colorant Decreasing wrinkles Face mask Foot care products Makeup remover Mouthwash, Toothpaste Shaving products	SiO_2	Body or face paint Concealer Face mask Eye makeup products Lip care products Hand care products Eye contour products Hair care products Nail care products
Au	Colorant Eye contour products Face mask Hair conditioner Makeup remover Nanocarrier Shampoo Sun protection Wrinkle treatment	Fe_2O_3	Eye makeup Foundation Lipstick Mascara Nail varnish
		Al_2O_3	Face mask Nail varnish Sun protection
Cu	Colorant Facial spray Foundation Mouthwash Nail sculpting products Shampoo	$Fe_4[Fe(CN)_6]_3$	Eye makeup Face makeup Nail makeup
		$LiMgNaO_6Si_2$	Chemical depilatories Cuticle remover Eye contour products Face masks Foundation Mascara Nail varnish
Pt	Face care products Foundation		
TiO_2	Anticaking agent Colorant Eye contour products Face mask Hand care products Lip care products Shaving products Skin lightening products Sunscreen, UV-filter		
		$MgNa_2O_6Si_2$	Eye contour products Face care products Hand care products Nail varnish
ZnO	Concealer Eye contour products Face mask Foot care products Hand care products Lip care products Physical epilation products Skin lightening products Sunscreen, UV-filter		

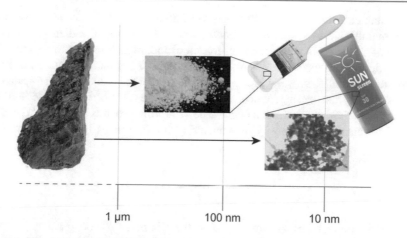

Fig. 3.4 Classical submicron-sized TiO_2 pigments ($\sim$0.2 μm) as e.g. used in paints and as opacifiers in various applications are prepared from ilmenite (a titania ore) using chemical processes (dry or wet). Further reducing the particle size to nano-sized particles ($<$50 nm) recently allowed increasing light absorbance in the UV region while reducing (undesired) skin whitening in high-quality sunscreen. Reprinted with permission from [1]. Copyright (2015) The Royal Society of Chemistry

Unlike molecular UV absorbers, nanoparticles of inorganic oxides such as TiO_2, ZnO, CeO_2, and ZrO_2 are used as physical UV filters in cosmetic sunscreen formulations due to their ability to scatter and reflect UV radiations. Even though these metal oxides mainly operate through the physical filtering of UV light, they can absorb a portion of the UV light incident on them [13]. These nanoscale materials have the capability to attenuate UV radiation, and they effectuate an improvement in the capacity to withstand UV radiation and widen the range of UV radiation that is blocked [14]. Applicability of these particles in efficient UV screening formulations is an outcome of their nanoscale dimensions and is not observed in their bulk state forms (Fig. 3.4).

Disadvantage of TiO_2 and ZnO in the bulk form is that they create a visible opaque layer on the skin. This visual effect can be avoided by decreasing the particle size of metal oxides to nanoscale. Hence micron-sized TiO_2 and ZnO have been replaced by TiO_2 and ZnO nanoparticles (radius $<$100 nm) in UV filters and sunscreens. These phenomena can be explained through the interaction of radiation with solid materials. Attenuation of incident radiation on solid particles occurs through reflection, scattering, and absorption processes. The intensity of transmitted light in such processes is given by equation [15]:

$$I_t = I_0 - (I_r + I_s + I_a)$$

In which I_t is the intensity of transmitted radiation, I_0 is the intensity of the incident radiation; I_r, I_s, and I_a are the intensities of the reflected, scattered, and absorbed radiation, respectively. In case of particles with submicron sizes, the effect of specular reflection is negligible and the attenuation of the incident radiation occurs mainly due to scattering and absorbance. Scattering of radiation is mainly dependent on the particle size, and absorption of radiation is related to the chemical composition of the material. When the dimensions of the particles are an order of magnitude smaller than the wavelength of the interacting radiation, the scattering follows Rayleigh scattering. Whereas when the dimension of the particle is of the same order as the radiation wavelength, Mic scattering prevails [15, 16]. According to Mie's equation, the intensity of scattered radiation relates to several factors as follows:

$$I_s \propto \frac{Nd^6}{\lambda^4} \left[\frac{m^2 - 1}{m^2 + 2}\right] I_0$$

where N is the number of particles, d is the diameter of the particle, λ is the wavelength of the radiation used, and m is the relative refractive index, which is defined as the ratio of refractive index of the particle to that of the medium in which the particle is present. In general, it is desirable to have more scattering of the UV light and less scattering of the visible light. Scattering of visible light causes a whitening effect, which needs to be avoided. It is clear from the equation that the scattering depends on the particle size and the scattering of the light can be minimized by decreasing the particle size. An optimal particle size should be reached because reducing the particle beyond a certain limit would enhance UV absorption. For sunscreen applications, TiO_2 and ZnO nanoparticles with size less than 100 nm are generally used in this regard. Also, Mie's equation provides information about how the scattering can be controlled by modulating the refractive index. Intrinsic refractive indices of TiO_2 and ZnO are fixed quantities, but the refractive index of the medium can be customized to deliver optimal results.

Absorption of UV light by the nanoparticles can also contribute to the attenuation of radiation. Rutile phase of TiO_2 has an absorption peak around 420 nm (band gap: 3.06 eV), anatase phase of TiO_2 has a peak around 390 nm (band gap: 3.20 eV), and ZnO shows an absorption maximum at 380 nm (band gap: 3.35 eV). In principle, particle size cannot influence the absorption of single photons. Nevertheless, when the particle size is small, each photon would interact with a greater number of particles thereby increasing absorption.

Suitable size of nanoparticles in the formulation of UV filters is selected based on the aforementioned theoretical considerations. The as-synthesized nanoparticles have very high surface-to-volume ratios due to their small sizes, which result in high surface energy leading to aggregation. Binding forces such as van der Waals interactions, electrostatic interactions, hydrogen bonding, and water bridging hold the particles together in their aggregates. These aggregates can group together to form larger agglomerates also, which generally are not as tightly bound as the aggregates. In order to disperse the larger agglomerates, processes such as milling are utilized. In practical terms, measure of the size of the primary nanoparticle (as-synthesized particles) is less informative because of the high propensity of them to get aggregated. The photophysical performance of the nanoparticle system is thus dependent on the size distribution of the aggregates.

In the case of TiO_2 nanoparticles, attenuation of UVB (~ 300 nm) radiation is mainly due to absorption and this increases as the particle size decreases [17, 18]. Attenuation of UVA (~ 350 nm) in this case is mainly due to the scattering of the radiation [17, 18]. In order to optimize this performance, particle size should be tuned such that there is maximum scattering while avoiding the whitening effect. In case of ZnO, the dispersions of nanoparticles are more transparent in comparison with those of TiO_2 even for bigger particle sizes. Smaller particle sizes result in lesser quantity of UV radiation transmitted. The capability to withstand the harmful effect of UV radiation increases with decrease in particle size.

Sun protection factor (SPF) is a universally accepted parameter to indicate the performance of sunscreens. SPF is defined as the ratio of the length of time the skin can be exposed to UV (mainly UVB) radiation with the sunscreen divided by the time for same amount of radiation without the sunscreen. Efficiency of a sunscreen to block UVA radiation is indicated by another parameter called protection factor of UVA (PFA). This is measured from an experimental observation called minimal persistent pigment darkening dose (MPPD), which is obtained by observing color change of skins with definite dose of UVA radiation. PFA is the ratio of MPPD obtained with protection to that obtained without protection of the sunscreen. Research on cosmetic formulations composed on inorganic nanoparticles aims to obtain the best SPF and PFA values with least compromise in the aesthetic and safety aspects.

Top-down and *bottom-up* methods can be used for the synthesis of TiO_2 and ZnO nanoparticles. However, *bottom-up* chemical synthesis through solution phase synthesis is the most commonly used method for the synthesis of these nanomaterials. TiO_2 nanoparticles can be obtained through sol-gel synthetic route in which titanium isopropoxide is hydrolyzed in the presence of water and alcohol. $TiCl_4$ also can be used as a titanium precursor for the synthesis of TiO_2. Green chemistry and hydrothermal methods also could be used for the synthesis of TiO_2.

One of the major concerns regarding the usage of TiO_2 and ZnO is their ability to generate free radicals upon exposure to radiation [19]. Absorption of UV light by both these compounds can lead to excitation of electrons from valence band to conduction band, which will subsequently lead to electron-hole pairs. These electrons and holes will react with dissolved oxygen, surface hydroxyl groups, and water to form superoxide radicals (O_2^-) and hydroxide radicals (OH$^-$). In the absence of oxygen, Ti^{4+} ions can be reduced to Ti^{3+} ions by the electrons and this can cause color change and graying. The UV-induced radical species can cause cell damage, photocarcinogenesis, and skin aging [20]. In order to mitigate this, various types of coatings are applied over the TiO_2 and ZnO nanoparticle surfaces [21]. These coatings impart properties of photostability, compatibility in the formulation, and mechanical properties suitable for the application. Inorganic coatings such as alumina, silica, and silicates; organosilicon compounds such as dimethicone (polydimethyl siloxane, PDMS), and silanes; and organic compounds like stearic acid are used as coatings for TiO_2 and ZnO nanoparticles used in sunscreens.

Combination of inorganic solar filters with organic solar filters such as avobenzone (methoxydibenzoylmethane) can perform effectively than the individual components [22]. But this is not allowed in certain markets due to concerns about the possibility of metal ions reacting with avobenzone in the presence of light. This possibility can be avoided by using coated metal oxide nanoparticles. Composites with nano-sized TiO_2 particles dispersed on microsized ZnO particles were recently reported to have impressive SPF value due to the syncrgism of the components present [23].

Inorganic nanomaterials other than TiO_2 and ZnO also were studied for use as UV filters. It was shown, for example, that CeO_2 nanoparticles can offer UV blocking properties comparable with that of TiO_2 and it does not produce free radicals like TiO_2 does in presence of UV light [24, 25]. Calcium-doped CeO_2 was reported to show better screening of short UVA radiations [26]. Cerium phosphates $(CePO_4)$ [27] and cerium-titanium pyrophosphates $(Ce_{1-x} Ti_x P_2O_7; x = 0 - 1)$ [28] are also studied as potential substitutes for TiO_2 and ZnO nanoparticles in sunscreen applications.

3.4.1.2 Inorganic Nanoparticles as Antimicrobial Agents

Silver nanoparticles (Ag NPs) are probably the most widely used synthetic nanoscale material in practical applications and commercial products. Silver has antibacterial and antiseptic properties which help to empower the immune system in human beings. Ag NPs can cause inhibition of cell growth and cell death in microbial organisms which are harmful to health [29]. There have been reports showing that Ag NPs are more effective than silver ions in antibacterial activity [30, 31]. There are evidences in the literature showing Ag NPs-induced inhibition of growth of some of the bacteria strains which are resistant to silver ions [32]. Ag NPs have been shown as effective inhibitors of various microorganisms such as *S. aureus* MRSA, multidrug-resistant *Pseudomonas aeruginosa*, ampicillin-resistant *E. coli* O157:H7, and erythromycin-resistant *Streptococcus pyogenes,* and they were found effective against fungi species like *Candida albicans*, *Phoma glomerata*, and *Trichoderma* sp as well [30, 33].

The mechanism of antimicrobial activity of Ag NPs is not completely understood. Nevertheless, there are two prominent approaches toward the explanation of this potential of Ag NPs. One of them proposes that the antimicrobial activity is due to the interaction of nanoparticles with the cell membrane while the other approach proposes the nanoparticles are entering the cells of the microorganism and the antimicrobial action starts from there [32]. The first theory explains that the Ag NPs will attach to the cell membrane of the microorganism through electrostatic and other interactions. Subsequently, events such as bonding with amino acids having thiol group, inhibition of enzyme activity, affecting the movement of electrons in respiratory process, and interaction with DNA and RNA occur, which will inhibit the survival and growth of microorganism [30]. Different interactions of Ag NPs with bacterial cells are schematically represented in Fig. 3.5 [34]. Electron microscopy analysis has been used to study the interaction of Ag NPs with bacterial

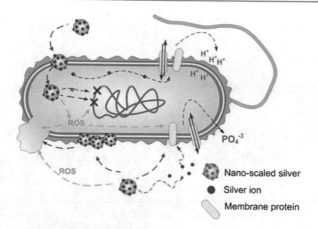

Fig. 3.5 Schematic illustration of the possible interactions of nano-scaled silver with bacterial cells. Nano-scaled silver may (1) release silver ions and generate reactive oxygen species (ROS); (2) interact with membrane proteins affecting their correct function; (3) accumulate in the cell membrane affecting membrane permeability; and (4) enter into the cell where it can generate ROS, release silver ions, and affect DNA. Generated ROS may also affect DNA, cell membrane, and membrane proteins, and silver ion release will likely affect DNA and membrane proteins. Reprinted with permission from [34]. Copyright (2010) Springer Science + Business Media B.V

cells such as *E. coli* (Fig. 3.6) [35]. Ag NPs have no effect on mammalian cells because of the difference in the structure of cell walls, thereby enabling the safe usage in cosmetic products.

Because of their potential as effective antimicrobial substance, Ag NPs are being used in various cosmetic formulations (Table 3.1). They are used in aftershave products to protect hair follicles from bacterial infection. Due to their ability to control sebaceous glands, they find use in matt creams. They can reduce *Propionibcaterium acnes,* thereby finding place in acne treatment formulations. Also, they are being used in products such as shampoos, antiperspirants, and deodorants, owing to their antibacterial activity. Ag NPs are used as a colorant substance in cosmetic products also. Apart from Ag NPs, inorganic nanoparticles such as gold nanoparticles and metal oxide nanoparticles are used in cosmetic products for antimicrobial applications.

3.4.2 Inorganic Nanoparticles as Nanocarriers

Nanocarriers are materials or assemblies in the nanosize dimensions which can facilitate the delivery of a drug or active substance to a targeted location in the body. These nanomaterials usually do not possess any biological activity and will act merely as a vehicle in the formulations. Nanoscale carriers are used in cosmetic products to deliver UV filters, antioxidants, vitamins, anti-acne substances, anti-aging substance, etc. to the target layer of the skin. They improve the

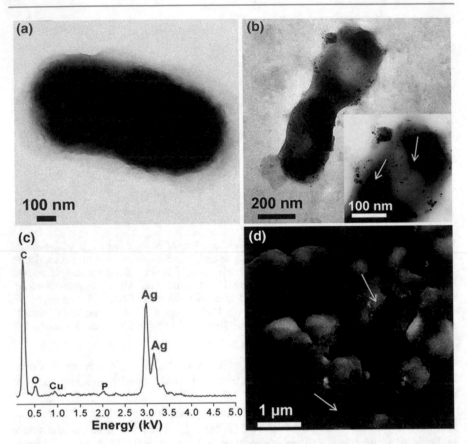

Fig. 3.6 TEM images of E. coli cells (**a**) untreated and (**b**) after treatment with silver nanoparticles. Inset shows the magnified image with the presence of silver nanoparticles, as shown by arrows. EDX analysis (**c**) of the bacteria shows the presence of silver. The SEM image (**d**) further confirms the presence of silver all over the bacterial surface. Reprinted with permission from [35]. Copyright (2014) The Royal Society of Chemistry

permeability, stability, and controlled delivery of the active substance. Nanoclays and mesoporous silica can deliver cosmetically active ingredients encapsulated in them through diffusion or erosion upon application to the skin [4].

Mesoporous silica is a porous form of silica consisting of hexagonal array of nano-sized pores. These pores or empty channels can accommodate biologically active molecules and can release them in controlled ways. Due to this capability, they are considered as good candidates for drug delivery applications [36]. Research works have shown that the usage of mesoporous silica in cosmetic products can improve the performance of the formulation. In a recent example from the literature, mesoporous silica was used to encapsulate the organic sunscreen material octal methoxycinnamate (MCX) and the resulting formulation showed 57% better UV protection than the free MCX [37]. Investigations on the encapsulation of other organic sunscreens like benzophenone-3 (BZP) with mesoporous silica also showed promising results [38].

3.4.3 Inorganic Nanoparticles as Formulation Aids in Cosmetics

Chemically inert inorganic nanoparticles are suitable as inert additives in various commercial products for specific applications [1]. Cosmetic products also use inorganic nanomaterials as formulation aids which can improve the rheology and optical properties of the cosmetic formulation [4]. Exclusive properties of the nanomaterials which are different from their bulk state counterparts are made use in these applications.

Nano-sized inorganic particles such as silica and clay are used as thickeners in cosmetic formulations [39, 40]. The thickening action of these particles results from their ability to form a percolating network of particles. Inorganic-organic hybrid nanomaterials also are used for these types of applications in cosmetic formulations. For instance, SiO_2 nanoparticles grafted to a perfluorocarbon hydrophobically modified alkali-soluble emulsion (HASE) polymer facilitate the formation of a nanostructured polymeric network barrier when applied to skin, which protects the skin from the toxicity of organophosphates [41].

3.5 Summary and Outlook

Owing to their unique characteristics different from their bulk state or molecular form counterparts, inorganic nanoparticles are appropriate for several applications related to cosmetic products. Commercial products in various categories containing these materials are already in the market, and they are showing improved performances compared to similar products which do not use nanoparticle components. One of the most prominent uses of inorganic nanoparticles is their application in sunscreens and UV filters. Active research is going on in this area for finding new materials and composition with improved performance and safety. Concerns about the health hazards and environmental impact are addressed in these studies. Inorganic nanomaterials are used as antimicrobial components in cosmetic formulations, and the use of nanoparticles instead of elemental or ionic state of the same material has been proved to be more efficient. Several physical and chemical methods can be used for the synthesis of inorganic nanomaterials relevant to cosmetic applications. Search for better methods for the synthesis of nanomaterials is also going on in related academic disciplines. Characterization of nanomaterials with current day techniques gives a complete picture about the structure and composition of the material, which enables the detailed understanding and prediction of their properties and functionality.

Recent trends in the scientific publications and patent applications related to the use of inorganic nanoparticles in cosmetics show that the use of inorganic nanoparticles in these products will expand and effectuate further in the forthcoming years.

References

1. Stark WJ, Stoessel PR, Wohlleben W, Hafner A. Industrial applications of nanoparticles. Chem Soc Rev. 2015;44(16):5793–805.
2. Vance ME, Kuiken T, Vejerano EP, McGinnis SP, Hochella MF Jr, Hull DR. Nanotechnology in the real world: redeveloping the nanomaterial consumer products inventory. Beilstein J Nanotechnol. 2015;6(1):1769–80.
3. Heiligtag FJ, Niederberger M. The fascinating world of nanoparticle research. Mater Today. 2013;16(7–8):262–71.
4. Mihranyan A, Ferraz N, Strømme M. Current status and future prospects of nanotechnology in cosmetics. Prog Mater Sci. 2012;57(5):875–910.
5. Catalogue of nanomaterials used in cosmetic products placed on the market. http://ec.europa.eu/docsroom/documents/24521. Accessed 03 Nov 2017.
6. The Project on Emerging Nanotechnologies. http://www.nanotechproject.org/cpi/browse/categories/health-and-fitness/cosmetics/. Accessed 03 Nov 2017.
7. Burda C, Chen X, Narayanan R, El-Sayed MA. Chemistry and properties of nanocrystals of different shapes. Chem Rev. 2005;105(4):1025–102.
8. Albanese A, Tang PS, Chan WCW. The effect of nanoparticle size, shape, and surface chemistry on biological systems. Annu Rev Biomed Eng. 2012;14:1–16.
9. Walter P, Welcomme E, Hallégot P, Zaluzec NJ, Deeb C, Castaing J, Veyssière P, Bréniaux R, Lévêque JL, Tsoucaris G. Early use of PbS nanotechnology for an ancient hair dyeing formula. Nano Lett. 2006;6(10):2215–9.
10. Epstein HA, Kielbassa A. Nanotechnology in cosmetic products. In: Bagchi, D, editor. Bio-nanotechnology: a revolution in food, biomedical and health sciences. United States: John Wiley & Sons; 2013; vol. 1, pp 414–23.
11. Singh TG, Sharma N. Nanobiomaterials in cosmetics: current status and future prospects A2 —Grumezescu, Alexandru Mihai. In: Nanobiomaterials in galenic formulations and cosmetics, William Andrew Publishing; 2016; pp 149–74.
12. Petrazzuoli M. Advances in sunscreens. Curr Probl Dermatol. 2000;12(6):287–90.
13. Levy SB. UV filters. In: Handbook of cosmetic science and technology, 4th ed. London: CRC Press; 2014.
14. Dransfield GP. Inorganic sunscreens. Radiat Prot Dosim. 2000;91(1–3):271–3.
15. Schlossman D, Shao Y. Inorganic ultraviolet filters. in sunscreens: regulations and commercial development. London: CRC Press; 2005.
16. Horvath H. Gustav Mie and the scattering and absorption of light by particles: historic developments and basics. J Quant Spectrosc Radiat Transfer. 2009;110(11):787–99.
17. Sakamoto M, Okuda H, Futamata H, Sakai A, Iida M. Influence of particle size of titanium dioxide on UV-ray shielding property. J Jpn Soc Colour Mater. 1995;68(4):203–10.
18. Stamatakis P, Palmer BR, Salzman GC, Bohren CF, Allen TB. Optimum particle size of titanium dioxide and zinc oxide for attenuation of ultraviolet radiation. J Coatings Technol. 1990;62(789):95–8.
19. Serpone N, Dondi D, Albini A. Inorganic and organic UV filters: their role and efficacy in sunscreens and suncare products. Inorg Chim Acta. 2007;360(3):794–802.
20. Shen C, Turney TW, Piva TJ, Feltis BN, Wright PFA. Comparison of UVA-induced ROS and sunscreen nanoparticle-generated ROS in human immune cells. Photochem Photobiol Sci. 2014;13(5):781–8.

21. González S, Fernández-Lorente M, Gilaberte-Calzada Y. The latest on skin photoprotection. Clin Dermatol. 2008;26(6):614–26.
22. Couteau C, Chammas R, Alami-El Boury S, Choquenet B, Paparis E, Coiffard LJM. Combination of UVA-filters and UVB-filters or inorganic UV filters—influence on the sun protection factor (SPF) and the PF-UVA determined by in vitro method. J Dermatol Sci. 2008;50(2):159–61.
23. Jiménez Reinosa J, Leret P, Álvarez-Docio CM, del Campo A, Fernández JF. Enhancement of UV absorption behavior in ZnO–TiO2 composites. Boletín de la Sociedad Española de Cerámica y Vidrio. 2016;55(2):55–62.
24. Yabe S, Sato T. Cerium oxide for sunscreen cosmetics. J Solid State Chem. 2003;171(1–2): 7–11.
25. Herrling T, Seifert M, Jung K. Cerium dioxide: Future UV-filter in sunscreen? SOFW J. 2013;139(5):10–4.
26. Truffault L, Winton B, Choquenet B, Andreazza C, Simmonard C, Devers T, Konstantinov K, Couteau C, Coiffard LJM. Cerium oxide based particles as possible alternative to ZnO in sunscreens: effect of the synthesis method on the photoprotection results. Mater Lett. 2012;68:357–60.
27. De Lima JF, Serra OA. Cerium phosphate nanoparticles with low photocatalytic activity for UV light absorption application in photoprotection. Dyes Pigm. 2013;97(2):291–6.
28. Wu W, Fan Y, Wu X, Liao S, Huang X, Li X. Preparation of nano-sized cerium and titanium pyrophosphates via solid-state reaction at room temperature. Rare Met. 2009;28(1):33–8.
29. Eckhardt S, Brunetto PS, Gagnon J, Priebe M, Giese B, Fromm KM. Nanobio silver: its interactions with peptides and bacteria, and its uses in medicine. Chem Rev. 2013;113 (7):4708–54.
30. Kedziora A, Gorzelańczyk K, Bugla-Płoskońska G. Positive and negative aspects of silver nanoparticles usage. Biol Int. 2013;53:67–76.
31. Guzman M, Dille J, Godet S. Synthesis and antibacterial activity of silver nanoparticles against gram-positive and gram-negative bacteria. Nanomed Nanotechnol Biol Med. 2012;8 (1):37–45.
32. Kapuccinska, A.; Nowak, I., Silver nanoparticles as a challenge for modern cosmetology and pharmacology. In: Nanobiomaterials in galenic formulations and cosmetics: applications of nanobiomaterials; 2016; pp 391–417.
33. Lara HH, Ayala-Núñez NV, del Turrent LCI, Padilla CR. Bactericidal effect of silver nanoparticles against multidrug-resistant bacteria. World J Microbiol Biotechnol. 2010;26 (4):615–21.
34. Marambio-Jones C, Hoek EMV. A review of the antibacterial effects of silver nanomaterials and potential implications for human health and the environment. J Nanopart Res. 2010;12 (5):1531–51.
35. Agnihotri S, Mukherji S, Mukherji S. Size-controlled silver nanoparticles synthesized over the range 5–100 nm using the same protocol and their antibacterial efficacy. RSC Adv. 2014;4 (8):3974–83.
36. Li Z, Barnes JC, Bosoy A, Stoddart JF, Zink JI. Mesoporous silica nanoparticles in biomedical applications. Chem Soc Rev. 2012;41(7):2590–605.
37. Chen-Yang YW, Chen YT, Li CC, Yu HC, Chuang YC, Su JH, Lin YT. Preparation of UV-filter encapsulated mesoporous silica with high sunscreen ability. Mater Lett. 2011;65 (6):1060–2.
38. Li CC, Chen YT, Lin YT, Sie SF, Chen-Yang YW. Mesoporous silica aerogel as a drug carrier for the enhancement of the sunscreen ability of benzophenone-3. Colloids Surf B Biointerfaces. 2014;115(Supplement C):191–6.
39. Bolzinger MA, Briançon S, Chevalier Y. Nanoparticles through the skin: managing conflicting results of inorganic and organic particles in cosmetics and pharmaceutics. Wiley Interdisc Rev Nanomed Nanobiotechnol. 2011;3(5):463–78.

40. Choy J-H, Choi S-J, Oh J-M, Park T. Clay minerals and layered double hydroxides for novel biological applications. Appl Clay Sci. 2007;36(1):122–32.
41. Cécile B, Sonia A, Frédéric G. Silica- and perfluoro-based nanoparticular polymeric network for the skin protection against organophosphates. Mater Res Express. 2016;3(6):065019.

Micelles and Nanoemulsions

4

Yves Chevalier and Marie Alexandrine Bolzinger

Abstract

Nanoemulsions and block copolymer micelles have specific features that are make them quite attractive for their application in cosmetic products. Both of them are organic nanoparticles having a liquid core. They are made from the same ingredients as classical emulsions, and their overall organization is also the same. Only the droplet size is smaller by a factor of 10 to 100. The chapter first gives a presentation of the physical chemistry of nanoemulsions and block copolymer micelles, then a discussion of the consequences of the small size to properties focusing on skin absorption. The sub-micron size brings about new properties such as accelerated skin delivery of active substances, absence of creaming in fluid products, immediate skin occlusion, transparency, and gloss after spreading. On the one hand, they are used as drug carriers for topical administration. On the other hand, fast occlusion of skin provides an immediate feeling which is a benefit in moisturizing products. Such properties are discussed with some help of studies carried out for pharmaceutical applications. Their current and prospective implementations in cosmetic technologies are addressed. Finally, the possible safety issues related to the size in the nano-scale are discussed on technical grounds.

Keywords

Nanoemulsions · Block copolymer micelles · Spontaneous emulsification · Phase inversion · Skin absorption · Topical drug delivery · Nanotoxicology · Formulation · Cosmetic cream · Cosmetic lotion

Y. Chevalier (✉) · M.-A. Bolzinger
Laboratoire d'Automatique, de Génie des Procédés et de Génie Pharmaceutique, LAGEPP, University Lyon 1, CNRS UMR 5007, 43 bd 11 Novembre, 69622 Villeurbanne Cedex, France
e-mail: yves.chevalier@univ-lyon1.fr

© Springer Nature Switzerland AG 2019
J. Cornier et al. (eds.), *Nanocosmetics*,
https://doi.org/10.1007/978-3-030-16573-4_4

4.1   Introduction

Nanoemulsions and block copolymer micelles are organic nanoparticles having a
liquid core. They are primarily considered as drug carriers for topical administration. Other interesting properties also bring about definite benefits regarding
applications to cosmetic science. Specific properties are coming from their size
smaller than a micrometer and the large interfacial area that is a geometrical consequence of the small size. The present chapter addresses such dispersed systems
that are either already used in cosmetic formulation or considered quite promising
to that end use.

The chapter first gives a presentation of the physical chemistry of nanoemulsions
and block copolymer micelles, then a short discussion of the consequences of the
small size to properties requirements of cosmetic products focusing on skin
absorption, and finally their current or prospective implementation in cosmetic
technologies.

4.2   Micelles and Nanoemulsions in the World of Nanoparticles

A large variety of nanoparticles is used in cosmetic products [1, 2]. The consumer
first thinks of inorganic nanoparticles used in sunscreens and possibly in makeup
products. In this chapter, we will address organic nanoparticles. Nanoemulsions
being simply emulsions with droplet sizes smaller than one micrometer [3], they
belong to the world of emulsions (Fig. 4.1). Their main feature is the liquid state of
their droplets. Solid nanoparticles made of crystalline fats are called solid lipid
nanoparticles (SLN). However, there are emulsions containing waxes that contain
part of their core as solid crystalline waxes, and they are still referred to as
emulsions. Polymer nanoparticles are also considered as another different class of
nanoparticles. But there are emulsions where oil is made of molten polymer; as
example, silicone oil is a polymer. The distinction between nanoemulsion, SLN and
polymer nanoparticles is rather a gray area.

Block copolymer micelles are made of a core containing the hydrophobic block
surrounded by a shell made of the hydrophilic block swollen by water [4, 5]
(Fig. 4.1). They correspond to nanoemulsions when their core is in liquid state, but
some of them have their core in solid glassy state.

Most nanoemulsions used so far are of the oil-in-water (o/w) type. Considering
the liquid state of nanodroplets core as a main feature, the specific property of
nanoemulsions is their high loading capacity for hydrophobic active ingredients
(even oil can be the pure active substance itself). It provides an easier transport to a
biological target thanks to the small droplet sizes.

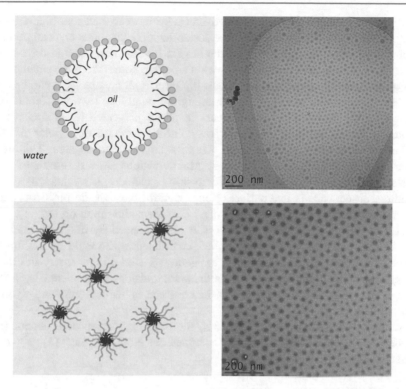

Fig. 4.1 Sketches of an o/w nanoemulsion droplet stabilized by emulsifier molecules (top left) and a block copolymer micelle (bottom left). Hydrophobic materials are in red, and hydrophilic materials are in green. The size of emulsifier molecules is enlarged for they can be distinguished at the surface of the oil nanodroplets; the scales are ~50–100 nm for nanoemulsion oil droplets and ~2 nm for emulsifier molecules. The typical size of block copolymer micelles is 30–50 nm. CryoTEM pictures of nanoemulsion (top right) and block copolymer micelles (bottom right) are given on the right-hand side

4.3 Nanoemulsions Versus Emulsions

Emulsions of o/w type are made of oil droplets dispersed in a continuous water phase. Nanoemulsions are not different from emulsions with that regard. The droplet size range of nanoemulsions is between 100 nm and 1 μm; their size is seldom below 50 nm because emulsification processes do not allow for smaller sizes. The droplet size of classical emulsions is larger than 1 μm.

An important feature of nanoemulsions is their slow creaming rate. Upon creaming, oil droplets migrate upward because of buoyancy and finally accumulate on top of the product as a layer of concentrated emulsion. Stokes law teaches that creaming rate increases as the square of droplet diameter, so that decreasing droplet size from 1 μm to 100 nm slows down creaming by a factor of 100. Thermal agitation by Brownian motion is also much more powerful for droplets of small sizes. In most instances, combination of these two effects keeps nanoemulsions homogeneous all along the lifetime of the product. One destabilization phenomenon is migration of oil molecules from small droplets to the largest by diffusion through the aqueous phase. This phenomenon known as Ostwald ripening comes from the small droplet size of nanoemulsion droplets that makes the interfacial area large. It can be a fast process, especially when oil has non-negligible solubility in water (polar oils). Destabilization by Ostwald ripening can be partly prevented by mixing a small amount of a supplementary oil of very low solubility with the main polar oil [3].

Another consequence of the small size is a large interfacial area. This makes materials transfer phenomena quite fast. Manifestations are fast release of active substances from oil droplets, fast transfer to skin as nanodroplets contact its surface. Another consequence is that surface groups control most of the physicochemical properties: stability, rheological properties, film formation upon drying, etc.

Their recognition by the immune system is enhanced by the large interfacial area. Unless the immune system is targeted, an efficient hydrophilic layer acting as a protective shell against recognition as exogenous material is necessary in order to prevent capture by macrophages. Stealth nanoemulsions are covered by a thick enough layer of hydrophilic polymer at their surface, either poly(ethylene glycol) or polysaccharide [6].

Nanoemulsions can be manufactured by high-power mechanical dispersion or by physicochemical processes known as "spontaneous emulsification" [7] as summarized in Table 4.1.

Table 4.1 Summary of the main manufacture processes for preparation of nanoemulsions

Mixing technique	Process	Benefits	Constraints
High-power mixing	High-pressure homogenizer Microfluidizer Ultrasounds	Technology is available for large productions, possibly under continuous process	High-energy input Dissipation of much heat
Medium-power mixing with an impeller	Catastrophic phase inversion	Can easily be implemented into classical batch wise emulsification plants	Phase inversion takes place through a highly viscous intermediate double emulsion
Gentle stirring while cooling	Thermal phase inversion (PIT emulsions)	Emulsification requires mild stirring, so it looks spontaneous	Select emulsifiers for which phase behavior against temperature is favorable
Gentle stirring as a "self-emulsifying preconcentrate" is diluted with water	Concentration driven phase inversion (PIC emulsions)	Emulsification looks spontaneous and can be done by the customer	Select emulsifiers for which the isothermal ternary phase diagram (oil/water/emulsifier) is favorable Large concentration of emulsifier is required

More details are given in the body of the text

Achieving sub-micron droplet sizes by mechanical means require high-power emulsification processes using high-pressure homogenizer or microfluidizer [8, 9], or ultrasounds.

Processes relying on "spontaneous emulsification" phenomenon do not require high-energy input. They consist in shifting a monophasic composition into a diphasic region of the phase diagram. As entering the two-phase region, the new phase appears as droplets of nanometric size that do not grow if coagulation is prevented. The two-phase region can be entered by means of lowering the temperature or by dilution with water. This is a matter of selecting ingredients such that phase behavior is favorable.

The thermal "phase inversion emulsification" is one of the various "spontaneous emulsification" processes. This is more precisely called "transitional" phase inversion emulsification in order to specify its distinctive feature with respect to "catastrophic" phase inversion emulsification (Fig. 4.2). This process relies on the temperature sensitivity of nonionic surfactants: They are soluble in water at low temperature, yielding o/w emulsions, and soluble in oil at high temperature above the "Phase Inversion Temperature" (PIT), yielding w/o emulsions. The process by phase inversion consists in preparing a w/o emulsion above the PIT and cooling it down below the PIT such that the w/o emulsion undergoes phase inversion into o/w. There is no need for strong high-shear mixing during the cooling process; just gentle stirring required for temperature is homogeneous throughout the whole

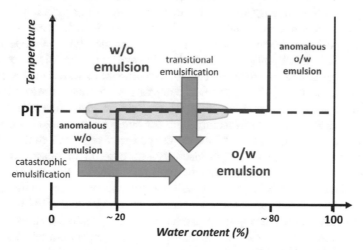

Fig. 4.2 Pathways followed through phase diagrams in emulsification processes relying on phase inversion phenomena. Horizontal arrow: Catastrophic phase inversion to an o/w emulsion is performed under strong stirring at constant temperature below the phase inversion temperature by means of progressive addition of water into an "anomalous" w/o emulsion. Phase inversion to the normal o/w emulsion takes place through intermediate double o/w/o emulsion that finally breaks into an o/w emulsion. Vertical arrow: Transitional phase inversion to an o/w emulsion is performed by cooling a w/o emulsion to temperature below the PIT. Formation of an o/w emulsion is successful when a monophasic area (microemulsion represented in yellow) is passed through as the temperature is decreased

product. Nanoemulsions (PIT emulsions) form in favorable cases when a monophasic region is passed through as temperature is lowered [10, 11]. Successful formation of nanoemulsion by thermal phase inversion requires a temperature-composition phase diagram where there is a monophasic region at the PIT that separates the two two-phase regions at higher and lower temperatures. Monophasic compositions containing oil, water and surfactant are microemulsions; they often contain quite high concentrations of surfactant. The concentrations of emulsifier in the final emulsion are then higher than those usually required for classical emulsification by mechanical stirring.

Spontaneous emulsification by dilution has been mainly studied in details in the framework of pharmaceutical applications [12]. The process starts with a monophasic self-emulsifying preconcentrate that contains oil and a suitable emulsifier mixture (and possibly a small amount of water). Dilution with water shifts the chemical composition into a two-phase region of the phase diagram such that the new phase appearing as droplets is made of oil [13, 14]. The resulting o/w PIC emulsion can be a nanoemulsion when droplet growth is limited by using a large emulsifier/oil ratio [15]. This often requires dilution with a large amount of water, which is a limitation since the concentration of the final emulsion is low [16]. Droplet sizes can reach diameters significantly lower than 100 nm.

"Catastrophic" phase inversion emulsification is another process yielding nanoemulsions. The physicochemical process of phase inversion and strong mechanical shearing are mixed together in that process. O/w emulsions (taken as example) form with hydrophilic emulsifier according to Bancroft rule. Although this is an empirical rule that a rationale did not receive definite consensus yet, the emulsion type is governed by the hydrophilic character (HLB number) of the emulsifier better than by the oil/water ratio. However, w/o emulsions can be prepared against Bancroft rule when dispersing low amount of water into an oil phase in presence of a hydrophilic emulsifier. Progressive addition of water into such "anomalous" w/o emulsions under strong stirring causes phase inversion when the amount of water is enough to accommodate for being the continuous phase of an o/w emulsion. Phase inversion into o/w occurs progressively through the formation of o/w/o double emulsions of high viscosity. As a result of phase inversion at microscopic scale, nanoemulsions of o/w type form. Such emulsions can be concentrated since catastrophic phase inversion proceeds for water amounts around 20% and can reach completion for water volume fraction of 40% (60% oil as a final dispersed phase). This looks quite an attractive process; there are cases of failure, however, where destabilization and macroscopic phase separation of oil and water happen instead of phase inversion.

Some authors confuse nanoemulsions and microemulsions. Indeed, the definitions are not that clear for applying them to formulation practice and the terminology is misleading [17]. Specifically, nanoemulsions are emulsions where the droplet size is smaller than 1 μm; they are made of two phases (dispersed and dispersing/continuous), so that their chemical composition is in a two-phase region of the equilibrium phase diagram. Microemulsions are solutions where some components are associated as droplets; they are made of one single phase, so that their composition is in a one-phase region of the equilibrium phase diagram. Both of them can be transparent or translucent, and they can have similar droplet sizes. The way to assess the difference between them consists in looking at whether they are at thermodynamic equilibrium. Microemulsions are at thermodynamic equilibrium, which means that they form spontaneously by simple mixing the ingredients and their formation does not depend on the way it has been prepared (strong or gentle stirring, order of addition of ingredients). Nanoemulsions are out of thermodynamic equilibrium, which means that their formation requires implementation of an emulsification process, most often strong stirring. Discrimination might reveal difficult in the case of "spontaneous emulsification" that yields nanoemulsions by changing the temperature or by dilution under gentle stirring. The spontaneous emulsification process requires following a specific pathway in the phase diagram for being successful, and the final droplet size depends on the detailed parameters of the process. On the contrary, microemulsions can be formed whatever the way ingredients are mixed and the final droplet sizes are the same whatever the mixing process.

Nanoemulsions are often referred to as "sub-micron emulsion" because the wording "nanoemulsion" that includes "nano" might suggest that they could be harmful owing to their small size. The suspicion regarding nanomaterials is not supported by clear scientific foundation. Indeed, their small size and large external area make penetration through biological barriers and materials exchanges possible and/or faster, so that any effects of a hazardous material are enhanced by the nano-ranged size. However, nanoparticles made of safe materials cannot be toxic. Nanoemulsions for cosmetic applications are manufactured from safe ingredients as it is the case for all other cosmetic products. Emulsions being essentially liquid, they cannot penetrate skin as intact particles [18], so that their size does not matter with that regard and only their chemical composition should be deserving attention. Nevertheless, use of the term "nano" is mostly avoided by cosmetic companies. The European regulation No 1223/2009 includes a specific article (Article 16) regarding solid nanomaterials specifying that they must be declared together with their detailed characterization and labeled in the list of ingredients with the word "nano" in brackets following the name of the substance. Precise definition of the term "nano" and standardization of measurement methods are still matters of dispute between scientific experts and lawyers.

4.4 Block Copolymer "Micelles"

Amphiphilic block copolymers are made of hydrophilic and hydrophobic blocks; their diblock structure is quite similar to that of surfactants. They share properties with classical surfactants, but there are also important differences. They self-associate in water in the same way as classical surfactants as micelles [4, 5], bilayers (liposomes) [19] or reverse micelles, depending on the balance between their hydrophilic and hydrophobic parts. Most applications deal with micelles of hydrophilic block copolymers. The internal structure of micelles is core-shell; it consists in a central core containing the hydrophobic blocks surrounded by a shell made of the hydrophilic blocks swollen by water [20]. The typical size of block copolymer micelles is 30–60 nm. Such size measured by dynamic light scattering includes the thickness of the water-swollen hydrophilic shell; the size of the hydrophobic core is much smaller. The major difference with classical water-soluble surfactants is that kinetics of self-association and dissociation of micelles is very slow. Even, some block copolymers are not soluble in water, so that their self-association process is not spontaneous. As consequence, a specific kind of an emulsification process is required for the preparation of micelles, and their final structure depends on the specificities of the process. In that regard, block copolymer micelles are closer to polymer nanoparticles than classical surfactant-based micelles. The properties of surfactant micelles are characterized by dynamic equilibrium between free and micellized surfactant molecules; surfactant adsorption to surfaces (oil droplets, skin surface) can operate through adsorption of free molecules and re-equilibration with micelles in solution. Such fast dynamic processes are not allowed with block copolymer micelles [21]. The critical micelle concentration (*CMC*) of block copolymers is very low [5], unless the hydrophobic block is very short [22]. Even the *CMC* may be difficult to measure with confidence. The term "block copolymer micelles" is misleading; the term "block copolymer nanoparticles" should be preferred.

The most popular block copolymers used so far for their application to life sciences are poly(ester)-*block*-poly(ethylene oxide) because the polyester part is biocompatible and biodegradable [23]. They are mainly poly(lactide)-*block*-poly

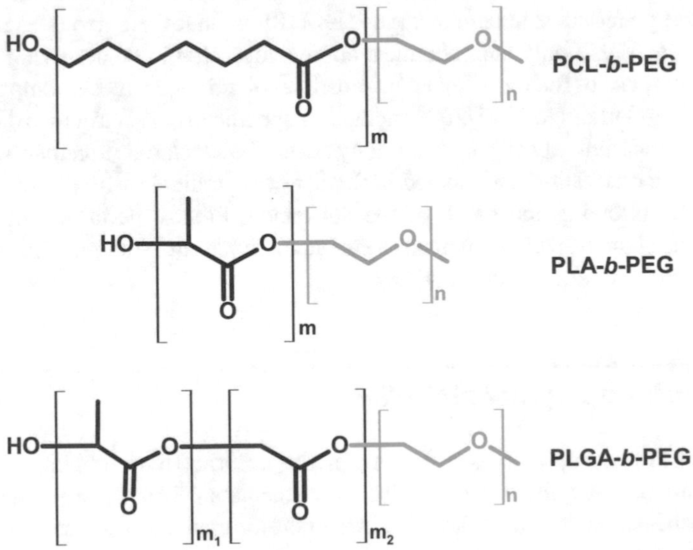

Fig. 4.3 Chemical structure of amphiphilic block copolymers PCL-*b*-PEG, PLA-*b*-PEG and PLGA-*b*-PEG. The hydrophilic and hydrophobic parts are drawn in green and red, respectively

(ethylene glycol) (PLA-*b*-PEG), poly(caprolactone)-*block*-poly(ethylene glycol) (PCL-*b*-PEG) and poly(lactide-*co*-glycolide)-*block*-poly(ethylene glycol) (PLGA-*b*-PEG) (Fig. 4.3). The state of the hydrophobic core is liquid or solid depending of the chemical nature of the hydrophobic polymer block. Semi-crystalline polymers cannot crystallize inside the very small core of micelles because of confinement; the only possible solid state is glassy. The glass transition temperature (T_G) of PCL being −60 °C, the core of PCL-*b*-PEG nanoparticles is liquid. The core of PLA-*b*-PEG is glassy [24] because T_G of PLA is +60 °C. Drugs solubilize inside block copolymer nanoparticles in two possible solubilization sites, in the hydrophobic core or in the shell by means of interactions with the hydrophilic part of the block copolymer [25]. That the core is in glassy state does not prevent drug loading inside it; slow drug release is expected because molecular motions in a glassy medium are very slow.

4.5 Nanocarriers and Skin

The behavior of drug carriers deposited at the surface of skin is quite a complex issue that has been most considered through the skin penetration of active ingredients, considering skin as a simple barrier against passive diffusion. This is justified since the outermost layer of skin, *stratum corneum*, is not a living tissue. Skin absorption is quite a complex issue because there are several paths for penetration [26]. The small size of nanoemulsions and block copolymer nanoparticles provides them specific properties when they are in contact with the skin. One can think that skin absorption of nanocarriers is made easier by their small size. However, there is a large body of evidence showing that nanoparticles actually do not penetrate deeply the healthy skin. This has been well documented regarding inorganic nanoparticles used in sunscreens in order to address fear that titanium dioxide particles could penetrate the viable layers of the skin and accumulate there. Indeed, they are neither biodegradable nor soluble in aqueous medium. Inorganic nanoparticles penetrate the most external part of the *stratum corneum* but do not pass through it during normal exposure time corresponding to utilization of cosmetic products by customers [27–29]. In an instance, where the *stratum corneum* barrier has been damaged for allowing silica nanoparticles to reach the viable epidermis, the main cellular uptake was into Langerhans dendritic cells although keratinocytes are in great majority [30]. The same conclusion has been reached for polystyrene nanoparticles that are penetrating through hair follicles pathway [31]. The same should be true for all organic particles [32]. Additionally, organic nanocarriers for cosmetic applications are made of materials that national and European regulatory bodies evaluated as safe. On the one hand, such organic nanoparticles made of biodegradable materials cannot accumulate for long times in living media and in environment in general. On the other hand, they are made of soft materials that cannot retain their particulate morphology once they have been spread on skin. Emulsions (and nanoemulsions) break in contact of skin because their thin layer spread on skin dries quite fast in open air; they release ingredients that can penetrate skin.

Specificities of nanoemulsions and block copolymer nanoparticles pertain to their higher ability to transfer active ingredients to skin and to target specific loci of the skin surface, namely skin appendages such as hair follicles and sweat ducts [33–35]. Fluorescence pictures of skin cryo-sections have been clearly shown the influence of particle size of fluorescent solid polymer particles. Thus, particles of 40 nm enter deep inside hair follicles and even penetrate the adjacent tissues through the epithelium, whereas larger particles of 750 or 1500 nm remained accumulated at the follicle entry (infundibulum) [31] (Fig. 4.4).

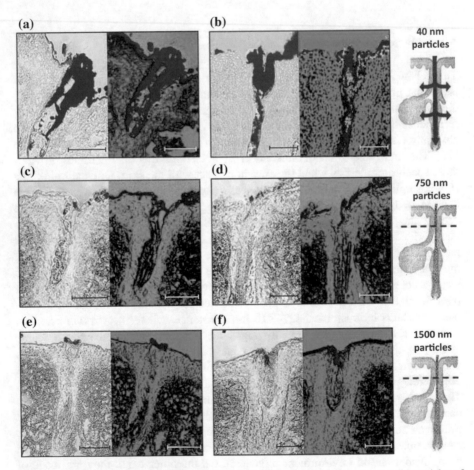

Fig. 4.4 Cryo-sections of human skin showing hair follicles where fluorescent solid particles of different sizes of 40 nm (**a**, **b**), 750 nm (**c**, **d**) and 1500 nm (**e**, **f**) have penetrated. The fluorescent regions are marked in red. The scale bars are 100 μm. Reproduced from ref [31] with permission

Hydrophobic active ingredients hardly can be released into the continuous aqueous phase and reach the skin by diffusion through it because their solubility in water is low. Transfer to skin is better achieved when nanocarriers contact the skin surface or slightly penetrate *stratum corneum* between corneocytes undergoing desquamation. This is more effective under non-occlusive conditions of skin exposure where the cosmetic product dries on the skin surface. This idea has been put forward as a rationale for higher skin absorption from Pickering emulsions than from classical surfactant-based emulsions [36, 37]. The measured 10-fold higher adhesion energy of such emulsion droplets to the skin surface makes the contact time longer and the transfer efficiency higher [36]. The small size of nanoemulsions can obviously enhance such adhesion owing to the large surface area of nanodroplets. Accordingly, the drug release rate that actually matters is not that into the aqueous phase but that into the intercellular lipid medium of *stratum corneum*.

Accumulation of nanocarriers in hair follicles has been mostly addressed in the framework of pharmaceutical application, either in dermatology or in transdermal delivery. These applications share many common features with skin care cosmetic applications since the active substance has to reach its site of action for it exerts its biological activity. Hair follicles and sweat ducts are "shunt routes" for skin penetration as they allow by-passing the *stratum corneum* barrier. On the one hand, a hair follicle enters deep inside the skin allowing by-passing *stratum corneum* and dermo-epidermal barriers; on the other hand, the epithelium inside hair follicles is much thinner than at the outer skin surface. Nevertheless, hair follicles can be filled (obstructed) with fats coming from sebaceous glands. The experimental determination of the contributions of skin absorption through *stratum corneum* and hair follicles is a difficult task. This methodological issue did not receive a satisfactory answer yet. The follicular pathway can make a large contribution, even when the active ingredient is not loaded inside nanoparticles. As example, it has been measured that absorption of caffeine to excised human female breast skin from an ethanol/propylene glycol solution took place half by hair follicles pathway and half by passive diffusion through *stratum corneum* [38].

4.6 Nanoemulsions in Cosmetics

There are a number of cosmetic products that include nanoemulsions. The "nanoemulsion" or "nanocream" technology has been used and claimed in many skin care or hair care products and still remains widely used. Currently with the mistrust of consumer organizations toward nanotechnology, this is difficult to sort them out because neither droplet sizes nor fabrication processes are given in products technical claims from the cosmetic manufacturers. Information about commercial cosmetic products is difficult to obtain because regulation only asks for the INCI list of ingredients to be written on packaging. The ingredients used in formulations of coarse and nanoemulsions can be the same, and the manufacture process is not disclosed. Nevertheless, few cosmetic products are declared containing nanoemulsions as marketing claim, and there are patents from cosmetic companies [39, 40] that let think that nanoemulsions are incorporated into cosmetic products supplied by these companies. The patent review of 2010 published in *Cosmetics and Toiletries* [39] shows that most major cosmetic companies have applied patents claiming for nanoemulsions. Production of nanoemulsions using high-pressure homogenizer requires a specific organization for maintenance and cleaning (dismantlement and reassembly) the equipment; such equipment is better suited for continuous production of large quantities. Conversely, spontaneous emulsification processes (PIT or PIC) can be implemented for the batch production of small quantities in SMEs that keep their manufacture process as know-how and do not apply patents.

Nanoemulsions are used in cosmetic products where benefits can be expected from the small droplet size. These are mainly: (i) stronger occlusive effect that keeps skin moisture, (ii) faster skin absorption of active ingredients, (iii) better penetration in narrow gaps such as pilosebaceous follicles and hair scales spacings, (iv) high fluidity with no creaming, (v) visual appearance of the product (clear to slightly turbid, pearlescent) and glossy coating on skin, (vi) cosmetic wet wipes and (vii) sprayable products (mist).

L'Oréal claims the advantageous use of a nanoemulsion (Aqua-Oléum®) containing avocado and jojoba seed oils in a serum dedicated to repair and nourish dry hairs. Aqua-Oléum® hair care serum from Kérastase is an o/w nanoemulsion which has a composition that does not differ from that of a classical emulsion (Table 4.2); the nanoemulsion form comes from the manufacturing process. It is claimed for fast penetration of hair fibers as oil nanodroplets are 100 times smaller than spacing between hair scales.

La Maison Chanel claimed some years ago for the nanoemulsions that improve the efficacy of several products. The Fresh Body Moisturizing Mist "Brume Fraîche pour le Corps" Coco Mademoiselle from Chanel is an o/w nanoemulsion used as a spray. This is a fluid sprayable nanoemulsion which remains homogeneous (no creaming in the container) though it is fluid, thanks to the nano-size of droplets. In addition, the nanoemulsion provides a fast and strong moisturizing effect by means of skin occlusion.

Chanel Précision "Solution Déstressante" Calming Emulsion (Table 4.3) contains centella asiatica extract (Gotu Kola or Indian Pennywort) as a soothing active ingredient that provides a de-stressing feeling upon application onto skin. Immediacy of perceived effect comes from fast penetration of the active ingredient brought about by the formulation in a nanoemulsion.

Table 4.2 INCI list of ingredients of Kérastase Aqua-Oléum® hair care serum	Kérastase Aqua-Oléum®
	Water • Alcohol Denat • Jojoba Seed Oil • Avocado Oil • Glycerin • Cyclopentasiloxane • PEG-8 Isostearate • Behentrimonium Chloride • Amodimethicone • C11-15 Pareth-7 • C12-16 Pareth-9 • Trideceth-12 • Butylphenyl Methylpropional • Cinnamyl Alcohol • Citronellol • Fragrance

Table 4.3 INCI list of ingredients of Chanel Précision® Solution Destressante calming emulsion	Précision® Solution Déstressante Calming Emulsion
	Water • Lauroyl Lysine • Dimethicone, Butylene Glycol • PPG-15 Stearyl Ether • Prunus Amygdalus Dulcis Seed Extract • Propylene Glycol • PEG-5 Rapeseed Sterol • Alphaglucan Oligosaccharide • Phenoxyethanol • Cetyl Alcohol Chlorphenesin • Glycerin • Palmitic Acid • Hydrogenated Lecithin • Thiethanolamine • Acrylates/C10-30 Alkyl Acrylate Crosspolymer • Laureth-3 • Pisum Sativum Extract • Tetrasodium EDTA • Hydroxyethycelulose • Centella Asiatica Extract • Methylparaben • Acetyl Dipeptide-1 Cetyl Ester • Glycine Soja Seed Extract • Ceteth-16 • Propylparaben • Potassium Sorbate • Cholesterol • BHT • Lecithin • Hydrolyzed Yeast • Ethylparaben • Butylparaben • Isobutylparaben • Lysine HCL • Sodium Dehydroacetate

Another type of product where nanoemulsions are intensively used is wet cosmetic wipes.

Evonik company has developed a self-emulsifying system containing polyglyceryl-4 laurate and dilauryl citrate which allows the preparation of nanoemulsions by the PIC technology [41, 42]. Heunnemann et al. [43] studied the phase behavior of this self-emulsifying system under dilution taking diethylhexyl carbonate as oil phase. The oil/surfactant composition before dilution consists of 66 wt% diethylhexyl carbonate, 12 wt% of preservative mixture consisting of phenoxyethanol (main component, 72 wt%) and various paraben esters (methyl- (16 wt%), ethyl- (4 wt%), propyl- (2 wt%), butyl- (4 wt%) and isobutyl- (2 wt%)), and 22 wt% polyglyceryl-4 laurate admixed with a small amount (0.9 wt%) of dilauryl citrate as surfactants. A nanoemulsion forms upon dilution with water. Small addition of water (<10%) to the dry mixture causes molecular dissolution forming a clear slightly bluish inverse microemulsion. As the water fraction is increased, a biphasic system appears containing the microemulsion as the upper phase (becoming turbid as the water fraction increases) and a lower phase corresponding to a clear excess of water. Around 50% of water, a single-phase microemulsion appears and finally the phase inversion concentration (PIC) is reached, which yields a single bluish milky nanoemulsion above 75% water. In between 50 and 75%, phase inversion takes place through a wide area of bluish microemulsions exhibiting flow birefringence.

Sinerga has also developed a combination of emulsifiers called Nanocream® [44, 45] (Potassium Lauroyl Wheat Amino Acids, Palm Glycerides, Capryloyl Glycine) which allows the formulation of o/w nanoemulsions. This product is intended to be used in sprayable emulsions, hyperfluid emulsions and wet wipes.

Most available data can be found in articles from academic research institutes published in the open literature. The great majority of them did not find an industrial application in a marketed product. Literature reports dealing with cosmetic applications are very scarce because cosmetic industry does not publish results from their research. Papers from academic research may claim for potential

Table 4.4 Reports on skin absorption from nanoemulsions for cosmetic and pharmaceutical applications

Author (year)	Cosmetic or pharma	Purpose: (skin absorption or transdermal)	Active substance	Emulsification process	Skin model	Comparison to control	References
Brownlow et al. (2015)	Cosmetic and pharma	Skin absorption and photoprotection	Vitamin E and genistein	US	Mixed cellulose esters membrane and fibroblast cells culture	Solution	[47]
Campani et al. (2016)	Pharma	Skin absorption	Vitamin K1	PIC	Excised pig ear skin	Solution in methanol	[48]
Calderilla-Fajardo et al. (2006)	Cosmetic	Absorption to *stratum corneum*	Octylmethoxy cinnamate	EmEvap	In vivo human skin	Coarse emulsion Nanocapsules	[49]
Mou et al. (2008)	Pharma	Skin absorption	Camphor + menthol + methyl salicylate	HPH	Excised rat abdominal skin	Coarse emulsion	[50]
Zhou et al. (2010)	None	Skin absorption	Nile Red dye	HPH	In vivo rat abdominal skin	Coarse emulsion	[51]
Rachmawati et al. (2015)	Pharma	Transdermal	Curcumin	PIC	Shed snake skin	Coarse emulsion	[52]
Alves et al. (2007)	Pharma	Skin absorption	Nimesulide	PIC	Excised human abdominal skin	Nanocapsules Polymer nanospheres	[53]
Nam et al. (2012)	Cosmetic	Skin absorption	Tocopheryl acetate	HPH	Excised abdominal skin of hairless guinea pigs	No	[54]
Lu et al. (2014)			D-limonene	US	Excised rat abdominal skin	No	[55]

(continued)

Table 4.4 (continued)

Author (year)	Cosmetic or pharma	Purpose: (skin absorption or transdermal)	Active substance	Emulsification process	Skin model	Comparison to control	References
Isailović et al. (2016)	Pharma	Absorption to *stratum corneum*	Aceclofenac	HPH	In vivo human skin	No	[56]
Klang et al. (2010)	Pharma	Skin absorption	Progesterone	HPH	Excised pig abdominal skin	No	[57]
Klang et al. (2012)	Pharma	Skin absorption	Fludrocortisone acetate	HPH	Excised pig abdominal and ear skin	No	[58]
Hoeller et al. (2009)	Cosmetic	Skin absorption	Fludrocortisone acetate and flumethasone pivalate	HPH	Excised pig abdominal skin	Placebo	[59]
Baspinar et al. (2012)	Pharma	Skin absorption	Prednicarbate	HPH	Excised human abdominal skin	No	[60]
Yilmaz et al. (2006)	Cosmetic	Skin humidity and elasticity	Ceramide	HPH	In vivo human skin	Placebo	[61]
Youenang Piemi et al. (1999)	Dermato	Skin absorption	Econazole and miconazole nitrate	PIT	Excised abdominal skin of hairless rats	Placebo	[62]
Kim et al. (2008)	Dermato	Skin absorption	Ketoprofen	PIC	Excised mouse skin	No	[82]
Sakeena et al. (2010)	Dermato	Skin absorption Anti-inflammatory activity from œdema	Ketoprofen	PIC	Cellulose acetate membrane Excised rat skin In vivo on rats	No	[83] [84] [85]
Elrashid et al. (2011)	Cosmetic	Skin absorption	*P. urinaria* extract	PIC	Cellulose acetate membrane	No	[86]

(continued)

Table 4.4 (continued)

Author (year)	Cosmetic or pharma	Purpose: (skin absorption or transdermal)	Active substance	Emulsification process	Skin model	Comparison to control	References
Fontana et al. (2011)	Dermato	Skin absorption	Clobetasol propionate	PIC	Cellulose acetate membrane	Nanocapsule	[87]
Yu et al. (2014)	Dermato	Skin absorption	Metronidazole	PIC	Excised pigskin and mouse ear	No	[88]
Ngan et al. (2015)	Cosmetic	Collagen protection	Fullerene	US	In vivo human skin	No	[89]
Su et al. (2017)	Pharma		Dyes	HPH	In vivo rat skin	No	[90]

HPH high-pressure homogenizer; *US* ultrasound; *PIT* thermal phase inversion; *PIC* spontaneous emulsification by dilution; *EmEvap* emulsification + solvent removal cosmetic application, but it is not possible to know whether an actual application has been developed. Finally, most of published data address pharmaceutical applications that are mainly of two types: dermatology or transdermal administration of drugs. They differ from cosmetic applications: Transdermal administration aims at reaching the systemic circulation whereas a cosmetic active substance should not pass to the bloodstream; dermatological products are used to cure pathologic skins whereas cosmetic products aim at skin care of healthy skin. Data resources dealing with pharmaceutical applications are of interest because they provide experimental results on skin absorption of molecules that are either the same or close to cosmetic ingredients, and they provide mechanistic information of the skin absorption processes. Table 4.4 gives a list (not intended to be exhaustive) of such skin absorption data taken in the open literature.

Published information mainly focuses on skin absorption. There are several technical benefits of nanoemulsions that deserve being considered. Firstly, one obvious benefit is slow creaming (or no creaming); the cosmetic product remains homogeneous, and it does not need being thickened for that. Fluid nanoemulsions can be stable and homogeneous over long time. This is a definite advantage when fluid emulsions are needed such as for sprayable products. It should be noted that sprays of aqueous suspensions of nanoparticles are made of liquid droplets much larger than the size of nanoparticles contained inside, so that there is no hazard related to pulmonary entry as in the case of breathing dry powders of nanoparticles. The viscosity of o/w emulsions is primarily determined by the volume fraction of oil, so that droplet size does not matter so much with that regard. Nanoemulsions are not intrinsically more fluid than classical emulsions; classical emulsions are often thicker because thickener has been added in order to prevent against creaming. Secondly, nanoemulsions bring about new textures and new sensorial properties. In principle, the same moisturizing activity is expected as for classical emulsions after their destabilization at the skin surface left a continuous layer of oil with occlusive behavior retaining skin hydration. Actually, nanoemulsions show a better moisturizing effect [8]. The small size of nanoemulsion droplets allows the formation of a continuous occlusive film even if coalescence of oil droplets at the skin surface does not occur [46]. Upon application of a coarse o/w emulsion, skin moisturizing coming from occlusion is delayed by the time required for droplets coalescence releases a continuous oil film at the skin surface. Close packing of nanoemulsion droplets makes a continuous film that shows up an immediate occlusive effect; an immediate sensory effect upon spreading the cream is felt that a coarse emulsion cannot provide (unless it is "quick breaking").

Nanoemulsions have been evaluated for their ability to improve drug delivery to skin; this is presented in Table 4.4 from which studies dealing with microemulsions have been discarded. Almost all studies conclude for "improved" skin delivery of hydrophobic drugs from nanoemulsions. Claim for "improvement" is supported by experiments when delivery from nanoemulsions has been compared to control samples, either solutions [47, 48] or classical emulsions of droplet size larger than 1 μm [49–52]. Most studies barely concluded for faster drug absorption or permeation; some of them addressed mechanistic issues, however. Interestingly, definitely different mechanisms of skin delivery of curcumin to skin have been shown [52]: zero-order kinetics from a coarse emulsion which was typical of an insoluble drug dispersed in water and a Higuchi-type kinetic profile from a nanoemulsion which was typical of a diffusion controlled release. With the aim at entering detailed parameters of formulations, it is difficult to figure out information of general breadth from many studies focusing on optimization of specific nanoemulsion composition. Beyond the influence of droplet size, an interesting comparative study of several nanocarriers concluded that skin delivery of nimesulide was faster from nanoemulsion than nanocapsules and nanospheres [53], which disclosed a definite difference between nanodroplets covered by surfactants and polymer nanoparticles or, in other words, between liquid and solid nanoparticles.

Droplet size has often been put forward as a main parameter that controlled drug absorption. The type of emulsifier has a definite influence [49, 54–56] because some of them have intrinsic skin penetration enhancer properties. It is difficult to ascribe variations of skin absorption to the effect of droplet size and/or penetration enhancer properties of the emulsifier because changing droplet size usually requires changing the emulsifier. Other ingredients of the formulation may also matter; as example, added cyclodextrins bring about their strong penetration enhancer effect [57, 58].

The surface charge of nanodroplets contributes to drug absorption. It is postulated that positively charged carriers strongly bind to negatively charged biological surfaces. Most epithelia are indeed negatively charged; so are corneocytes. This idea has been confirmed by in vitro absorption experiments of neutral hydrophobic drugs, fludrocortisone acetate and flumethasone pivalate into pigskin explants [59] and prednicarbate absorption into human skin explants [60], showing that a cationic nanoemulsion made delivery of those drugs faster than an anionic one. The same kind of nanoemulsion stabilized by cationic emulsifier improves skin hydration and elasticity [61], but it is not clear whether the origin is occlusive effect of the nanoemulsion or reinforcement of the *stratum corneum* barrier by penetration of the cationic ceramide emulsifier. The situation further gains complexity when the drug itself is charged. Skin delivery of cationic econazole nitrate was identical from cationic and anionic nanoemulsions when it has been measured on excised abdominal skin of hairless rats; anionic nanoemulsions improved skin delivery of the similar drug miconazole nitrate under the same conditions [62].

Nanoemulsions have been used for skin delivery of photosensitizers (or precursors of them) for the purpose of photodynamic therapy. The application is really far from cosmetics, but it has been again given evidence of faster skin penetration of various ingredients from nanoemulsions than coarse emulsions [63–67].

Interesting information comes from studies devoted to skin decontamination, some of these formulations involving nanoemulsions. Such works have not been listed in Table 4.2 because they are really far from topical administration of cosmetic ingredients. As example, extraction of radionuclides (uranyl acetate) from the skin has been achieved by using nanoemulsions containing a complexing agent for uranyl cations [68, 69]. O/w nanoemulsions containing a calixarene prepared by the thermal phase inversion process have been assessed in vitro by measurements of uranyl uptake from solution [70, 71] and extraction from excised pig ear skin [77]. It has been disclosed that the calixarene complexing agent was adsorbed at the surface of nanodroplets, so that nanoemulsions having large interfacial area between oil and water looked suitable for the purpose of extraction. The decontamination process has been optimized to reach high levels of uranyl removal; however, an experimental comparison with related formulations and controls has not been reported. As a prospect, skin decontamination strategies provide good clues for designing cosmetic products claimed "anti-pollution" as they recently appeared on the market.

4.7 Block Copolymer Nanoparticles in Cosmetics

Cosmetic chemists working in R&D are active in the field of block copolymer nanoparticles as they know about their promising specific properties. To our knowledge, developments are still at the stage of R&D. There is no current application in actual cosmetic products. The most studied block copolymers are made of two blocks: hydrophobic biodegradable polyesters and hydrophilic poly (ethylene oxide). Neither have they received a CAS number nor a REACH registration yet. The Sigma-Aldrich company supplies PCL-*b*-PEG, PLA-*b*-PEG and PLGA-*b*-PEG under the respective names "Poly(ethylene glycol)-*block*-poly (ε-caprolactone) methyl ether," "Poly(ethylene glycol)-*block*-polylactide methyl ether" and "Poly(ethylene glycol) methyl ether-*block*-poly(L-lactide-*co*-glycolide").

Research in the field of amphiphilic block copolymers is mainly concerned with pharmaceutical drug delivery applications. The first report on topical administration aiming at transdermal delivery is from year 2004 [72]. Micelles of polycaprolactone-*block*-poly(ethylene glycol) copolymer loaded with the antihypertensive vasodilator minoxidil were evaluated on guinea pigskin. The higher skin permeation through hairy than hairless skin indicated a large contribution of penetration through hair follicles. A significant influence of nanoparticle size was observed on hairy skin but not on hairless skin. This interesting set of observations suggested that penetration by the follicular pathway depended on size and that the size-independent penetration through the *stratum corneum* came from a release of the drug at the external surface of skin with no penetration of the carriers. In the field of dermatology, where drug should penetrate skin and stay there for long, poly (dihexyl lactide)-*block*-poly(ethylene glycol), nanoparticles improved skin delivery of several antifungal drugs compared to the commercial o/w emulsion Pevaryl® cream [73]. Skin delivery of hydrophobic active substances not only depends on nanoparticle size but on their ability to release drugs into skin. The nature of the hydrophobic block is of high relevance through its physical state, either liquid or solid, and through its interactions with the drug. As example, in vitro evaluation of absorption of vitamin A on excised pigskin [74] disclosed the enhanced skin absorption from block copolymer nanoparticles, by one order of magnitude compared to Polysorbate 80 surfactant micelles and by two orders of magnitude with

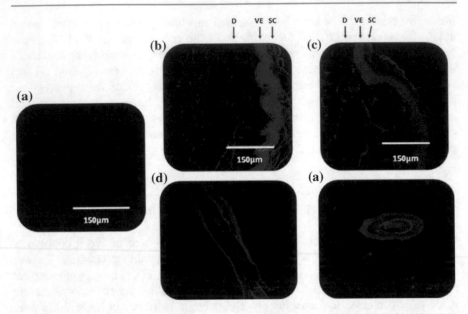

Fig. 4.5 Fluorescence microscopy pictures of histological section sections of pigskin after 24 h exposure to PLA-*b*-PEG (**b**, **d**) and PCL-*b*-PEG (**c**, **e**) nanoparticles loaded with Nile Red. Pictures show hairless parts of pigskin (**b** and **cc** and hair follicles (**d** and **e**). Picture A shows the very weak self-fluorescence of skin caused by endogenous fluorescent compounds of the control (24 h exposure to pure water). SC = *stratum corneum*, VE = viable epidermis, D = dermis. Reproduced from ref [83] with permission

respect to an oil solution. More interesting was disclosure of the influence of the nature of the block copolymer: a much higher absorption from PLA-*b*-PEG than from PCL-*b*-PEG nanoparticles. Fluorescence imaging histological skin sections provides a picture of this huge effect on penetration of the Nile Red fluorescent dye (Fig. 4.5). This was rationalized by considering interactions between retinol and the hydrophobic polyester blocks, PLA and PCL. The distance of solubility parameters according to the Hansen framework [75] revealed that PCL was a good solvent medium for retinol; conversely, PLA was quite a poor solvent. Therefore, retinol was retained inside the hydrophobic core of PCL-*b*-PEG nanoparticles whereas PLA-*b*-PEG nanoparticles released retinol as they contacted favorable media for transfer of retinol, namely the lipid medium of *stratum corneum*. Tuning drug delivery to skin can be achieved by several parameters: the nanoparticles size, their adhesion to different parts of the skin and their ability to release the loaded drug molecules.

Amphiphilic block copolymers with longer hydrophobic block can self-assemble liposomes called "Polymersomes." Their behavior is close to classical phospholipid liposomes [76]. Indeed, the difference of in-out dynamic exchange behavior between "dynamic" classical surfactant micelles and "frozen" block copolymers micelles does not pertain to classical and block copolymer-based liposomes which both have very slow dynamics.

4.8 Combination of Nanoemulsions and Block Copolymer Nanoparticles

Block copolymer nanoparticles can be used as stabilizers of o/w emulsions [77]. If block copolymers are "dissolved" in the aqueous phase as it is done with usual emulsifiers, they self-assemble as nanoparticles and adsorb at the surface of oil droplets as nanoparticles. Emulsions stabilized by adsorbed nanoparticles are called Pickering emulsions [78]. Examples are o/w emulsions of medium chain triglycerides stabilized by PCL-*b*-PEG or by PLA-*b*-PEG nanoparticles [79]. There are two sites for drug solubilization in such formulations: oil droplets and the part of block copolymer nanoparticles that is not adsorbed to oil droplets. Each vehicle can be loaded (or not) by active substances, and it delivers them to skin at its own rate [80, 81]. Delivery by block copolymer nanoparticles is much faster than by emulsion droplets. This double level of skin delivery provides much versatility in terms of drug delivery profiles. A requirement of cosmetic applications is to provide the consumer with an immediate perceived effect and to achieve sustained delivery that ensures the long-term activity according to product claims.

4.9 Conclusion

Cosmetic science aims at producing products with improved performance or new properties, mainly pertaining to efficiency and sensoriality. To that end, innovation is the guideline.

Though physiological activity is the main property claimed on the packaging (anti-aging, moisturizing, sun protection), the consumer generally does not feel it easily when it is a long-term activity (anti-aging care). Immediate feeling relies on sensorial properties.

Nanoemulsions bring about such new features. They are made from the same ingredients as classical emulsions, and their overall organization is also the same. Only the droplet size is smaller by a factor of 10–100. Of course, this brings about new properties such as accelerated skin delivery of active substances, which is a trivial consequence of larger interfacial area. Absence of creaming in fluid products, immediate skin occlusion, transparency, and gloss after spreading, are specific consequences of the sub-micron size of nanoemulsions. Block copolymer micelles are not used in current cosmetic products. For sure, they will find applications to cosmetics in the near future.

All of these products are made of nanoparticles that the consumer is anxious about their potentially harmful side effects. Hazard of nanoparticles in cosmetics is not supported by definite evidence, especially when nanoparticles are made of organic biodegradable compounds. All experimental investigations on nanoparticles used in cosmetic products concluded for there is no deep skin penetration. In addition, nanoemulsion droplets do not penetrate skin as intact objects: Emulsion is broken at the surface of skin as water evaporates. At a whole, possible hazards of nanoemulsions might only come from the presence of harmful ingredients (regulated by REACH and the Cosmetic Regulation); in no way, it can come from the small droplet size.

New properties related to the small size provide new textures of cosmetic products and open new types of utilization. Improved delivery of active substances allows decreasing their concentrations, while keeping the same effects. With regard to skin care, the high occlusive effect of nanoemulsion is a definite improvement for moisturizing products; gloss is an immediate sensory effect felt by consumers. In hair repair, permanent shaping and hair dye applications, small droplets penetrate easier between hair scales, allowing more efficient transport of active ingredients to the cortex.

The development of more applications of nanoemulsions and related nano-objects such as block copolymer micelles faces two main challenges that are current research prospects: (i) have a convincing presentation concerning safety of nanoparticles based on scientific facts and the demonstration of a good mastery over production and utilization of nanoparticles and (ii) develop efficient processes for large-scale production of nanoemulsions.

References

1. Wu X, Guy RH. Applications of nanoparticles in topical drug delivery and in cosmetics. J Drug Deliv Sci Technol. 2009;19:371–84.
2. Briançon S, Chevalier Y, Bolzinger M-A. Biopharmaceutical evaluation of various dosage forms intended for caffeine topical delivery. In: Chilcott R, Brain K editors. Advances in dermatological sciences. Issues in Toxicology No 20, RSC Publishing, Cambridge; 2013, Chap 8. pp 88–100.
3. Solans C, Izquierdo P, Nolla J, Azemar N, Garcia-Celma MJ. Nano-emulsions. Curr Opin Colloid Interface Sci. 2005;10:102–10.
4. Riess G. Micellization of block copolymers. Prog Polym Sci. 2003;28:1107–70.
5. Gohy J-F. Block copolymer micelles. Adv Polym Sci. 2005;190:65–136.
6. Torchilin VP, Trubetskoy VS. Which polymers can make nanoparticulate drug carriers long-circulating? Adv Drug Deliv Rev. 1995;16:141–55.
7. Yukuyama MN, Ghisleni DDM, Pinto TJA, Bou-Chacra NA. Nanoemulsion: process selection and application in cosmetics—a review. Int J Cosmet Sci. 2016;38:13–24.
8. Sonneville-Aubrun O, Simonnet J-T, L'Alloret F. Nanoemulsions: a new vehicle for skincare products. Adv Colloid Interface Sci. 2004;108–109:145–9.
9. Baspinar Y, Keck CM, Borchert H-H. Development of a positively charged prednicarbate nanoemulsion. Int J Pharm. 2010;383:201–8.
10. Förster T, von Rybinski W, Wadle A. Influence of microemulsion phases on the preparation of fine-disperse emulsions. Adv Colloid Interface Sci. 1995;158:119–49.
11. Izquierdo P, Esquena J, Tadros TF, Dederen JC. Phase behavior and nano-emulsion formation by the phase inversion temperature method. Langmuir. 2004;20:6594–8.
12. Date AA, Desai N, Dixit R, Nagarsenker M. Self-nanoemulsifying drug delivery systems: formulation insights, applications and advances. Nanomedicine. 2010;5:1595–616.
13. Pouton CW. Formulation of self-emulsifying drug delivery systems. Adv Drug Deliv Rev. 1997;25:47–58.
14. Meyer J, Scheuermann R, Wenk HH. Combining convenience and sustainability: Simple processing of PEG-free nanoemulsions and classical emulsions. SOFW J. 2008;134:58–64.

15. Date AA, Nagarsenker MS. Design and evaluation of self-nanoemulsifying drug delivery systems (SNEDDS) for cefpodoxime proxetil. Int J Pharm. 2007;329:166–72.
16. Sadurní N, Solans C, Azemar N, García-Celma MJ. Studies on the formation of O/W nano-emulsions, by low-energy emulsification methods, suitable for pharmaceutical applications. Eur J Pharm Sci. 2005;26:438–45.
17. Klang V, Valenta C. Lecithin-based nanoemulsions. J Drug Deliv Technol. 2011;21:55–76.
18. Katz LM, Dewan K, Bronaugh RL. Nanotechnology in cosmetics. Food Chem Toxicol. 2015;85:127–37.
19. Massignani M, Lomas H, Battaglia G. Polymersomes: a synthetic biological approach to encapsulation and delivery. Adv Polym Sci. 2010;229:115–54.
20. Riley T, Heald CR, Stolnik S, Garnett MC, Illum L, Davis SS, King SM, Heenan RK, Purkiss SC, Barlow RJ, Gellert PR, Washington C. Core-shell structure of PLA-PEG nanoparticles used for drug delivery. Langmuir. 2003;19:8428–35.
21. Zana R, Marques C, Johner A. Dynamics of micelles of the triblock copolymers poly(ethylene oxide)-poly(propylene oxide)-poly(ethylene oxide) in aqueous solution. Adv Colloid Interface Sci. 2006;123–126:345–51.
22. Letchford K, Liggins R, Burt H. Solubilization of hydrophobic drugs by methoxy poly (ethylene glycol)-block-polycaprolactone diblock copolymer micelles: Theoretical and experimental data and correlations. J Pharm Sci. 2008;97:1179–90.
23. Kumar N, Ravikumar MNV, Domb AJ. Biodegradable block copolymers. Adv Drug Deliv Rev. 2001;53:23–44.
24. Heald CR, Stolnik S, Kujawinski KS, De Matteis C, Garnett MC, Illum L, Davis SS, Purkiss SC, Barlow RJ, Gellert PR. Poly(lactic acid)-poly(ethylene oxide) (PLA-PEG) nanoparticles: NMR studies of the central solidlike PLA core and the liquid PEG corona. Langmuir. 2002;18:3669–75.
25. Patel SK, Lavasanifar A, Choi P. Roles of nonpolar and polar intermolecular interactions in the improvement of the drug loading capacity of PEO-b-PCL with increasing PCL content for two hydrophobic cucurbitacin drugs. Biomacromol. 2009;10:2584–91.
26. Bolzinger M-A, Briançon S, Pelletier J, Chevalier Y. Penetration of drugs through skin, a complex rate controlling membrane. Curr Opin Colloid Interface Sci. 2012;17:156–65.
27. Pflücker F, Hohenberg H, Hölzle E, Will T, Pfeiffer S, Wepf R, Diembeck W, Wenck H, Gers-Barlag H. The outermost stratum corneum layer is an effective barrier against dermal uptake of topically applied micronized titanium dioxide. Int J Cosmet Sci. 1999;21:399–411.
28. Monteiro-Riviere NA, Riviere JE. Interaction of nanomaterials with skin: aspects of absorption and biodistribution. Nanotoxicology. 2009;3:188–93.
29. Monteiro-Riviere NA, Baroli B. Nanomaterial penetration. In: Monteiro-Riviere NA. editor. Toxicology of the skin, target organ toxicology series. New York: Informa Healthcare; 2010, Ch. 22. pp. 333–46.
30. Rancan F, Gao Q, Graf C, Troppens S, Hadam S, Hackbarth S, Kembuan C, Blume-Peytavi U, Rühl E, Lademann J, Vogt A. Skin penetration and cellular uptake of amorphous silica nanoparticles with variable size, surface functionalization, and colloidal stability. ACS Nano. 2012;6:6829–42.
31. Vogt A, Combadiere B, Hadam S, Stieler KM, Lademann J, Schaefer H, Autran B, Sterry W, Blume-Peytavi U. 40 nm, but not 750 or 1,500 nm, nanoparticles enter epidermal CD1a + cells after transcutaneous application on human skin. J Investigative Dermatol. 2006;126:1316–22.
32. Bolzinger M-A, Briançon S, Chevalier Y. Nanoparticles through the skin: managing conflicting results of inorganic and organic particles in cosmetics and pharmaceutics. WIREs Nanomed Nanobiotechnol. 2011;3:463–78.
33. Schaefer H, Watts F, Brod J, Illel B. Follicular penetration. In: Scott RC, Guy RH, editors. Prediction of percutaneous penetration, methods, measurements, modelling. London: IBC Technical Services; 1990. p. 163–73.
34. Illel B. Formulation for transfollicular drug administration: some recent advances. Crit Rev Therap Drug Carrier Syst. 1997;14:207–19.

35. Lekki J, Stachura Z, Dąbroś W, Stachura J, Menzel F, Reinert T, Butz T, Pallon J, Gontier E, Ynsa MD, Moretto P, Kertesz Z, Szikszai Z, Kiss AZ. On the follicular pathway of percutaneous uptake of nanoparticles: Ion microscopy and autoradiography studies. Nucl Instrum Methods Phys Res B. 2007;260:174–7.
36. Frelichowska J, Bolzinger M-A, Valour J-P, Mouaziz H, Pelletier J, Chevalier Y. Pickering w/o emulsions: drug release and topical delivery. Int J Pharm. 2009;368:7–15.
37. Marku D, Wahlgren M, Rayner M, Sjöö M, Timgren A. Characterization of starch Pickering emulsions for potential applications in topical formulations. Int J Pharm. 2012;428:1–7.
38. Trauer S, Patzelt A, Otberg N, Knorr F, Rozycki C, Balizs G, Büttemeyer R, Linscheid M, Liebsch M, Lademann J. Permeation of topically applied caffeine through human skin—a comparison of *in vivo* and *in vitro* data. Br J Clin Pharmacol. 2009;68:181–6.
39. Fox C. Stable o/w nanoemulsions for skin and other topics: literature findings. Cosmet Toiletries. 2010;125(3):30–8.
40. Rigano L, Lionetti N. In: Grumezescu A, editor. Nanobiomaterials in galenic formulations and cosmetics. Applications of nanobiomaterials, vol 10. Oxford: William Andrew— Applications of nanobiomaterialsElsevier; 2016. Chap 6. p. 121–48.
41. Meyer J, Polak G, Scheuermann R. Preparing PIC emulsions with a very fine particle size. Cosmet Toiletries. 2007;122(1):61–70.
42. Meyer J, Scheuermann R, Wenk HH. Combining convenience and sustainability: simple processing of PEG-free nanoemulsions and classical emulsions. SOFW J. 2008;6:58–64.
43. Heunnemann P, Prévost S, Grillo I, Marino CM, Meyer J, Gradzielski M. Formation and structure of slightly anionically charged nanoemulsions obtained by the phase inversion concentration (PIC) method. Soft Matter. 2011;7:5697–710.
44. https://www.sinerga.it/files/materie-prime/nanocream/nanocream-flyer.pdf.
45. Comini M, Lenzini M, Guglielmini G. Nanoemulsions comprising lipoaminoacids and monoglycerides, diglycerides and polyglycerides of fatty acids. Patent WO. 2006;2006087156:A1.
46. Tsutsumi H, Utsugi T, Hayashi S. Study on the occlusivity of oil films. J Soc Cosmet Chem. 1979;30:345–56.
47. Brownlow B, Nagaraj VJ, Nayel A, Joshi M, Elbayoumi T. Development and *in vitro* evaluation of vitamin E-enriched nanoemulsion vehicles loaded with genistein for chemo-prevention against UV$_B$-induced skin damage. J Pharm Sci. 2015;104:3510–23.
48. Campani V, Biondi M, Mayol L, Cilurzo F, Pitaro M, De Rosa G. Development of nanoemulsions for topical delivery of vitamin K1. Int J Pharm. 2016;511:170–7.
49. Calderilla-Fajardo SB, Cazares-Delgadillo J, Villalobos-García R, Quintanar-Guerrero D, Ganem-Quintanar A, Robles R. Influence of sucrose esters in the in vivo penetration of octyl methoxycinnamate formulated in nanocapsules, nanoemulsions and emulsions. Drug Develop Ind Pharm. 2006;32:107–13.
50. Mou D, Chen H, Du D, Mao C, Wan J, Xu H, Yang X. Hydrogel-thickened nanoemulsion system for topical delivery of lipophilic drugs. Int J Pharm. 2008;353:270–6.
51. Zhou H, Yue Y, Liu G, Li Y, Zhang J, Gong Q, Yan Z, Duan M. Preparation and characterization of a lecithin nanoemulsion as a topical delivery system. Nanoscale Res Lett. 2010;5:224–30.
52. Rachmawati H, Budiputra DK, Mauludin R. Curcumin nanoemulsion for transdermal application: formulation and evaluation. Drug Dev Ind Pharm. 2015;41:560–6.
53. Alves MP, Escarrone AL, Santos M, Pohlmann AR, Guterres SS. Human skin penetration and distribution of nimesulide from hydrophilic gels containing nanocarriers. Int J Pharm. 2007;341:215–20.
54. Nam YS, Kim J-W, Park JY, Shim J, Lee JS, Han SH. Tocopheryl acetate nanoemulsions stabilized with lipid–polymer hybrid emulsifiers for effective skin delivery. Colloids Surf B. 2012;94:51–7.
55. Lu W-C, Chiang B-H, Huang D-W, Li P-H. Skin permeation of D-limonene-based nanoemulsions as a transdermal carrier prepared by ultrasonic emulsification. Ultrason Sonochem. 2014;21:826–32.

56. Isailović T, Đorđević S, Marković B, Ranđelović D, Cekić N, Lukić M, Pantelić I, Daniels R, Savić S. Biocompatible nanoemulsions for improved aceclofenac skin delivery: formulation approach using combined mixture-process experimental design. J Pharm Sci. 2016;105:308–23.

57. Klang V, Matsko N, Zimmermann A-M, Vojnikovic E, Valenta C. Enhancement of stability and skin permeation by sucrose stearate and cyclodextrins in progesterone nanoemulsions. Int J Pharm. 2010;393:152–60.

58. Klang V, Haberfeld S, Hartl A, Valenta C. Effect of γ-cyclodextrin on the in vitro skin permeation of a steroidal drug from nanoemulsions: Impact of experimental setup. Int J Pharm. 2012;423:535–42.

59. Hoeller S, Sperger A, Valenta C. Lecithin based nanoemulsions: A comparative study of the influence of non-ionic surfactants and the cationic phytosphingosine on physicochemical behaviour and skin permeation. Int J Pharm. 2009;370:181–6.

60. Baspinar Y, Borchert H-H. Penetration and release studies of positively and negatively charged nanoemulsions—is there a benefit of the positive charge? Int J Pharm. 2012;430:247–52.

61. Yilmaz E, Borchert H-H. Effect of lipid-containing, positively charged nanoemulsions on skin hydration, elasticity and erythema—an in vivo study. Int J Pharm. 2006;307:232–8.

62. Youenang Piemi MP, Korner D, Benita S, Marty J-P. Positively and negatively charged submicron emulsions for enhanced topical delivery of antifungal drugs. J Control Release. 1999;58:177–87.

63. Reinhold U. A review of BF-200 ALA for the photodynamic treatment of mild-to-moderate actinic keratosis. Future Oncol. 2017;13:2413–28.

64. Schmitz L, Novak B, Hoeh A-K, Luebbert H, Dirschka T. Epidermal penetration and protoporphyrin IX formation of two different 5-aminolevulinic acid formulations in ex vivo human skin. Photodiagn Photodyn Ther. 2016;14:40–6.

65. Maisch T, Santarelli F, Schreml S, Babilas P, Szeimies R-M. Fluorescence induction of protoporphyrin IX by a new 5-aminolevulinic acid nanoemulsion used for photodynamic therapy in a full-thickness ex vivo skin model. Exper Dermatol. 2010;19:e302–5.

66. Primo FL, Rodrigues MA, Simioni AR, Bentley MVLD, Morais PC, Tedesco AC. In vitro studies of cutaneous retention of magnetic nanoemulsion loaded with zinc phthalocyanine for synergic use in skin cancer treatment. J Magn Magn Mater. 2008;320:e211–4.

67. Primo FL, Michieleto L, Rodrigues MAM, Macaroff PP, Morais PC, Lacava ZGM, Bentley MVLB, Tedesco AC. Magnetic nanoemulsions as drug delivery system for Foscan®: Skin permeation and retention in vitro assays for topical application in photodynamic therapy (PDT) of skin cancer. J Magn Magn Mater. 2007;311:354–7.

68. Spagnul A, Bouvier-Capely C, Phan G, Landon G, Tessier C, Suhard D, Rebière F, Agarande M, Fattal E. Ex vivo decrease in uranium diffusion through intact and excoriated pig ear skin by a calixarene nanoemulsion. Eur J Pharm Biopharm. 2011;79:258–67.

69. Spagnul A, Bouvier-Capely C, Phan G, Rebière F, Fattal E. A new formulation containing calixarene molecules as an emergency treatment of uranium skin contamination. Health Phys. 2010;99:430–4.

70. Spagnul A, Bouvier-Capely C, Phan G, Rebière F, Fattal E. Calixarene-entrapped nanoemulsion for uranium extraction from contaminated solutions. J Pharm Sci. 2010;99:1375–83.

71. Spagnul A, Bouvier-Capely C, Adam M, Phan G, Rebière F, Fattal E. Quick and efficient extraction of uranium from a contaminated solution by a calixarene nanoemulsion. Int J Pharm. 2010;398:179–84.

72. Shim J, Kang HS, Park W-S, Han S-H, Kim J, Chang I-S. Transdermal delivery of mixnoxidil with block copolymer nanoparticles. J Control Release. 2004;97:477–84.

73. Bachhav YG, Mondon K, Kalia YN, Gurny R, Möller M. Novel micelle formulations to increase cutaneous bioavailability of azole antifungals. J Control Release. 2011;153:126–32.

74. Laredj-Bourezg F, Bolzinger M-A, Pelletier J, Valour J-P, Rovère M-R, Smatti B, Chevalier Y. Skin delivery by block copolymer micelles (block copolymer nanoparticles). Int J Pharm. 2015;496:1034–46.

75. Hansen CM. Hansen solubility parameters. Boca Raton, FLA: A User's Handbook. CRC Press; 2000.
76. Rastogi R, Anand S, Koul V. Flexible polymerosomes—an alternative vehicle for topical delivery. Colloids Surf B. 2009;72:161–6.
77. Chausson M, Fluchère A-S, Landreau E, Aguni Y, Chevalier Y, Hamaide T, Abdul-Malak N, Bonnet I. Block copolymers of the type poly(caprolactone)-*b*-poly(ethylene oxide) for the preparation and stabilization of nanoemulsions. Int J Pharm. 2008;362:153–62.
78. Chevalier Y, Bolzinger M-A. Emulsions stabilized with solid nanoparticles: pickering emulsions. Colloids Surf A. 2013;439:23–34.
79. Laredj-Bourezg F, Chevalier Y, Boyron O, Bolzinger M-A. Emulsions stabilized with solid organic particles. Colloids Surf A. 2012;413:252–9.
80. Laredj-Bourezg F, Bolzinger M-A, Pelletier J, Chevalier Y. Pickering emulsions stabilized by biodegradable block copolymer micelles for controlled topical drug delivery. Int J Pharm. 2017;531:134–42.
81. Laredj-Bourezg F, Bolzinger M-A, Pelletier J, Rovère M-R, Smatti B, Chevalier Y. Pickering emulsions stabilised by biodegradable particles offer a double level of controlled delivery of hydrophobic drugs. In: Chilcott R, Brain K, editors. Advances in dermatological sciences. Issues in Toxicology No 20. Cambridge: RSC Publishing; 2013, Chap 12. pp 143–56.
82. Kim BS, Yang MW, Lee KM, Kim CS. In vitro permeation studies of nanoemulsions containing ketoprofen as a model drug. Drug Deliv. 2008;15:465–9.
83. Sakeena MHF, Muthanna FA, Ghassan ZA, Kanakal MM, Elrashid SM, Munavvar AS, Azmin MN. Formulation and *in vitro* evaluation of ketoprofen in palm oil esters nanoemulsion for topical delivery. J Oleo Sci. 2010;5:223–8.
84. Sakeena MHF, Elrashid SM, Muthanna FA, Ghassan ZA, Kanakal MM, Laila L, Munavvar AS, Azmin MN. Effect of limonene on permeation enhancement of ketoprofen in palm oil esters nanoemulsion. J Oleo Sci. 2010;5:395–400.
85. Sakeena MHF, Yam MF, Elrashid SM, Munavvar AS, Azmin MN. Anti-inflammatory and analgesic effects of ketoprofen in palm oil esters nanoemulsion. J Oleo Sci. 2010;5:667–71.
86. Elrashid SM, Azmin MN, Sakeena MH, Ghassan ZA, Muthanna FA, Munavvar AS. Formulation and in vitro release evaluation of newly synthesized palm kernel oil esters-based nanoemulsion delivery system for 30% ethanolic dried extract derived from local *Phyllanthus urinaria* for skin antiaging. Int J Nanomed. 2011;6:2499–512.
87. Fontana MC, Rezer JFP, Coradini K, Leal DBR, Beck RCR. Improved efficacy in the treatment of contact dermatitis in rats by a dermatological nanomedicine containing clobetasol propionate. Eur J Pharm Biopharm. 2011;79:241–9.
88. Yu M, Ma H, Lei M, Li N, Tan F. *In vitro/in vivo* characterization of nanoemulsion formulation of metronidazole with improved skin targeting and anti-rosacea properties. Eur J Pharm Biopharm. 2014;88:92–103.
89. Ngan CL, Basri M, Tripathy M, Karjiban RA, Abdul-Malek E. Skin intervention of fullerene-integrated nanoemulsion in structural and collagen regeneration against skin aging. Eur J Pharm Sci. 2015;70:22–8.
90. Su R, Fan W, Yu Q, Dong X, Qi J, Zhu Q, Zhao W, Wu W, Chen Z, Li Y, Lu Y. Size-dependent penetration of nanoemulsions into epidermis and hair follicles: implications for transdermal delivery and immunization. Oncotarget. 2017;8:38214–26.

Polymeric Nanoparticles

5

Sílvia S. Guterres, Karina Paese and Adriana R. Pohlmann

Abstract

This chapter addresses recent advances and benefits of using polymeric nanoparticles to encapsulate organic sunscreens, antioxidants, anti-acne agents and fragrances. The nanoparticles can improve the stability of labile ingredients, offer efficient protection for the skin from harmful UV radiation, target the active ingredient to the desired layer of the skin or to the hair follicles, enhance photostability, antioxidant, and antimicrobial effects, and control the release or the volatilization for a prolonged time.

Keywords

Polymeric nanoparticles · Cosmetics · Sunscreens · Antioxidants · Anti-acne agents · Fragrances

S. S. Guterres (✉) · K. Paese · A. R. Pohlmann
Programa de Pós-Graduação em Ciências Farmacêuticas,
Faculdade de Farmácia, Av. Ipiranga 2752, Porto Alegre 90610-000, Brazil
e-mail: silvia.guterres@ufrgs.br

K. Paese
Departamento de Produção e Controle de Medicamentos,
Faculdade de Farmácia, Av. Ipiranga 2752, Porto Alegre 90610-000, Brazil

A. R. Pohlmann
Departamento de Química Orgânica, Instituto de Química,
Universidade Federal do Rio Grande do Sul, Av Bento Gonçalves 9500,
PBox 15003, Porto Alegre 91501-970, Brazil

© Springer Nature Switzerland AG 2019
J. Cornier et al. (eds.), *Nanocosmetics*,
https://doi.org/10.1007/978-3-030-16573-4_5

5.1   Introduction

Nanostructuration modifies chemical, physical, and biological properties of materials leading to a range of new applications in health, including novel delivery systems for pharmaceutics and cosmetics, advanced materials for implants and biosensors, tissue engineering, and new tools for diagnosis. Polymeric nanoparticles are a common term for nanospheres (matrix system) and nanocapsules (reservoir system), in general, used to encapsulate poorly water-soluble substances [20]. The most common polymers used to produce polymeric nanoparticles are biodegradable aliphatic polyesters, such as poly(lactide) (PLA), poly(lactide-*co*-glycolide) (PLGA), and poly(ε-caprolactone) (PCL), as well as biocompatible polymers like poly(methyl methacrylate) and poly-acrylates. This chapter summarizes the recent advances and benefits of using poly-meric nanoparticles to encapsulate organic sunscreens, antioxidants, anti-acne agents, and fragrances (Table 5.1).

Table 5.1 Type of particles, polymers used, and advantages of nanoencapsulation of sunscreens, antioxidants, antiacne agents, and fragrances

Application	Particles	Polymers	Advantages
Organic sunscreens	Lipid-core nanocapsules Nanocapsules Nanoparticles Nanostructured polymeric lipid carriers	Cellulose acetate phthalate; Chitosan; poly(ε-caprolactone); poly(D,L-lactide); poly(methyl methacrylate); and poly(vinyl alcohol) modified with different saturated fatty acids (C12-C20)	Increased photostability Retention of sunscreen in the upper layers of the skin Increased sun protection due to absorption and scattering of UV radiation
Antioxidants and anti-aging	Core-shell nanofibers Lipid-core nanocapsules Nanocapsules Nanoparticles	[poly(ethylene oxide)-*block*-poly(caprolactone)]; blend of ethylcellulose and methylcellulose; Eudragit® E100-poly(vinyl alcohol); Eudragit® RS100; poly(ε-caprolactone); poly(acrylonitrile); and poly(DL-lactide-co-glycolide)	Increased photostability and chemical stability Control of the release of antioxidants Retention of antioxidants in the superficial layers of the skin as a reservoir system Increased antioxidant activity in vitro Incorporation of lipophilic antioxydants into hydrogels
Acne vulgaris	Lipid-core nanocapsules Micelles Nanoparticles Nanospheres	Eudragit® RL100; PCL-PEG-PCL; Pluronic® F127; poly(ε-caprolactone); poly(D,L-lactic acid); and poly(DL-lactide-co-glycolide)	Retention of the active substance in hair follicles Increased antimicrobial activity Control of the release rate of the encapsulated substances Increased photostability of retinoids
Fragrances	Nanocapsules Nanoparticles Nanospheres	Blend of hydroxypropyl methylcellulose, poly(vinyl alcohol), and ethylcellulose; *N*-succinyl chitosan; poly(stearyl acrylate); and polyurea	Control the volatilization for a prolonged time

5.2 Organic Sunscreens

Moderate exposure to ultraviolet (UV) radiation exerts beneficial effects in health, leading to the production of melanin and vitamin D3. However, excessive sun exposure of unprotected skin induces short-term damage, inflammatory response and immunosuppression, as well as chronic effects, such as photoaging and photocarcinogenesis [32]. In view of this, the use of sunscreens on a daily basis is mandatory. An efficient formulation should have sunscreens against both UVB (290–320 nm) and UVA (320–400 nm) radiation, photostability [32], and acceptable sensory properties.

Generally, sunscreens can be classified into two major groups, organic and inorganic filters. Inorganic filters absorb, reflect, and disperse the UV photons [32]. Organic filters absorb UVA and/or UVB radiation due to their chemical structure presenting a chromophore (C = O) conjugated to aromatic rings substituted with electron donating groups (amine or methoxyl) at the *ortho-* or *para*-positions. The absorbed radiation is dissipated at longer wavelengths, usually in the form of heat. The chromophore absorbs UV radiation in the singlet ground state (S_0) promoting an electron to an available vibrational level of the first singlet excited state (S_1). The molecule is rapidly relaxed to the lowest vibrational level of S_1, then, the electron decays from S_1 to the ground state (S_0) or by a spin electron reversion occurring an intersystem crossing the lowest energy triplet state (T1), and then, the molecule decays to S_0. Chemical transformations can occur in S_1 or T_1.

The photostability of an organic filter is associated with its ability to release energy without undergoing changes in the chemical structure, ensuring the effectiveness of UV absorption by the filter during sun exposure. A reduction in the UV absorption capacity results in a greater amount of UV radiation affecting the skin and inducing damage. Moreover, the degradation products may induce allergic or toxic reactions [44].

Among the advantages of polymeric nanoparticles, it is worth highlighting their ability in protecting organic filters. In addition, these carriers are skilled in controlling the release of the encapsulated substance delivering it to the layer of the skin of interest and keeping it available for a longer period. Due to the small particle size and the use of polymers with adhesive properties, there is a high contact time of the formulations with the *stratum corneum*. Thus, nanoparticles are adhered by affinity to the horny layer gradually releasing the sunscreen and ensuring a prolonged effect and an improved photoprotection. Polymeric nanoparticles can also provide a synergistic effect with the encapsulated organic filters since the UV radiation is simultaneously absorbed and scattered [24].

The effectiveness of sunscreens depends on its maintenance on the surface of the skin forming a protective film. The percutaneous absorption of sunscreens is reduced by nanoencapsulation at the same time as the effect of the formulation is prolonged promoting a constant release of the filter. Octyl methoxycinnamate (OMC), an UVB filter, loaded in PCL nanocapsules is retained 3.4-fold in the *stratum corneum* compared to the nonencapsulated sunscreen, after 6 h of exposure, as demonstrated in vitro using porcine ear skin as a biological membrane. OMC (nonencapsulated or encapsulated) was not detected in the Franz cell acceptor compartment; this is the desired result since the permeability of this substance through the skin shall be minimal avoiding systemic toxicity [2].

In addition, the nanoparticle aqueous dispersions can be incorporated into different end formulations, such as emulsions and hydrogels. PCL nanocapsules containing OMC were incorporated in o/w and w/o emulsions in order to evaluate the in vitro penetration and permeation of the sunscreen in porcine skin [24]. After 3 and 24 h of exposure, OMC penetration into the *stratum corneum* and other skin layers, as well as the permeation through the skin were higher for the formulations containing nonencapsulated OMC. There was a greater accumulation of OMC on the surface of the skin when the o/w and w/o emulsions containing OMC nanocapsules were applied, corroborating previous findings [2]. The results point to the low diffusion of OMC and its high retention at the skin surface when it is encapsulated, demonstrating that the nanocapsules act as OMC reservoirs on the surface of the skin.

The supramolecular structure of nanocarriers affects OMC skin distribution. Tape stripping technique to determine the in vivo retention showed that OMC applied in a nanoemulsion (closed to 160 nm) penetrated more in the *stratum corneum,* after 1 h, than OMC incorporated in (cellulose acetate phthalate)-nanocapsules (about 400 nm). The reduced size and flexibility of the droplets in the nanoemulsion explain the results [39].

Hydroxyethylcellulose hydrogel containing OMC-PCL nanocapsules was applied to pigskin [62]. After 6 h, OMC was extracted using acetonitrile to dissolve the nanocapsules and OMC or *iso*propyl myristate to dissolve OMC. The approach allowed the quantification of the OMC content or the OMC released in the layers of the skin. Using one or another solvent, the concentrations of OMC in the dermis were similar. On the other hand, on the surface of the skin and in the *stratum corneum,* a significant difference between the content of OMC and the released OMC was observed, indicating that the nanocapsules are accumulated on the surface of the skin and gradually release the sunscreen to penetrate the viable skin layers, acting as reservoirs. Similarly, PLA nanocapsules containing OMC were evaluated [60]. The nanoencapsulated sunscreen remained mostly on the surface of the skin (about 80%), and the amount released that reached the viable epidermis was three times smaller than the amount measured when an emulgel containing nonencapsulated OMC was assayed [60].

The permeation of benzophenone-3 (BZ3), a broad-spectrum sunscreen, was evaluated in vitro. The filter was encapsulated using as structuring material poly (vinyl alcohol) modified with different saturated fatty acids (C12-C20) with two degrees of substitution (40 and 80%). Using as pendant groups myristic or palmitic acids, smaller particles have been obtained than using stearoyl or behenoyl substituents. The degree of substitution influenced more the skin permeation of the sunscreen than the nature of the substituent. Poly(vinyl alcohol) modified with high degrees of substitution prevents the percutaneous absorption of the filter, while the polymer with low degree of substituents enhances the sunscreen location in the epidermis [31].

Coating the nanoparticles with cationic polymers is a strategy to increase the electrostatic interaction between the particles and the negatively charged biological surfaces. Chitosan is a cationic bioadhesive polymer used in the biomedical area since it has biocompatibility, biodegradability, and low toxicity [65]. Chitosan-coated PCL nanocapsules containing BZ3 incorporated into a hydroxyethylcellulose hydrogel were evaluated for up to 8 h in terms of skin penetration/permeation using Franz diffusion cells and porcine skin as a membrane [52]. The chitosan coating contributed for the maintenance of BZ3 in the *stratum corneum* due to its bioadhesiveness. On the other hand, no difference was observed for BZ3 retained in the other layers of the skin when compared to a hydrogel formulation containing the free sunscreen. As regards the permeation, chitosan-coated nanocapsules were able to decrease the amount of BZ3 in the acceptor compartment [52].

Polymeric nanoparticles can modulate the penetration/permeation of nanoencapsulated organic sunscreens differently than the lipid systems. As previously mentioned, nanoemulsions may increase the flux of encapsulated substances because of their flexibility when compared to polymeric nanocapsules [39]. The skin permeation of BZ3 released from PCL nanoparticles was evaluated in vitro employing human skin during 24 h of exposure. Comparatively, the permeation of the sunscreen from the other two formulations, a conventional emulsion and solid lipid nanoparticles consisting of cetyl palmitate, was also determined. Nanoencapsulation in polymeric nanoparticles was able to decrease the flux of BZ3 through the skin in about 70% in the epidermis and dermis and 80% in the receptor fluid. In another study, a diverse result was observed; the solid lipid nanoparticles did not decrease the permeation of BZ3 when compared to a conventional emulsion [33]. The percutaneous absorption and bioavailability of BZ3 through pig ear skin was also evaluated for 24 h. Similarly to the previous study, polymeric systems were compared to lipid nanoparticles using as control BZ3 dissolved in an aqueous solution of albumin [18]. Solid lipid nanoparticles (SLN), nanostructured lipid carriers (NLC), nanostructured polymeric lipid carriers (NPLC), and nanocapsules (NC) were studied. The results showed that NPLC and NC reduced the skin permeation of BZ3. A higher flux through the skin was observed for the control (nonencapsulated filter) due to the characteristics of BZ3 (moderate lipophilicity and low molecular weight). The formulations without polymer (SLN and NLC) did not differ significantly from the control, while the polymeric formulations (NPLC and NC) reduced the flow. The result can be explained in the case of NPLC by the high association of BZ3 to the carrier (about 50%), compared to other systems which showed encapsulation efficiency of around 30%. In the case of NC, the presence of poloxamer in the formulation might increase BZ3 affinity for the dispersed phase decreasing the permeation of the sunscreen through the skin [18]. Beyond modulating the skin permeation/penetration, the polymeric systems have other advantages over the lipid formulations, such as higher encapsulation efficiencies. For instance, Paiva et al. [43] have encapsulated BZ3 and avobenzone in liposomes and in poly(methyl methacrylate)-nanoparticles. Chloroform was used for the production of the liposomes; on the other hand, for the polymeric nanoparticles, produced by miniemulsion polymerization, no organic solvent was used. The encapsulation efficiency was close to 100% for the polymeric nanoparticles, while for the liposomes, after filtration, the value was less than 20%.

OMC photostability was evaluated by comparing PCL nanocapsules produced with two different surfactant systems (sorbitan monostearate/polysorbate 80 or lecithin) [61]. Quercetin has been co-encapsulated with OMC to act as a scavenger of reactive oxygen species. The formulations were exposed to UVA radiation, and the degradation of OMC and quercetin was monitored. The nanocapsules were able partially to protect OMC from photodegradation. It was observed recovery from 17 to 35% of nanoencapsulated OMC after 15 days of exposure, while for the control, an OMC methanol solution, the recovery was less than 1%. Co-encapsulation was effective in further delaying both quercetin and OMC photodegradation when compared to the formulations containing the isolated substances. The surfactant system that better stabilized OMC was sorbitan monostearate/polysorbate 80, likely because the use of sorbitan monostearate provided a formulation with a greater number of nanocapsules dispersed in water since it is located at the lipid core, while lecithin is located at the nanocapsule–water interface. Thus, light scattering was higher for the former causing higher protection to OMC. In parallel, the polymeric coating may help to protect the encapsulated substance, and then, the use of different polymers has been evaluated [59]. PLA nanoparticles containing OMC (*trans*-isomer) were incorporated into an emulgel formulation and exposed to a solar simulator for 2 h. The degree of photodegradation was evaluated quantifying the OMC *cis*- and *trans*-isomers. The not desired *cis*-isomer, which absorbs less ultraviolet radiation, is formed by photoisomerization of *trans*-OMC. When the filter was encapsulated, the *cis-trans* ratio was decreased indicating the reduction in the OMC degradation compared to the control formulation (nonencapsulated OMC in emulgel).

The light scattering from polymeric nanoparticles prevents the radiation-induced degradation of nanoencapsulated molecules being probably one of the mechanisms involved in the photostabilization of organic filters. In addition, the isolation of the molecules from contact with other reactive molecules, provided by the polymeric wall or matrix, may also be responsible for increasing the photostability of organic filters. The stability of BZ3 loaded in PCL nanoencapsules incorporated in hydrogels against exposure to UVA is higher than that observed for hydrogels containing the nonencapsulated sunscreen (control). After 13 h of exposure, about 56% of BZ3 was recovered from the formulation containing nanocapsules and only 29% was recovered from the control [42].

Studies have been dedicated to describe the improvement of the sun protection using polymeric nanocarriers. A hydrogel containing OMC encapsulated in PCL nanocapsules was able to decrease the development of erythema. This formulation was compared to a hydrogel containing nonencapsulated OMC by an in vivo methodology, which evaluated the development of erythema upon exposure to UV

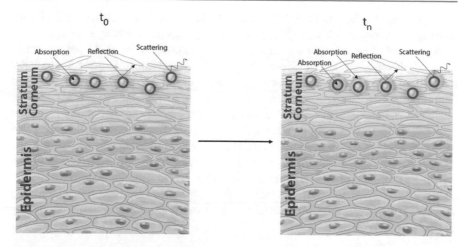

Fig. 5.1 Illustration showing the absorption, reflection and scattering of UV photons by sunscreen-loaded nanocapsules, as reservoirs, located at the *stratum corneum* (artwork by Max Saito). *Note* The nanocapsules are oversized to better demonstrate the mechanisms

radiation of photoprotected pigs. The improved protection of the encapsulated OMC indicates that the formulation adhered to the skin, forming a protective film, resulting in an erythema of lower intensity [1]. Additionally, a hydrogel containing BZ3-loaded polymeric nanocapsules was able to simultaneously absorb and scatter UV radiation (Fig. 5.1) [42].

The in vitro solar protection factor (SPF) was determined for hydrogels containing BZ3 encapsulated in polymeric nanocapsules or in SLN. Both formulations were compared to a control (hydrogel containing nonencapsulated sunscreen). The SPF values were 18 for the polymeric nanocapsules, 21 for SLN, and 16 for the control formulation. The highest SPF for SLN might be due to a higher light scattering, consequence of the high crystallinity of cetyl palmitate (78.4%) compared to the slightly lower crystallinity of PCL (71.2%) [33]. Cosmetic emulsion containing a combination of nanoencapsulated organic sunscreens (OMC, octocrylene, and BZ3) also showed high in vitro SPF (22) when evaluated using a spectrophotometer comprising an integrating sphere to determine the intensity of UV light transmittance through a substrate of poly(methyl methacrylate) coated by the formulation. In this study, the SPF of an emulsion containing the same combination of nonencapsulated sunscreens was 12 [37]. Diverse structures used to nanoencapsulate BZ3 (nanostructured polymeric lipid carriers, poly(ε-caprolactone)-nanocapsules, solid lipid nanoparticles, and nanostructured lipid carriers) showed higher in vitro SPF than the control (nonencapsulated BZ3), showing the ability of the systems, in addition to absorbing UV radiation, scattering the light, acting as physical filters [18].

Regarding the safety of the cutaneous application of nanoencapsulated organic sunscreens in polymeric nanoparticles, few studies have been conducted so far. The cytotoxicity and phototoxicity of polymeric nanocapsules and SLN containing BZ3 were evaluated using BALB/c 3T3 mouse embryo fibroblast cell line and HaCaT human keratinocyte cell line [33]. Both formulations were not cytotoxic to these strains. Regarding the phototoxicity, BZ3 encapsulated in PCL nanoparticles showed phototoxic potential against HaCaT cells. In addition, a cosmetic emulsion containing OMC, octocrylene, and nanoencapsulated BZ3 was considered as a minor irritant in the hen's egg test–chorioallantoic membrane assay (HET-CAM), which assesses vascular phenomena and indirectly predicts the possibility of ocular irritation [37]. However, in the chorioallantoic membrane–trypan blue staining test (CAM-TBS), which also evaluates ocular irritation using trypan blue absorption as an indicator of CAM lesion, the result demonstrates that the formulation is nonirritating. Furthermore, this same formulation showed no irritant effect using the red blood cell test for irritation potential (RBC), which quantifies the photometric absorption of oxyhemoglobin, hemolysis, and denaturation of proteins caused by surfactants.

5.3   Antioxidants and Anti-aging

The generation of reactive oxygen species (ROS) is one of the main causes of oxidative stress. ROS can interact with the skin macromolecules (lipids, proteins, and DNA) inducing damage [35]. Skin exposure to the sun is the main reason that triggers the production of ROS, responsible for accelerating the process of aging skin, which is characterized by wrinkles and atypical pigmentation.

Antioxidants interact with free radicals shutting down the chain reaction before the skin macromolecules are damaged. In the skin, antioxidants can prevent the lipid peroxidation by reaction with lipid radicals and peroxides, converting them into more stable molecules. However, antioxidants have some limitations with regard to their use in cosmetics, such as low stability (light, pH, temperature, and oxygen), low ability to penetrate the different layers of the skin, and poor solubility in water. All these limitations compromise the antioxidant incorporation into hydrophilic formulations such as gels [11].

Semisolid vehicles have been described for the incorporation of antioxidant-loaded polymeric nanoparticles in cosmetics. Spray-dried-coenzyme Q10-loaded nanocapsules were incorporated into a Carbopol hydrogel [55], nanocapsules of rosehip oil in a chitosan gel [9], curcumin-loaded nanoparticles either in a sodium carboxymethylcellulose hydrogel [38] or in w/o or o/w lotions [54], and lipoic acid-loaded nanocapsules either in a silicone emulsion or in an Aristoflex AVC® gel [27].

Fixed oils of plant origin contain antioxidants useful for cosmetic formulations. Grape seed, sunflower, rose, and carrot oils were used as oily cores in the production of Eudragit® RS100 nanocapsules (mean diameter of about 280 nm and physically stable for 90 days) [7]. This cationic polymer has the advantage of increasing the interaction of the particles with the skin surface, which has anionic residual charge. The nanoencapsulation of vegetable oils can be advantageous to

prevent their oxidation induced by UV light. A hydrogel containing Eudragit®
RS100 nanocapsules of rosehip oil showed the protection of the oil from lipid
peroxidation after 24 h of exposure to UVA and UVC radiations. The malondialdehyde concentration was about three times lower when the oil was nanoencapsulated compared to the control, a dispersion of this oil in water [9].

Nanoencapsulation can improve the skin bioavailability of antioxidants for a
prolonged period. Core-shell nanoparticles produced with amphiphilic and thermosensitive triblock copolymers (PNIPAM-*b*-PCL-*b*-PNIPAM; 108:384:108,
17:38:17, and 29:88:26) controlled the release of α-tocopherol. Dialysis method
showed a burst effect in the first 20 h and a sustained release profile within 140 h
[46]. The lipophilicity of antioxidants may result in different release profiles when
conanoencapsulated. Resveratrol and curcumin co-encapsulated in lipid-core
nanocapsules showed that 70% of resveratrol and only 10% of curcumin were
released after 24 h [16]. The result is attributed to a higher solubility of curcumin
compared to that of resveratrol in the oily core of the nanocapsules, retarding the
curcumin diffusion to the aqueous outer phase where it is less soluble.

Core-shell-poly(acrylonitrile) nanofibers controlled the release of derivatives of
vitamins A and E, magnesium L-ascorbic acid 2-phosphate, and α-tocopherol
acetate compared to their released from blend nanofibers. In the first 6 h, the blend
nanofibers released 35–40% of the antioxidants, while the core-shell-poly(acrylonitrile) nanofibers released only 10–12%. After 76 h, about 80% of the antioxidants were released from the blend nanofibers and 67% from the core-shell-poly
(acrylonitrile) nanofibers. Highest burst release for the blend nanofibers was due to
the antioxidants location at fibers surface, whereas the gradual release from the
core-shell nanofibers was due to their encapsulation [63].

The release profiles are also influenced by the degradation rate of the polymers.
PLGA nanoparticles containing different concentrations of ascorbic acid (15, 30,
and 50% relative to the polymer) were maintained for 2 months in saline at 37 °C.
The samples were collected weekly to observe the particles by scanning electron
microscopy and to quantify the ascorbic acid released. In the first 24 days, less than
10% of ascorbic acid was released from all samples, but within 8 weeks, the total
amount of ascorbic acid was released. At the beginning of the study, the particles
were spherically shaped, but after 24 days, aggregates were observed. After
2 months, there was no trace of nanoparticles in the solution. The PLGA degradation was confirmed by observing the decrease in pH values from 7 to 3
(2 months) due to the presence of glycolic and lactic acids [53].

A cosmetic formulation design must consider at which skin layer the effect is
expected avoiding absorption and systemic distribution of the active substance.
Polymeric nanocarriers have been designed for this purpose since they can act as
reservoirs in the *stratum corneum* and epidermis, slowly releasing the substances
with minimal absorption by the systemic circulation. Lipid-core nanocapsules and
nanostructured lipid carriers were able to maintain a greater amount of resveratrol in
the *stratum corneum* and in the epidermis when compared to the antioxidant
administered as a solution. In this study, the permeation was evaluated for 8 h and
the skin exposed to UVA radiation [12]. Furthermore, Eudragit® E100-poly(vinyl

alcohol) nanoparticles containing 7,3',4'-trihydroxyisoflavone were evaluated
in vitro for 8 h using porcine skin. When nanoencapsulated, the amounts of
antioxidant retained in the *stratum corneum* and in the viable epidermis/dermis
were about five times greater compared to a solution of 7.3',4'-trihydroxyisoflavone
[23].

The cutaneous permeation of curcumin-loaded nanoparticles produced with
ethylcellulose or a blend of ethylcellulose and methylccllulose incorporated into
o/w or w/o lotions was compared [54]. Laser scanning microscopy was used to
measure curcumin after application of the products on pigskin. Nanoencapsulation
improved the curcumin penetration through the hair follicle and the type of
emulsion influenced its deposition. After 1 h, the lotions containing
curcumin-loaded nanoparticles showed deeper follicular curcumin penetration
compared to an aqueous suspension of curcumin, and the penetration was greater
when the w/o lotion containing curcumin-loaded nanoparticles was administered. In
addition, the polymer blend improved the penetration of curcumin into the skin due
to the presence of methylcellulose, which presents amphiphilic characteristics
allowing better penetration into the skin due to its water-soluble portion [54].
Moreover, the highest deposition of nanoencapsulated curcumin (from sodium
alginate nanocarriers) in the first 50 μm of skin, equivalent to the epidermis, was
visualized by confocal microscopy 24 h after application on human skin in an
in vitro experiment, indicating no transdermal delivery of curcumin [38]. Additionally, nanoencapsulated curcumin in solid lipid nanoparticles, nanoemulsion, and
PCL nanoparticles was compared in terms of in vitro skin penetration/permeation
for 24 h using Franz permeation cells and porcine skin [5]. After 24 h, the epidermis was separated from the dermis for quantification of curcumin, and confocal
microscopy was used to visualize the deposition of curcumin in the different layers
of the skin. PCL nanoparticles led to a greater accumulation in the epidermis with
2.5- and 3.3-times more curcumin than the solid lipid nanoparticles and the
nanoemulsion, respectively. Confocal microscopy confirmed the results showing
higher fluorescence emission in the epidermis compared to the dermis when
polymeric nanoparticles were applied.

The use of [poly(ethylene oxide)-*block*-poly(caprolactone)] (PEO-*b*-PCL)
copolymer to stabilize a nanoemulsion formulation containing tocopherol acetate
also limited the amount of antioxidant permeated through the skin as compared to a
nanoemulsion stabilized with only unsaturated phospholipids. The amount of
tocopherol acetate permeated from the PEO-*b*-PCL-stabilized nanoemulsion was
59% of the value observed for the phospholipid-stabilized nanoemulsion although
the droplet diameters are similar, closed to 100 nm [36]. These results indicate that
the diblock copolymer can act as a polymer wall for the nanoemulsion, which
behaves as a reservoir system on the skin surface.

The skin penetration/permeation of encapsulated substances may also be influenced by their release profiles and interactions with the different layers of the skin.
The permeation of curcumin and resveratrol co-encapsulated in lipid-core
nanocapsules was evaluated in Franz cells for 24 h using human skin [16]. The
low rate of curcumin released from nanocapsules correlated with its low penetration

into all layers of skin compared to the control. Resveratrol co-encapsulated with curcumin showed greater amounts in the epidermis and dermis when compared to the solutions or with the nanocapsules containing only resveratrol. The interaction of curcumin with the *stratum corneum* explains the results. Curcumin may reversibly alter the structure of the phospholipid bilayer in the *stratum corneum* due to its high lipophilicity, facilitating the transport of resveratrol that accumulates in the deeper layers of the skin.

A limitation in the development of cosmetic formulations containing antioxidants is related to their low chemical or photochemical stabilities. Lipoic acid loaded in PCL-lipid-core nanocapsules showed chemical stability for 28 days, compared to the control (lipoic acid in water), which showed a significant reduction of the lipoic acid content [26]. Beta-carotene encapsulated in micelles of chitosan-*graft*-poly(lactide) presented loss of only 4.5 and 8.1% when stored for 15 days at 4 and 25 °C, respectively [17]. Moreover, liposomes, nanostructured lipid carriers, nanocapsules or polymeric nanospheres decreased the photodegradation rate of resveratrol. The longer half-life was observed for resveratrol encapsulated in liposomes, but they lost physical stability after exposure to UVA radiation. Polymeric nanocapsules and nanostructured lipid carriers showed similar and significantly lower resveratrol photodegradation compared to the ethanol solution [12].

Coenzyme Q10-loaded lipid-core nanocapsules dispersed in water had mean diameter close to 250 nm [55]. To obtain a redispersible powder, the formulation was spray-dried using lactose as an adjuvant. After redispersion in water, the initial particle size was recovered. This process is useful since it reduces the chemical and microbiological instability characteristic of liquid products. Moreover, nanoparticles in aqueous dispersions can also be freeze-dried as demonstrated for curcumin-loaded calcium alginate nanoparticles using trehalose as a cryoprotectant, whose mean diameter increased only 30% after redispersion [38].

The antioxidant capacity can also be increased by nanoencapsulation. PNIPAM-*b*-PCL-*b*-PNIPAM nanoparticles containing α-tocopherol were able to increase its in vitro antioxidant activity in aqueous medium. Two phenomena can explain this increase, the diffusion of free radicals from the nanoparticles to the aqueous outer phase, and the surface characteristics of the particles that promote α-tocopherol migration to the aqueous phase [46]. In addition, nanoencapsulated beta-carotene had higher in vitro antioxidant activity being able to eliminate 100% free radicals, while the nonencapsulated antioxidant was able to eliminate only 60% of the free radicals after 10 min. [17]. Also, the in vitro antioxidant activity of resveratrol and curcumin conanoencapsulated against hydroxyl radicals was increased with a synergistic effect [10].

The incorporation of nanocarriers into cosmetic formulations may alter some sensory characteristics. Silicone emulsions containing lipoic acid-loaded PCL nanocapsules were evaluated in a discriminative and affective analysis performed with 88 individuals, men and women aged from 17 to 40 years old without previous training. The subjects applied the formulations on the forearm and evaluated the spreadability (ease of distributing the product on the surface of the skin), viscosity (intensity of finger-to-skin adhesion at the time of application), oiliness (greasy skin sensation), and residual odor of thiol. In the discriminative analysis, viscosity and oiliness were statistically identified as properties that differed between the nonencapsulated or nanoencapsulated lipoic acid silicon emulsions. Considering the affective analysis, nanoencapsulated lipoic acid silicon emulsions were preferred due to the lowest consistency, and consequently, lower viscosity added to the lower residual odor of thiol. In contrast, the nonencapsulated lipoic acid silicon emulsions were preferred in terms of oiliness and spreading [28].

Discriminative and affective analyses have also been used to optimize a chitosan hydrogel containing Eudragit® RS100-nanocapsule formulation [8]. In this study, 60 individuals evaluated chitosan hydrogels compared to hydroxyethylcellulose gels, as vehicles for the nanocapsules. Chitosan gel presented higher viscosity and higher film formation on the skin leading to few acceptances than the hydroxyethylcellulose gels. Besides increasing the consistency of the chitosan gel, the presence of nanocapsules also increased the perception of the formation of the film on the surface of the skin. To improve the sensory characteristics of the chitosan gel, Silicone DC245® and carboxylic pyrrolidone acid sodium salt were added in the formulation diminishing the film formation, which also presented higher homogeneity [8].

Polymeric nanocapsules have been commercialized for more than two decades. In 1995, L'Oreal introduced vitamin E nanocapsules in the cosmetic market providing powerful antioxidant protection. More recently, L'Oréal marketed Primordiale Intense and Hydra Zen Serum, among others encapsulating a diversity of substances.

5.4   Acne Vulgaris

Acne vulgaris is a chronic inflammatory cutaneous disorder considered the most prevalent disease of the pilosebaceous unit [51]. In the pathophysiology of acne, the genetic predisposition is involved, besides other factors: sebaceous gland hyperplasia and increased sebum production, hyperkeratinization of the pilosebaceous follicle, proliferation of *Propionibacterium acnes* (*P. acnes*), and inflammatory response initiated by bacterial antigens and cytokines [14]. Acne symptoms include the development of comedones, papules, pustules, nodules, abscesses, and scars depending on the severity of the disease [51].

The mainstay of acne therapy includes topical and/or systemic treatment aimed at reducing the sebum production, normalizing the hyperkeratinization of the pilosebaceous follicle, reducing *P. acnes* proliferation, and decreasing or preventing the release of inflammatory mediators [51]. Topical therapy with antimicrobials, retinoids, benzoyl peroxide, keratolytic, and α-hydroxyacids is widely employed [30]. Many of these substances have some drawbacks, like induction of irritating skin reactions and chemical instability against environmental factors, such as light, heat, and temperature [4, 56].

Benzoyl peroxide is unstable and has low solubility in water, which limits its incorporation into hydrophilic semi-solid formulations [25]. In addition, benzoyl peroxide triggers adverse effects at the beginning of treatment including erythema, dryness, scaling, and skin burning. In order to reduce these inconveniences, benzoyl peroxide has been encapsulated in liposomes, lipid nanoparticles, and polymeric nanoparticles modulating its release and reducing the skin irritation.

Nanoencapsulation enables the incorporation of benzoyl peroxide into hydrophilic formulations, for instance, gels, and provides greater accumulation in the hair follicle [6, 66]. The penetration of benzoyl peroxide-loaded Pluronic® F127 micelles (diameter of about 25 nm) through the porcine skin was compared to the penetration of a commercial gel. After 24 h, benzoyl peroxide was not detected in the acceptor compartment indicating no skin permeation. However, threefold higher amounts of benzoyl peroxide from the micelle formulation penetrated and were retained in the skin compared to the commercial formulation. This higher penetration of benzoyl peroxide was attributed to the increase in the thermodynamic activity. Furthermore, the small diameter of the micelles allows an increase in the contact surface with the skin acting as a reservoir [25].

PLA nanoparticles and lipid nanoparticles, both containing benzoyl peroxide, were incorporated into foams formed by heptafluoropropane as a propellant and Pluronic® L62D as a surfactant. The benzoyl peroxide permeation from the formulations (polymeric nanoparticles, lipid nanoparticles, and their respective foams) through silicone membranes was evaluated in comparison with a cream containing nonencapsulated benzoyl peroxide and a commercial gel. The lipid nanoparticles were able to increase the benzoyl peroxide permeation, presenting a flux of 43 μg cm^{-2} h^{-1}, while PLA nanoparticles promoted a lower flux (2 μg cm^{-2} h^{-1}), likely due to the higher surface area of the lipid nanocarriers (50 nm) compared to the polymeric ones (350 nm) [66].

Nanoencapsulation of triclosan, an acne-mitigating agent, in Eudragit® E100 nanoparticles was evaluated in vitro using pigskin, showing a flux of 10.7 μg cm^{-2} h^{-1} higher than those observed for the nonencapsulated drug (6.3 μg cm^{-2} h^{-1}) or for an emulsion w/o containing triclosan (6.3 μg cm^{-2} h^{-1}), within the first 2 h. However, after that time, the flux for the nanoencapsulated triclosan decreased to 2.1 μg cm^{-2} h^{-1}, whereas the flux of the nonencapsulated drug was maintained. The rapid penetration of nanoparticles into the hair follicles caused a saturation, which decreased the triclosan flux through the skin [13], explaining the result. In turn, cationic nanospheres produced with Eudragit® RL100 were able to increase 8.5-fold the retention of triclosan in pigskin compared to a triclosan solution, proving again the nanoparticles act as reservoirs in the hair follicles. In addition, the amount of triclosan retained in the skin using cationic nanospheres was ten times greater than the minimum inhibitory concentration of triclosan against *P. acnes* (7.8 ppm) (Rodríguez-Cruz et al. [49]). Additionally, nanoencapsulation is a strategy to control the release rate of azelaic acid. Azelaic acid-PLGA nanoparticles showed a "burst effect" in the first 12 h. After this period, a sustained release was observed reaching 100% in 75 h [48].

With the view of improving the antimicrobial activity against *P. acnes*, lauric acid (LA), a short chain fatty acid with antimicrobial activity, was encapsulated in PCL-PEG-PCL micelles, synthesized with different ratios of each block. The higher the PCL portion was, the greater the encapsulation efficiency of LA in the micelles. The lower and the greatest encapsulation efficiencies were 3.3 and 15.4%. The minimum inhibitory concentration (MIC) and minimum bactericidal concentration (MBC) were determined. The MIC of LA in solution was 20 $\mu g\ mL^{-1}$, and the MIC of LA-PCL-PEG-PCL micelles (highest PCL portion) was 10 $\mu g\ mL^{-1}$. MBC was 80 $\mu g\ mL^{-1}$ for LA in solution and half of that concentration (40 $\mu g\ mL^{-1}$) for the LA-PCL-PEG-PCL micelles [57].

In acne, besides acting as anti-inflammatory agents, retinoids modulate the proliferation and differentiation of keratinocytes inside the pilosebaceous unit preventing the formation of comedones. Tretinoin, isotretinoin, adapalene, and tazarotene are the most commonly used retinoids. Nevertheless, their cutaneous application is limited by the development of topical irritation, high chemical instability against oxygen, light, and heat, as well as low solubility in water, which makes difficult their incorporation in hydrogels [51]. The photostability against UVC radiation of tretinoin-PCL-lipid-core nanocapsules was studied. The half-life times of the tretinoin degradation were, respectively, 89 and 95 min for the nanocapsules prepared using sunflower oil or medium-chain triglycerides, as oily cores. Comparatively, the half-life time of tretinoin in methanol was 40 min. When evaluating the photostability of a nanoemulsion, a formulation in which the polymeric wall is omitted, the half-life was 69 min (nanocapsules prepared using sunflower oil) and 82 min (nanocapsules prepared using medium-chain triglycerides). The polymeric wall of the lipid-core nanocapsules was determinant for the stabilization likely due to the crystallinity of the polymer and the ability of the nanocapsules to scatter light [40]. Moreover, retinol acetate has also been encapsulated in polymeric nanoparticles of poly(ethylene glycol)-4-ethoxycinnamoylphthaloylchitosan and exposed for 1 h to UVA radiation showing lower degradation than the retinol acetate solution. The polymer used to obtain the nanoparticles was able to absorb the UV radiation and this mechanism explains the increase of the photostability [3].

The lipid-core nanocapsules dispersed in water can be incorporated into hydrogels, maintaining the drug stabilizing capacity. Carbopol hydrogels containing nanoencapsulated tretinoin exhibit lower photodegradation (about 25%) than the same semisolid formulation containing the nonencapsulated substance (about 70% degradation) after 8 h of UVA irradiation [41]. Additionally, tretinoin-loaded lipid-core nanocapsules were added to 10% (m/v) lactose and spray-dried. The photostability of the liquid dispersion and that of the resuspended powder has been compared after exposure to UVA radiation. After 20 min, the photodegradation of tretinoin was 6.5% for the nanocapsule aqueous dispersion and 8.2% for the resuspended powder of nanocapsules, values of about fourfold lower than the degradation of tretinoin in methanol (30.7%). After 60 min, the photodegradation of tretinoin in nanocapsule aqueous dispersion and in the resuspended powder of nanocapsules was about 30%, while the photodegradation of tretinoin in methanol was about 60% [34].

The decrease in skin permeation is also observed when tretinoin is nanoencapsulated. Using human heat separation epidermis as a diffusion membrane, the permeability coefficient (Kp) of the nanoencapsulated tretinoin across the membrane was around 0.30 cm s^{-1} and greater than 1.00 cm s^{-1} for the tretinoin solution [41]. Increased deposition of nanoencapsulated tretinoin in hair follicles was also verified. Moreover, polymeric micelles of methoxy-poly(ethylene glycol)-poly(hexylsubstituted lactic acid), a biocompatible and biodegradable polymer, were produced to incorporate tretinoin. The penetration/permeation of tretinoin was evaluated in vitro in Franz cells using porcine skin and human skin as membranes. After 12 h, biopsy of the hair follicles was performed to quantify tretinoin retained. The tretinoin micelles were compared to a tretinoin solution and to a commercial gel containing tretinoin microparticles. For the last two formulations, tretinoin was not detected in the acceptor compartment indicating no permeation through the skin. When encapsulated in micelles, tretinoin penetrated more in both pigskin and human skin (7.1 and 7.5% of the applied dose, respectively) compared to the microencapsulated substance in hydrogel (0.4 and 0.8% of the applied dose). The efficiency of the delivery in the pilosebaceous unit in human skin was also higher for nanoencapsulated tretinoin (10.4%) compared to the commercial formulation (0.6%) [29].

Adapalene is commercially available as gel and cream. To improve solubility, the commercial formulations contain alcohols acting as cosolvents and surfactants. Nevertheless, both alcohols and surfactants can also be irritating to the skin [19, 47]. Adapalene-containing acid-responsive polymer nanocarriers were developed using Eudragit® EPO. The release of adapalene from the nanocarriers was pH-dependent, being faster at pH 4.0 and slower at pH 5.0 and 6.0, due to the higher solubility of the polymer at lower pH values. Despite a difference in the release profile determined for adapalene when nanoencapsulated, its retention in *stratum corneum* of pigskin was similar to the control after 24 h [19]. In a different way, the nanoencapsulation of adapalene in tyrosine-derived nanospheres (Tyro-Spheres) was able to increase its accumulation in porcine hair follicles after 12 h, compared to a commercial formulation. The amount of adapalene in the hair follicle was 288 and 185 ng cm^{-2} for nanoencapsulated adapalene and for the commercial formulation, respectively [47]. Another retinoid, the retinyl acetate was encapsulated in poly(ethylene glycol)-4-methoxycinnamoylphthaloyl-chitosan nanoparticles showing slower penetration in mouse skin compared to retinyl acetate solution, and 100% of cutaneous retention after 24 h. Additionally, confocal microscopy analyzes demonstrated the presence of both polymer and retinyl acetate in the hair follicle region [3].

Adapalene nanoencapsulated in TyroSpheres was not irritating when evaluated by in vitro methodologies. The nanoencapsulated formulation was less irritating as the metabolic activity of the HaCaT cells (the human keratinocyte cell line) exposed to this formulation was significantly higher than that of the cells exposed to the commercial formulation. An in vitro skin model, fully differentiated human epidermis (EpiDerm™), has also been exposed to the formulations. The skin irritation of adapalene was assessed by determining the mitochondrial activity of the cells and the concentrations of inflammatory mediators (IL-1α and IL-8) in the medium. After 24 h, for nanoencapsulated adapalene, the tissue viability was higher with the lower secretion of IL-1α and IL-8 compared to the commercial formulation [47].

5.5 Fragrances

Fragrances are volatile substances acting as chemical messengers to the olfactory cells in the nasal cavity [15]. These substances may be lost by evaporation during the manufacture, storage, and use of perfumed products [22]. Encapsulation can be effective in protecting these compounds against rapid evaporation prolonging their release. Dementholized peppermint was encapsulated in nanocapsules prepared by interfacial polymerization of isophorone diisocyanate and hexamethylene diamine forming polyurea, in the presence or not of medium-chain triglycerides [64]. The thermal stability of the nanocapsules was evaluated by TGA from 30 to 600 °C. The volatile fragrance, evaluated isolated, presented a range of evaporation from 50 to 150 °C. The evaporation of medium-chain triglyceride occurred between 200 and 300 °C, and the curve of the polymeric wall indicated a thermal decomposition at 326 °C. In the case of fragrance nanocapsules, three stages were observed, evaporations at 110–200 °C, at 243 °C and decomposition of the polymer wall at 330 °C, characterized by weight loss. Whereas for the medium-chain triglyceride nanocapsules containing the fragrance, a significant weight loss occurred at 253 °C, indicating that the thermal stability was higher than in the nanocapsules without medium-chain triglycerides.

A blend of hydroxypropyl methylcellulose, poly(vinyl alcohol), and ethylcellulose (EC) (1:1:6) was used to obtain nanoparticles containing camphor, eucalyptol, citronellal, limonene, menthol, and 4-tert-butylcyclohexyl acetate. The formulation was able to control the release of these volatile compounds. The nanoparticles had a load capacity of more than 40% and an encapsulation efficiency of more than 80%. The release was determined by the electronic nose or by TGA. In both analyses, only the limonene presented a rapid release, and for the other fragrances, the release was slow when compared to the nonencapsulated controls. After the encapsulation, the release rate has been controlled by the diffusion of the fragrances through the polymer, and this diffusion was dependent on the chemical interactions between the polymer blend and the encapsulated substances [50].

Nanoparticles responsive to pH or temperature are an interesting strategy to control the release of encapsulated substances depending on the external conditions to which the particles are exposed. Nanocapsules containing α-pinene produced by miniemulsion polymerization using the monomers methyl methacrylate, butyl methacrylate, methacrylic acid, and 1,4-butanediol dimethacrylate were developed. This pH-triggered formulation can control the release of the fragrance since the solubility of the polymer is pH-dependent. The percentage of acrylic acid influenced the release of the encapsulated fragrance at alkaline pH, since, in the protonated form at acid pH, the glassy state of the polymeric wall acts as a barrier to the diffusion of the encapsulated fragrance. Increasing the pH, the carboxylic acid groups are deprotonated, and therefore, the increased wettability enhances hydrophilicity of the polymeric wall facilitating the diffusion of the fragrance to the outer phase. After 60 min at 60 °C, the release of α-pinene was 2% and 90%, respectively, when 2.5 wt% and 10 wt% of acrylic acid were used in the preparation [21].

Termoresponsive nanoparticles prepared by miniemulsion polymerization using stearyl acrylate and dipropylene glycol caprylate acrylate in the presence of dipropylene glycol diacrylate sebacate as the crosslinking agent are capable of reducing the fragrance evaporation at room temperature (25 °C), which is gradually released at 31, 35, and 39 °C. The transition of poly(stearyl acrylate) crystalline state to the amorphous state triggers the release of fragrances at elevated temperatures. At 25 °C, a slow release of encapsulated fragrance occurs ($10–28$ mg g^{-1}) depending on the composition of the copolymer, which is lower than the release of the control (an emulsion of sodium dodecyl sulfate) (50 mg g^{-1}). The increase in temperature (39 °C) accelerates the fragrance release reaching 379 mg g^{-1} in 48 h, depending on the composition of the formulation [45].

A strategy combining physical and chemical controls on the release of fragrances was proposed [58]. Nanospheres were produced by covalently bonding the aldehyde groups of different fragrances with the N-succinyl chitosan amine. The resulting imine portion is located in the core of the particles. The release profiles of the four fragrances (vanillin, cinnamaldehyde, citronellal, and citral) were evaluated for 16 days at 32 °C compared to a control (ethanol: water solution). For the nanospheres, the contents of fragrances ranged from 20 to 80%, and for the solution, the contents varied from 5 to 30%.

5.6 Concluding Remarks

Polymeric nanoparticles have been shown to be versatile for cosmetic applications. They find special place in the cosmetic arena because these nanoparticles can improve the stability of labile ingredients, offer efficient protection for the skin from harmful UV radiation, target the active ingredient to the desired layer of the skin or to the hair follicles, enhance photostability, antioxidant and antimicrobial effects, and control the release or the volatilization for prolonged time. As a rule, the penetration and transport extent of polymeric nanoparticles through the skin are dependent on the chemical composition of ingredients, on the particle size and flexibility of the particles, as well as on the encapsulation mechanism, which, by consequence, influences the drug release mechanism. Among the polymeric nanocarriers, the most outstanding are the polymeric nanocapsules due to their core-shell supramolecular structure, which guarantees a reservoir effect for the delivery of substances in the upper layers of the skin or in the hair follicles, in a very controlled way. Triggered nanoparticles composed of smart polymers able to release substances through specific stimuli (temperature, enzymes, etc.) appear as a promising trend for more selective targeting at the desired site. As a recommendation, it should be emphasized that nanotoxicological studies are always needed to ensure that innovative nanotechnological topical formulations are not only highly effective, but also safe.

References

1. Alvarez-Román R, Barré G, Guy RH, Fessi H. Biodegradable polymer nanocapsules containing a sunscreen agent: preparation and photoprotection. Eur J Pharm Biopharm. 2001;52:191–5. https://doi.org/10.1016/S0939-6411(01)00188-6.
2. Alvarez-Román R, Nai A, Kalia YN, Guy RH, Fessi H. Enhancement of topical delivery from biodegradable nanoparticles. Pharmaceut Res. 2004;21:1818–25. https://doi.org/10.1023/B:PHAM.0000045235.86197.
3. Arayachukeat S, Wanichwecharungruang SP, Tree-Udom T. Retinyl acetate-loaded nanoparticle: Dermal penetration and release of the retinyl acetate. 2011;404:281–8. https://doi.org/10.1016/j.ijpharm.2010.11.019.
4. Brisaert M, Plaizier-Vercammen J. Investigation on the photostability of a tretinoin lotion and stabilization with additives. Int J Pharm. 2000;199:49–57. https://doi.org/10.1016/S0378-5173(00)00366-5.
5. Caon T, Mazzarino L, Simões CMO, Senna EL, Silva MAS. Lipid- and polymer-based nanostructures for cutaneous delivery of curcumin. AAPS Pharm Sci Tech. 2017;18:920–5. https://doi.org/10.1208/s12249-016-0554-7.
6. Castro GA, Ferreira LAM. Novel vesicular and particulate drug delivery systems for topical treatment of acne. Expert Opin Drug Deliv. 2008;5:665–79. https://doi.org/10.1517/17425240802167092.
7. Contri RV, Ribeiro KLF, Fiel LA, Pohlmann AR, Guterres SS. Vegetable oils as core of cationic polymeric nanocapsules: influence on the physicochemical properties. J Exp Nanosci. 2013;8:913–24. https://doi.org/10.1080/17458080.2011.620019.
8. Contri RV, Külkamp-Guerreiro IC, Krieser K, Pohlmann AR, Guterres SS. Applying the sensory analysis in the development of chitosan hydrogel containing polymeric nanocapsules for cutaneous use. J Cosmet Sci. 2014;65:1–16.
9. Contri RV, Kulkamp-Guerreiro IC, Silva SJ, Frank LA, Pohlmann AR, Guterres SS. Nanoencapsulation of rose-hip oil prevents oil oxidation and allows obtainment of gel and film topical formulations. AAPS Pharm Sci Tech. 2016;17:863–71. https://doi.org/10.1208/s12249-015-0379-9.
10. Coradini K, Lima FO, Oliveira CM, Chaves OS, Athayde ML, Carvalho LM, Beck RCR. Co-encapsulation of resveratrol and curcumin in lipid-core nanocapsules improves their in vitro antioxidants effects. Eur J Pharm Biopharm. 2014;88:178–85. https://doi.org/10.1016/j.ejpb.2014.04.009.
11. Costa R, Santos L. Delivery systems for cosmetics—From manufacturing to the skin of natural antioxidants. Powder Technol. 2017;322:402–16. https://doi.org/10.1016/j.powtec.2017.07.086.
12. Detoni CB, Souto GD, Silva ALM, Pohlmann AR, Guterres SS. Photostability and skin penetration of different E-resveratrol-loaded supramolecular structures. Photochem Photobiol. 2012;88:913–21. https://doi.org/10.1111/j.1751-097.2012.01147.x.
13. Domínguez-Delgado CL, Rodríguez-Cruz IM, Escobar-Cháves JJ, Calderón-Lojero IO, Quintanar-Guerrero D, Ganem A. Preparation and characterization of triclosan nanoparticles

intended to be used for the treatment of acne. Eur J Pharm Biopharm. 2011;79:102–7. https://doi.org/10.1016/j.ejpb.2011.01.017.

14. Dréno B. What is new in the pathophysiology of acne, an overview. JEADV. 2017;31(Suppl 5):8–12. https://doi.org/10.1111/jdv.14374.

15. Firestein S. How the olfatory system makes sense of scents. Nature. 2001;413:211–8. https://doi.org/10.1038/35093026.

16. Friedrich RB, Kann B, Coradini K, Offerhaus HL, Beck RCR, Windbers M. Skin penetration behavior of lipid-core nanocapsules for simultaneous delivery of resveratrol and curcumin. Eur J Pharm Sci. 2015;78:204–13. https://doi.org/10.1016/j.ejps.2015.07.018.

17. Ge W, Li D, Chen M, Wang X, Liu S, Sun R. Characterization and antioxidant activity of & β-carotene loaded chitosan-graft-poly(lactide) nanomicelles. Carbohydr Polym. 2015;117:169–76. https://doi.org/10.1016/j.carbpol.2014.09.056.

18. Gilbert E, Roussel L, Serre C, Sandouk R, Salmon D, Kirilov P, et al. Percutaneous absorption of benzophenone-3 loaded lipid nanoparticles and polymeric nanocapsules: A comparative study. Int J Pharm. 2016;504:48–58. https://doi.org/10.1016/j.ijpharm.2016.03.018.

19. Guo C, Khengar RH, Sun M, Wang Z, Fan A, Zhao Y. Acid-responsive polymeric nanocarriers for topical adapalene delivery. Pharm Res. 2014;31:3051–9. https://doi.org/10.1007/s11095-014-1398.

20. Guterres SS, Alves MP, Pohlmann AR. Polymeric nanoparticles, nanospheres and nanocapsules, for cutaneous applications. Drug Target Insights. 2007;2:147–57.

21. Hofmeister I, Landfester K, Taden A. pH-sensitive nanocapsules with barrier properties: Fragrance encapsulation and controlled release. Macromolecules. 2014;46:5768–77. https://doi.org/10.1021/ma501388w.

22. Hu J, Xiao Z, Zhou R, Li Z, Wang M, Ma S. Synthesis and characterization of polybutylcyanoacrylate-encapsulated rose fragrance nanocapsules. Flavour Fragr J. 2011;26,162–73. https://doi.org/10.1002/ffj.2039.

23. Huang P, Hu SC, Lee C, Yeh A, Tseng C, Yen F. Design of acid-responsive polymeric nanoparticles for 7,3',4'-trihydroxyisoflavone topical. Int. J. Nanomed. 2016;11:1615–27. https://doi.org/10.2147/ijn.s100418.

24. Jiménez MM, Pelletier J, Bobin MF, Martini MC. Influence of encapsulation on the in vitro percutaneous absorption of octyl methoxycinnamate. Int J Pharm. 2004;272:45–55. https://doi.org/10.1016/j.ijpharm.2003.11.029.

25. Kahraman E, Özhan G, Özsoy Y, Güngör S. Polymeric micellar nanocarriers of benzoyl peroxide as potential follicular targeting approach for acne treatment. Colloids Surf B Biointerfaces. 2016;146:692–9. https://doi.org/10.1016/j.colsurfb.2016.07.029.

26. Külkamp IC, Paese K, Guterres SS, Pohlmann AR. Stabilization of lipoic acid by encapsulation in polymeric nanocapsules designed for cutaneous administration. Quim Nova. 2011;32:2078–84. https://doi.org/10.1590/S0100-40422009000800018.

27. Külkamp-Guerreiro IC, Terroso TF, Assumpção ER, Berlitz SJ, Contri RV, Pohlmann AR, et al. Development and stability of innovative semisolid formulations containing nanoencapsulated lipoic acid for topical use. J Nanosci Nanotechnol. 2012;12:7723–32. https://doi.org/10.1166/jnn.2012.6662.

28. Külkamp-Guerreiro IC, Berliz SJ, Contri RV, Alves LR, Henrique EG, Barreiros VRM, et al. Influence of nanoencapsulation on the sensory properties of cosmetic formulations containig lipoic acid. Int J Cosmet Sci. 2013;35:105–11. https://doi.org/10.1111/ics.12013.

29. Lapteva M, Möller M, Gurny R, Kalia YN. Self-assembled polymeric nanocarriers for the targeted delivery of retinoic acid to the hair follicle. Nanoscale. 2015;7:18651–62. https://doi.org/10.1039/c5nr04770f.

30. Leyden JJ. A review of the use of combination therapies for the treatment of acne vulgaris. J Am Acad Dermatol. 2003;49:S200–10. https://doi.org/10.1067/S0190-9622(03)01154-X.

31. Luppi B, Cerchiara T, Bigucci F, Basile R, Zecchi V. Polymeric nanoparticles composed of fatty acids and polyvinylalcohol for topical application os sunscreens. J Pharm Pharmacol. 2004;56:407–11. https://doi.org/10.1211/0022357022926.
32. Mancuso JB, Maruthi R, Wang SQ, Lim HW. Sunscreens: An Update. Am J Clin Dermatol. 2017;1–8. https://doi.org/10.1007/s40257-017-0290-0.
33. Marcato PD, Caversan J, Rossi-Bergmann B, Pinto EF, Machado D, Silva RA, et al. Nanostructured polymer and lipid carriers for sunscreen. Biological effects and skin permeation. J Nanosci Nanotechnol. 2011;11:1880–6. https://doi.org/10.1166/jnn.2011.3135.
34. Marchiori MCL, Ourique AF, Silva CB, Raffin RP, Pohlmann AR, Guterres ss, et al. Spray-dried powders containing tretinoin-loaded engineered lipid-core nanocapsules: Development and photostability study. J Nanosci Nanotechnol. 2012;12:2059–67. https://doi.org/10.1166/jnn.2012.5192.
35. Masaki H. Role of antioxidants in the skin: Anti-aging effects. J Dermatol Sci. 2010;58:85–90. https://doi.org/10.1016/j.jdermsci.2010.03.003.
36. Nam YS, Kim J, Park J, Shim J, Lee JS, Han SH. Tocopheryl acetate nanoemulsions stabilized with lipid-polymer hybrid emulsifiers for effective skin delivery. Colloids Surf B Biointerfaces. 2012;94:51–7. https://doi.org/10.1016/j.colsurfb.2012.01.016.
37. Nascimento DF, Silva AC, Mansur CRE, Presgrave RF, Alves EN, Silva ES. Characterization and evaluation of poli(ε-caprolactone) nanoparticles containing 2-ethylhexyl-p-methoxycinnamate, octocrylene and benzophenone-3 in anti-solar preparations. J Nanosci Nanotechnol. 2012;12:7155–66. https://doi.org/10.1166/jnn.2012.5832.
38. Nguyen HTP, Munnier E, Souce M, Perse X, David S, Bonnier F, et al. Novel alginate-based nanocarriers as a strategy to include high concentrations of hydrophobic compounds in hydrogels for topical application. Nanotechnology. 2015;26:1–13. https://doi.org/10.1088/0957-4484/26/25/255101.
39. Olvera-Martínez BI, Cázares-Delgadillo J, Calderilla-Fajardo SB, Villalobos-García R, Ganem-Quintanar A, Quintanar-Guerreo D. Preparation of polymeric nanocapsules containing octyl methoxycinnamate by the Emulsification-diffusion technique: Penetration across the Stratum Corneum. J Pharm Sci. 2005;94:1552–9. https://doi.org/10.1002/jps.20352.
40. Ourique AF, Polhlmann AR, Guterres SS, Beck RCR. Tretinoi-loaded nanocapsules: Preparation, physicochemical characterization, and photostability study. Int J Pharm. 2008;352:1–4. https://doi.org/10.1016/j.ijpharm.2007.12.035.
41. Ourique AF, Melero A, Silva CB, Schaefer UF, Pohlmann AR, Guterres SS, et al. Improved photostability and reduced skin permeation of tretinoin: Development of a semisolid nanomedicine. 2011;79:95–101. https://doi.org/10.1016/j.ejpb.2011.03.008.
42. Paese K, Jäger A, Poletto FS, Pinto EF, Rossi-Bergmann B, Pohlmann AR, et al. Semisolid formulation containing a nanoencapsulated sunscreen: effectiveness, in vitro photostability and immune response. J Biomed Nanotechnol. 2009;5:240–6. https://doi.org/10.1166/jbn.2009.1028.
43. Paiva T, Melo P Jr, Pinto JC. Comparative analysis of sunscreen nanoencapsulation processes. Macromol Symp. 2016;368:60–9. https://doi.org/10.1002/masy.201600028.
44. Perugini P, Simeoni S, Scalia S, Genta I, Modena T, Conti B, et al. Effect of nanoparticle encapsulation on the photostability of the sunscreen agente, 2-ethylhexyl-p-methoxycinnamate. Int J Pharm. 2002;246:37–45. https://doi.org/10.1016/S0378-5173(02)00356-3.
45. Popadyuk N, Popadyuk A, Kohut A, Voronov A. Thermoresponsive latexes for fragrance encapsulation and release. Int J Cosmet Sci. 2016;38:139–47. https://doi.org/10.1111/ics.12267.
46. Quintero C, Vera R, Perez LD. α-Tocopherol loaded thermosensitive polymer nanoparticles: preparation, in vitro release and antioxidant properties. Polímeros. 2016;26:304–12. https://doi.org/10.1590/0104-1428.2324.

47. Ramezanli T, Zhang Z, Michniak-Kohn BB. Development and characterization of polymeric nanoparticle-based formulation of adapalene for topical acne therapy. Nanomed Nanotech Biol Med. 2017;13:143–52. https://doi.org/10.1016/j.nano.2016.08.008.

48. Reis CP, Gomes A, Rijo P, Candeias S, Pinto P, Baptista M, et al. Development and evaluation of a novel topical treatment for acne with azelaic acid-loaded nanoparticles. Microsc Microanal. 2013;19:1141–50. https://doi.org/10.1017/S1431927613000536.

49. Rodríguez-Cruz IM, Merino V, Merino M, Díez O, Nácher A, Quintanar-Guerrero D. Polymeric nanospheres as strategy to increase the amount of triclosan retained in the skin: passive diffusion vs. iontophoresis. J Microencapsul. 2013;30:72–80. https://doi.org/10.3109/02652048.2012.700956.

50. Sansukcharearnpon A, Wanichwecharungruang S, Leepipatpaiboon N, Kerdcharoen T, Arayachukeat S. High loading fragrance encapsulation based on a polymer-blend: Preparation and release behavior. Int J Pharm. 2010;391:267–73. https://doi.org/10.1016/j.ijpharm.2010.02.020.

51. Sathish D, Shayeda, Rao M. Acne and its treatment options—a review. Curr Drug Deliv. 2011;8:634–9. https://doi.org/10.2174/156720111797635540.

52. Siqueira NM, Contri RV, Paese K, Beck RCR, Pohlmann AR, Guterres SS. Innovative sunscreen formulation based on benzophenone-3-loaded chitosan-coated polymeric nanocapsules. Skin Pharmacol Physiol. 2011;24:166–74. https://doi.org/10.1159/000323273.

53. Stevanovic M, Savic J, Jordovic B, Uskokovic D. Fabrication, in vitro degradation and the release behaviours of poly(DL-lactide-co-glycolide) nanospheres containing ascorbic acid. Colloids Surf B Biointerfaces. 2007;59:215–23. https://doi.org/10.1016/j.colsurfb.2007.05.011.

54. Suwannateep N, Wanichwecharungruang S, Fluhr J, Patzelt A, Lademann J, Meinke MC. Comparison of two encapsulated curcumin particular systems contained in different formulations with regard to in vitro skin penetration. Skin Res Technol. 2013;19:1–9. https://doi.org/10.1111/j.1600-0846.2011.00600.x.

55. Terroso T, Külkamp IC, Jornada DS, Pohlmann AR, Guterres SS. Development of semi-solid cosmetic formulations containing coenzyme Q10-loaded nanocapsules. Lat Am J Pharm. 2009;28:819–26.

56. Thielitz A, Krautheim A, Gollnick H. Update in retinoid therapy of acne. Dermatol Ther. 2006;19:272–9. https://doi.org/10.1111/j.1529-8019.2006.00084.x.

57. Tran T, Hsieh M, Chang K, Pho Q, Nguyen V, Cheng C, et al. Bactericidal effect of lauric acid-loaded PCL-PEG-PCL nano-sized micelleson skin commensal Propionibaterium acnes. Polymers. 2016;8:1–19. https://doi.org/10.3390/polym8090321.

58. Tree-udom T, Wanichwecharunruang SP, Seemork J, Arayachukeat S. Fragrant chitosan nanospheres: Controlled release systems with physical and chemical barriers. Carbohydr Polym. 2011;86:1602–9. https://doi.org/10.1016/j.carbpol.2011.06.074.

59. Vettor M, Perugini P, Scalia S, Conti B, Genta I, Modena T, et al. Poly(D, L-lactide) nanoencapsulation to reduce photoinactivation of a sunscrenn agent. Int J Cosmet Sci. 2008;30:219–27. https://doi.org/10.1111/j.1468-2494.2008.00443.x.

60. Vettor M, Bourgeois S, Fessi H, Pelletier J, Perugini P, Pavanetto F, et al. Skin absorption studies of octyl-methoxycinnamate loaded poly(D, L-lactide) nanoparticles: Estimation of the UV filter distribution and release behavior in skin layers. J Microencapsul. 2010;27:253–62. https://doi.org/10.3109/10717540903097770.

61. Weiss-Angeli V, Poletto FS, Zancan LR, Baldasso F, Pohlmann AR, Guterres SS. Nanocapsules of octyl methoxycinnamate containing quercetin delayed the photodegradation of both components under ultraviolet A radiation. J Biomed Nanotechnol. 2008;4:80–9. https://doi.org/10.1166/jbn.2008.004.

62. Weiss-Angeli V, Bourgeois S, Pelletier J, Guterres SS, Fessi H, Bolzinger MA. Development of an original method to study drug release from polymeric nanocapsules in the skin. J Pharm Pharmacol. 2010;62:35–45. https://doi.org/10.1211/jpp/62.01.0003.

63. Wu X, Branford-White CJ, Yu D, Chatterton NP, Zhu L. Preparation of core-shell PAN nanofibers encapsulated α-tocopherol acetate and ascorbic acid 2-phosphate for photoprotection. Colloids Surf B Biointerfaces. 2011;82:247–52. https://doi.org/10.1016/j.colsurfb.2010.08.049.

64. Ye K, Zhao D, Shi X, Lu X. Use of caprylic/capric triglyceride in the encapsulation of dementholized peppermint fragrance leading to smaller and better distributed nanocapsules. RSC Adv. 2016;6:84119–26. https://doi.org/10.1039/c6ra19003k.

65. Zargar V, Asgharu M, Dashti A. A review on chitin and chitosan polymers: structure, chemistry, solubility, derivates and applications. ChemBioEng Rev. 2015;2:204–26. https://doi.org/10.1002/cben.201400025.

66. Zhao Y, Brown MB, Jones SA. The topical delivery of benzoyl peroxide using elegant dynamic hydrofluoroalkane foams. J Pharm Sci. 2012;99:1384–98. https://doi.org/10.1002/jps.21933.

Phospholipids in Cosmetic Carriers

6

Peter van Hoogevest and Alfred Fahr

Abstract

Phospholipids are attractive components in cosmetic products because of their natural origin and multifunctional properties. They are used, technically as surface-active compounds, but also as cosmetic actives and modulator of skin penetration. Three classes of phospholipids are being used in cosmetics: saturated and unsaturated phospholipids and monoacylphospholipids (lyso-phospholipids). Whereas the use of monoacylphospholipids is limited to technical use as emulsifiers, the cosmetic use of saturated and unsaturated phospholipids is much broader. The latter are used as actives; additionally saturated phospholipids are used as surface-active compounds, whereas unsaturated phospholipids are suitable to enhance skin penetration. The relation of phospholipids to nanotechnology is apparent when they are, e.g., used as an emulsifier in "nano-emulsions" and as component of "nano-liposomes", having particle sizes in the nano-range. Regulatory authorities discriminate, however, between nanoparticles comprising solid inorganic non-biodegradable particles, with potential (systemic) toxicity, and flexible lipid biodegradable carriers, respectively, comprising, e.g., phospholipids which are non-toxic [1] and fall apart in the skin. Since basically the particle characteristics in the formulation and the component of the particles are believed to cause the desired beneficial cosmetic skin effects, in this chapter the

P. van Hoogevest (✉)
Phospholipid Research Center, Im Neuenheimer Feld 515, 69120 Heidelberg, Germany
e-mail: pvanhoogevest@phospholipid-institute.com

A. Fahr
Professor Emeritus, Pharmaceutical Technology,
Friedrich-Schiller-University Jena, Jena, Germany

© The Author(s) 2019
J. Cornier et al. (eds.), *Nanocosmetics*,
https://doi.org/10.1007/978-3-030-16573-4_6

95

use and degree of skin interaction of especially natural phospholipids as cosmetics ingredient in traditional and nano-cosmetic carriers are described and demonstrated by means of illustrative products.

Keywords

Phospholipids · Liposomes · Emulsions · Lamellar structures · Solubilizers · Biocompatibility

Abbreviations

ATR-FTIR	Attenuated total reflection—Fourier-transform infrared spectroscopy
CLE	Corneocyte lipid envelope
COSMOS	Cosmetics to optimize safety
DHA	Docosahexaenoic acid (DHA, 22:6n-3)
DiI	1,1′-dioctadecyl-3,3,3′,3′-tetramethylindocarbocyanine perchlorate
DMEM	Dulbecco's modified Eagle's medium
DNA	Deoxyribonucleic acid
DPPC	1,2-dipalmitoylphosphaitdylcholine
DSPC	1,2-distearoylphosphatidylcholine
EPA	Eicosapentaenoic acid (20:5n-3)
GMO	Genetically modified organism
GPC	Glycerophosphocholine
HA	Hyaluronic acid
HET-CAM	Hen's egg test on chorioallantoic membrane
HLB	Hydrophilic–lipophilic balance
HPC	Hydroxypropylcelullose
HSPC	Hydrogenated soybean phosphatidylcholine
IFN	Interferon
IP	Identity preserved (i.e. non-GMO)
LUV	Large unilamellar vesicle
MCT	Medium-chain triglyceride
MLV	Multilamellar vesicle
MMP	Matrix metalloprotease
NaFl	Sodium fluorescein
NHDF	Normal human dermal fibroblasts
NHEK	Normal human epidermal keratinocytes
PA	Phosphatidic acid
PC	Phosphatidylcholine
PE	Phosphatidylethanolamine
PG	Phosphatidylglycerol
PI	Phosphatidylinositol
PLA	Phospholipase A
PLD	Phospholipase D

PMA   Phorbol myristate acetate
PS   Phosphatidylserine
RH   Relative humidity
RNA   Ribonucleic acid
SDS   Sodium dodecylsulphate
SLS   Sodium lauryl sulphate
sPLA2   Secretory phospholipase A2
SD   Standard deviation
SLM   Skin lipid matrix
SUV   Small unilamellar vesicle
TEWL   Transepidermal water loss
TGF   Transforming growth factor
TNF   Tumor necrosis factor
US-FDA   US Food and Drug Administration
UV   Ultraviolet

6.1 Introduction

Since a few decades, material scientists developed new materials which could be prepared in particle form at the nanoscale. By introducing these materials in research in natural sciences, it was overlooked that nanoparticles were already known since a much longer time (e.g. micelles, liposomes) and that they also occur naturally (in milk, blood, duodenum, etc.).

Since the newly developed nano-materials were unknown and the use of nanoparticles was launched as a hype, raising unrealistic expectations regarding new medical therapies, soon warnings on the medical, dietetic and cosmetic use of such particles came up, simply because of the lack of information on the toxicology of these unknown materials in the nano-size range. Unfortunately, also nanoparticles with decade-long records of safe use and proven existence in nature and food were linked with this way of thinking and consideration, primarily by not very well-informed scientists, as potentially toxic. Nowadays, neither the EU nor the US-FDA gives a precise definition of nanotechnology. They assess new products containing particles in the nano-range from a scientifically sound objective and a case-by-case risk-benefit evaluation, based on careful and when needed extensive characterization and toxicity testing of the nano-sized particles. The US-FDA makes very clear that (nano-sized) lipid carriers belong to the class of flexible, biodegradable particles which upon administration onto the skin penetrate into the skin and fall apart in the skin [1]. Compared to the other class of nanoparticles comprising solid nano-sized particles, which do not fall apart, there is no risk of systemic absorption and systemic toxicity of these lipid carriers.

In cosmetic products, phospholipids are popular as biodegradable natural ingredients. They can be used in classical formulations like suspensions, oil-in-water and water-in-oil emulsions and mixed micelles. These formulation types can be made or have particle sizes which lie in the range of nanoparticles. Based on the point of view of the US-FDA, the question can be raised whether the particle size of such lipid carriers, having a particle size in the nano-range (liposomes, oil-in-water nano-emulsions), will influence the degree of skin interaction. The skin distribution and penetration of lipid particles varying in size and composition have been studied primarily in pharmaceutical research. Mainly, liposomes or liposomes with commercially attractive names like "invasomes" [2] and "transfersomes" [3] suggesting powerful permeation properties were investigated. Liposomes were primarily of interest due to their intriguing cell-like structure, enabling the encapsulation of water-soluble drugs which needed to be transported through the skin to achieve transdermal delivery and systemic therapy. It is now generally understood that intact liposomes (flexible or not) are not able to pass the skin to any significant extent [4] but are excellent carriers to enhance the skin interaction of co-formulated compounds in cosmetic products. The results and status of this (pharmaceutical) research and the relevance for cosmetic use and resulting knowledge on the skin interaction of the phospholipids and delivery of compounds of cosmetic interest to the skin are described and discussed below.

6.1.1 Phospholipid Properties

The phospholipid molecule comprises a glycerol backbone which is esterified in positions 1 and 2 with fatty acids and in position 3 with phosphate. The systematic designation of, e.g., phosphatidic acid (PA) is 1,2-diacyl-*sn*-glycero-3-phosphate (where *sn* means 'stereospecific numbering'). The specific and non-random distribution of substituents over the positions 1, 2 and 3 of the glycerol molecule introduces chirality. In typical membrane phospholipids, the phosphate group is further esterified with additional alcohol, for instance in phosphatidylcholine (PC) with choline (Fig. 6.1), in phosphatidylethanolamine (PE) with ethanolamine, in phosphatidylglycerol (PG) with glycerol and in phosphatidylinositol (PI) with inositol. The

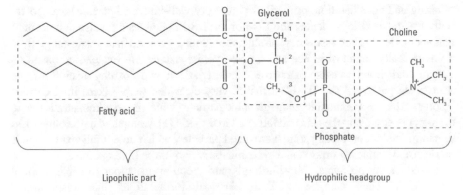

Fig. 6.1 Molecular structure of phosphatidylcholine

phospholipid without esterified alcohol is phosphatidic acid. Depending upon the structure of the polar region and pH of the medium, PE and PC are zwitterionic and have a neutral charge at pH values of about 7, whereas, e.g., PG is negatively charged.

The molecular structure of phospholipids comprises a hydrophilic part and a lipophilic part and has therefore a particular amphiphilic character. After mixing with an aqueous phase, they are able to form various structures depending on the number and type of fatty acids esterified to the glycerol backbone and the ratio of the surface areas occupied by the hydrophilic and lipophilic part of the phospholipid molecule. Diacylphospholipids having a cylindrical shape are organized as lipid bilayers (lamellar phase) with the hydrophobic tails lined up against one another and the hydrophilic head group facing the water on both sides (Fig. 6.2). The structures then formed are called liposomes. The bilayer membrane of such a liposome resembles the basic structures of cellular membranes. Because of this similarity, it is therefore evident that phospholipids are biocompatible and may have a beneficial interaction with skin cells.

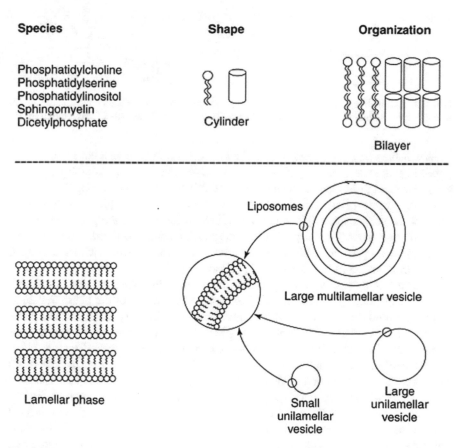

Fig. 6.2 Phospholipid molecules with a cylindrical shape which form upon hydration phospholipid bilayers and liposomal structures. Adapted from Lasic [5]

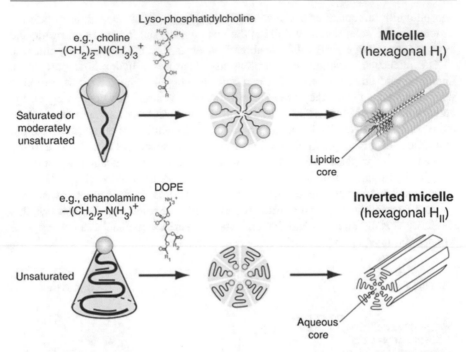

Fig. 6.3 Phospholipid molecules with either cone shape (above) or inverted cone (below) shape, forming upon hydration micelles or inverted micelles, respectively. Adapted from Kraft et al. [6]

When only one fatty acid is esterified to the glycerol backbone of the phospholipid molecule (monoacylphospholipids, also called lyso-phospholipids), and the polar head group are relatively large, the molecules are cone-shaped, and they can form micelles (also called hexagonal H_I phase). An example of such a phospholipid is lysophosphatidylcholine. When the surface area of the polar head group is small (e.g. in case of phosphatidylethanolamine) compared to the surface area of the fatty acid part, then so-called inverted cones are formed which are upon hydration arranged in the so-called H_{II} phase (Fig. 6.3).

The fatty acid composition of phospholipids determines the temperature at which the fatty acids change their mobility. Below the temperature of the phase transition from the liquid crystalline to the so-called gel state, the fatty acids and the phospholipid molecule are rigid (gel state), whereas above this phase transition temperature the fatty acids and the phospholipid molecule are mobile (Fig. 6.4).

Phospholipids with polyunsaturated fatty acids have a very low (below 0 °C) phase transition temperatures. This means that at skin temperatures of around 32 ° C, these lipids are in the liquid crystalline state and form, upon hydration, structures/liposomes with a flexible membrane. Phospholipids with unsaturated fatty acids can be converted by means of hydrogenation to phospholipids containing saturated fatty acids. Hydrogenated soybean PC (HSPC) is an example of a saturated phospholipid, which contains mainly saturated fatty acids. The fatty acid

Gel State   Liquid Crystalline State

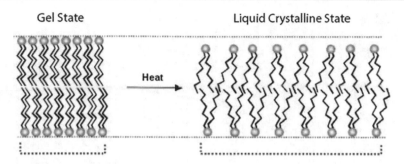

Heat

Fig. 6.4 Structures of phospholipid in the gel or liquid crystalline state. Adapted from Blaber [7]

composition is then approximately 85% stearic acid, 14% palmitic acid, and 1% other saturated fatty acids. The phospholipids with these saturated fatty acids have a high phase transition temperature of approximately 55 °C. At skin temperatures, liposomal dispersions with hydrogenated lipids are therefore in the gel state and rigid in nature.

The moisturizer effect of phospholipids on the skin is explained by their excellent skin interaction and their hygroscopic properties, as demonstrated in the following experiments. Figure 6.5 shows the hygroscopicity of unsaturated soybean phosphatidylcholine and saturated soybean phosphatidylcholine (data kindly provided by Lipoid GmbH, Germany).

Figure 6.5 shows that unsaturated soybean phosphatidylcholine is slightly more hygroscopic than saturated soybean phosphatidylcholine. At 25% RH and 30 °C, the mass increase of unsaturated soybean phosphatidylcholine by hydration is limited to only 3% after 5 h. The saturated soybean phosphatidylcholine is under these conditions not hygroscopic. When soybean phospholipids are processed in dry form or into dry formulations, the humidity and temperature of the production environment should therefore be controlled, and preferably the humidity should be 25% RH at 30 °C or lower. In general, dry phospholipids can absorb up to 12% by weight water; this represents about five molecules of water bound to one molecule of phospholipid. In case phospholipids are fully hydrated in the form of liposomes, they can bind 20 molecules of water per phospholipid molecule [9].

Unsaturated phospholipids may have a yellowish/brownish color, especially when considering lower purity grade fractions. Higher grade fractionated phospholipids of, e.g., soybean phosphatidylcholine are yellowish in color. Hydrogenated (saturated) phospholipids have a white color. The odour of phospholipids is slightly nutty. When needed, addition of fragrances can easily mask this odour.

Natural phospholipids used in cosmetics are mainly coming from soybeans, sunflower seed or rapeseed. Sunflower phospholipids are considered to be non-GMO; soybean and rapeseed (canola seed) phospholipids can be supplied in a non-GMO quality as well. Synthetic phospholipids are hardly being used by the cosmetic industry [10].

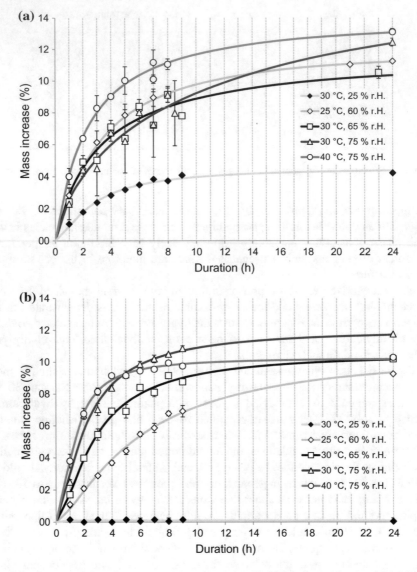

Fig. 6.5 Hygroscopicity of **a** unsaturated soybean phosphatidylcholine with n.l.t. 94% PC and **b** saturated soybean phosphatidylcholine with n.l.t 90% PC, as determined according to Yee and Davis [8]

In the literature on the application of phospholipids in cosmetics, there is some confusion on the use of the terms lecithin and phosphatidylcholine. This is because in the American literature, "lecithin" is used as a synonym for phosphatidylcholine, whereas according to generally accepted pharmacopeial definitions lecithin is the complex mixture of acetone-insoluble phosphatides (i.e. phospholipids), from vegetable source. This "consists chiefly of phosphatidylcholine, phosphatidylethanolamine, phosphatidylinositol and phosphatidic acid, present in conjunction with various amounts of other substances such as triglycerides, fatty acids, and carbohydrates, as separated from the crude vegetable oil source" [11]. For this reason, it is proposed to use "lecithin" to describe a product when it contains less than 80% by total weight phospholipids (and phosphatidylcholine as main component). Products containing, e.g., 80–90% of phospholipids from soybean can be called soybean phospholipids. Purified phospholipid fractions containing more than 90% by weight of phosphatidylcholine shall be designated as "phosphatidylcholine". In the description of the discussed phospholipid products in this manuscript, the weight percentage of phosphatidylcholine in these products is indicated.

6.1.2 Phospholipids Raw Materials

In cosmetic products, lecithins containing varying amounts of phosphatidylcholine and unsaturated and saturated diacylphosphatidylcholines, and to a lesser extent monoacylphospholipids (lysolecithins), are used. The phosphatidylcholine content of the lecithins can be enriched using fractionation methods like solvent extraction and chromatography. After fractionation, several grades can be produced starting to form crude lecithin containing ca. 15% PC still comprising significant amounts of the plant oil raw material from which it has been isolated, and de-oiled or fractionated lecithin to obtain higher contents of PC from 25 to 96%. Phospholipids are essential natural components of the membrane of all living cells; they are non-toxic and possess very high skin tolerability.

The purified phospholipids have a typical fatty acid composition related to the plant source used. The fatty acid composition of soybean phosphatidylcholine can be found in Table 6.1.

As described above, phospholipids containing polyunsaturated fatty acids have a very low (below 0 °C) phase transition temperature. At skin temperatures of around 32 °C, these lipids are therefore in the liquid crystalline state and form, upon hydration, very flexible structures/liposomes.

Table 6.1 Fatty acid composition (mole %) of soybean phosphatidylcholine determined by enzymatic hydrolysis followed by gas chromatography [12]

Fatty acid	In 1-position (%)	In 2-position (%)	Total (%)
Palmitic acid (C16:0)	24.0	1.7	12.9
Stearic acid (C18:0)	7.9	1.0	4.4
Oleic acid (C18:1)	10.9	10.0	10.5
Linoleic acid (C18:2)	52.4	80.6	66.5
Linolenic acid (C18:3)	4.7	6.7	5.7

6.2   Skin Structure

In order to understand the interaction of phospholipids with the skin, the physiology
of the epidermis, including the stratum corneum, the skin layer relevant for cosmetic treatment and the presence of (endogenous) phospholipids in these layers will
be reviewed in the first instance (Fig. 6.6).

The stratum corneum is the target skin layer for cosmetic products. It is the
utmost significant barrier of the epidermis, even that it is not thicker than a sheet of
paper. Without stratum corneum, the human body would lose every day about 20 l
of water in contrast to 0.15 l per day with an intact stratum corneum. The structure
of the stratum corneum is best described in short with the classic "brick-and-mortar
model" [13]. The corneocytes represent the bricks and the intercellular lipids, in
which the corneocytes are embedded, the mortar. The thickness of this barrier varies
depending on variables like individual persons, gender and body region between 6
and 40 μm, a body "average" thickness for four areas (abdomen, flexor forearm,
thigh and back) being estimated as 8.7–12.9 μm [14]. The number of corneocyte
layers is correspondingly higher at thicker stratum corneum (14–23 layers) [14].
Some studies indicate that the "brick-and-mortar" wall indeed seems to have narrow
channels [15] or small pores varying from 0.59 nm [16], to about 3.6 nm [17] and
up to 20 nm [18]. Please note that these measurements were made with stratum
corneum of different species and most values are obtained by various and indirect

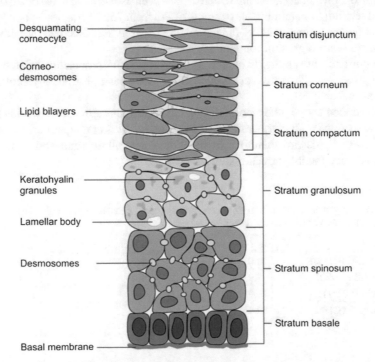

Fig. 6.6 Structure of the human skin layers, relevant for treatment with cosmetic products

methods. However, these structural data are crucial for the still ongoing discussion, if and how colloidal systems containing phospholipids may penetrate intact the stratum corneum (see Sects. 6.5.3 and 6.7).

The epidermis is a very dynamic system; in about four weeks, the epidermis is renewed in younger humans. However, at the age of 60 years this time is doubled. The stratum corneum itself is being considered as "dead", as the corneocytes do not show metabolism and have lost their nucleus.

The renewal of the stratum corneum starts at the border between epidermis and dermis, the stratum basale. The stratum basale consists of a layer of elongated basal cells which rest on the basement membrane. These basal cells are connected via desmosomes and are continuously proliferating. These newly produced cells are called keratinocytes (which can again divide themselves 4–5 times) and delaminate from the stratum basale still connected to their surrounding neighbours via desmosomes. They migrate into the stratum spinosum. The name was attributed to their spiny appearance.

The keratinocytes migrate from this point to the stratum granulosum (named after the granular appearance of the cytoplasm). These keratinocytes start to produce in their cytosol so-called lamellar bodies, which contain a variety of lipids and enzymes (see details below). Also keratohyalin granules are provided by the keratinocytes in the stratum granulosum, which contain precursor proteins for the keratin core of the finished corncocytes. At the transition between stratum granulosum and stratum corneum, the lamellar bodies (also called "Odland bodies") are excreted. This results not only in the formation of the intercellular lipid matrix, but also of a hydrophobic lipid envelope. This corneocyte lipid envelope (CLE) (see, e.g., [19]) is a monolayer of covalently bound ω-OH-ceramides to the surface of the developing corneocytes which allow the embedding of the hydrophilic corneocytes into the intercellular hydrophobic matrix. Also proteins are cross-linked in this envelope and make the corneocyte envelope even tighter [20]. This whole envelope structure reduces absorption of substances into the corneocytes. For these reasons, most of the active substances applied onto the skin are diffusing along the lipid lamellae in the intercellular regions. During migration closer to the stratum corneum region, the keratinocytes start to become non-viable corneocytes by losing DNA, RNA and almost all organelles.

The stratum compactum is the tightest part of the stratum corneum, which is also due to the corneodesmosomes (successor of the desmosomes) which keep the corneocytes together. These corneosomes are hydrolysed closer to the skin surface, as this is necessary for the desquamation at the surface and to be shed from the skin surface (exfoliation) to maintain homoeostasis.

There is a pronounced pH gradient across the stratum corneum, starting at the skin surface with 4.5 ± 0.2 for men and 5.3 ± 0.5 for women at the volar forearm [21]. After complete removal of the stratum corneum by stripping, a pH of 6.9 ± 0.4 is measured for men and 6.8 ± 0.5 for women [21]. This means a pH change of 2 units over a distance of about 10 µm! The acidified skin surface functions as a deterrent to skin infections by, e.g., inhibiting skin colonization with *Staphylococcus aureus* or *Streptococcus pyogenes* [22, 23].

6.3   Skin (Phospho)lipids

Lipids play a decisive role in creation and maintaining the barrier function of the stratum corneum. During the conversion of keratinocytes to corneocytes and the migration to the top of the skin, a tremendous activity of anabolism and catabolism of lipidic material is going on. For a better understanding, a schematic drawing of the prominent metabolic pathways is depicted in Fig. 6.7. There are many more reactions involved (see, e.g., [24]), but this is beyond this review.

These metabolic activities are in the healthy skin in steady state. The enzymes and the (phospho)lipid concentrations in the different regions of the epidermis have been determined in numerous studies.

Type I secretory phospholipase A_2 (sPLA$_2$) activity in tape strips of human skin could be detected [26]. In the tape strips taken from the skin surface (strips 1–10), activity of 1 mU/mg protein was obtained; in the lower region, a threefold higher activity was observed. Immuno-histochemical staining of skin slices with anti-porcine pancreatic sPLA$_2$ revealed the primary location of sPLA$_2$ at the stratum corneum-stratum granulosum junction. This supports the notion raised earlier [27] that sPLA$_2$ hydrolyses the phospholipids secreted from lamellar bodies and serves as producer of specific fatty acids improving thereby the barrier properties of the stratum corneum and acidifying the upper part of the stratum corneum. In this

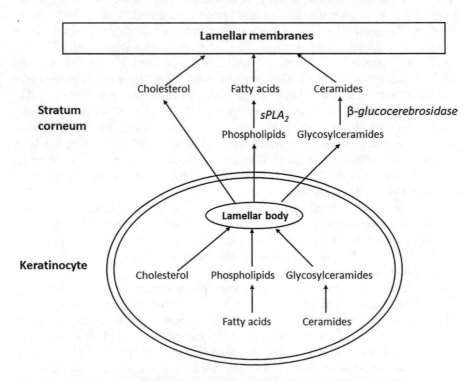

Fig. 6.7 Metabolic processes in the upper epidermis. Modified after Feingold [25]

earlier study [27] done in mice, the putative sPLA$_2$ was blocked by bromophenacyl bromide and MJ-33, which reduced the fatty acid content in the upper stratum corneum layers and caused thereby a perturbation of the barrier function, and epidermal hyperplasia. This defect could be reversed by topical application of free or esterified to phospholipids fatty acids. This also proved that the barrier impairment is not caused by the accumulation of phospholipids in the upper layer of the stratum corneum. Besides enzymatic degradation by means of PLA$_2$, phospholipase D may also play a role [9].

As one of the consequences of the phospholipase activity in the stratum corneum, the phospholipid percentage of the lipid composition decreases dramatically from the basal layer (about 50%) to the stratum corneum (<5%), as can be seen in Fig. 6.8, and the fatty acid content in the stratum corneum significantly increases. Similar distributions of phospholipids were found in pig skin [28], but here the phospholipid content gradient was not as pronounced between stratum corneum and viable epidermis (2.5 mg/g tissue vs. 10 mg/g tissue, respectively).

As a widely used rule of thumb, the major intercellular lipid composition of the stratum corneum is given as consisting of about 50% ceramides, 25% cholesterol, 10% fatty acids, cholesterol sulphate and esters [30]. These lipids form bilayers with a thickness of 4–5 nm. A particular group of ceramides contains a C36-side chain, which extends into the neighbouring bilayer, thereby "nailing" them together.

The overall lipid composition of the stratum corneum varies with regions of the body. The first investigation on human stratum corneum [31] showed a significant variation of the neutral lipid content (see Table 6.2). This neutral lipid group

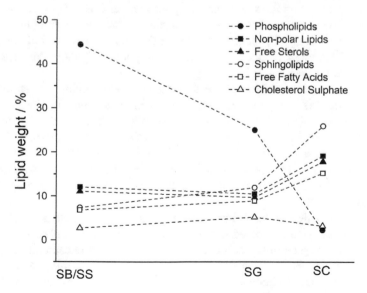

Fig. 6.8 Stratification of lipid species in human epidermis, redrawn after Lampe et al. [29]; SB/SS—stratum basale, stratum spinosum; SG—stratum granulosum; SC—stratum corneum

Table 6.2 Regional variations in lipid weight per cent and distribution of significant lipid species in the stratum corneum; values in weight percent

Lipid fraction	Face	Abdomen	Leg	Plantar
Phospholipids	3.3	4.9	5.2	3.2
Glycosylceramides	6.7	2.6	3.4	6.4
Cholesterol sulphate	2.7	1.5	6	3.4
Free fatty acids	19.7	19.3	13.9	9
Ceramides	19.9	15.5	22.6	28.4
Free sterols	17.3	14	20.1	32.8
Triacyl glycerols	13.5	25.2	20.1	5.9
Sterol esters	6.2	6.1	4.6	7.1
Squalene	6.9	6.5	3.6	2.9
n-alkanes	2.8	3.7	3	2.9

Adapted from Lampe et al. [31]

comprising free sterols, free fatty acids, triglycerides, sterol and wax esters, squalene and n-alkanes dominates the lipid content of the stratum corneum. The phospholipid content varied according to this study not statistically significant (3.2–5.2% weight of total lipid weight).

Concentration values for the gradient of lipid classes across the stratum corneum [32] were measured. The gradient was quite significant for free fatty acids, for which the fatty acid content decreased from 6.5 $\mu g/cm^2$ starting at the surface to 3.0 $\mu g/cm^2$ in the lower parts of the stratum corneum. A decrease in ceramides and cholesterol from the surface to the lower parts was statistically significant as well. A peculiar gradient profile was found for phospholipids, as the phospholipid content at the surface and the basal site of the stratum corneum are quite low (2.5 and 3 $\mu g/cm^2$ respectively), but at about the lower third a higher content of 5.5 $\mu g/cm^2$ was found. This might be due to the synthesis of phospholipids in this layer of the stratum corneum (see Fig. 6.7).

The fatty acid composition of the few detected phospholipids in stratum corneum comprised of 18:0, 18:1 and 18:2 fatty acid chains, which reflects the fatty acid composition of the viable epidermis phospholipids [29] coarsely. In contrast to that, 16:0 fatty acid chains that were present to 26% in the phospholipids of the viable epidermis were only present to 2.8% in the phospholipids of the stratum corneum. It has to be emphasized that neither phospholipids nor glycosphingolipids are present [29] in the outer layers of the human stratum corneum (stratum disjunctum).

To the knowledge of the authors, very little is known regarding which specific phospholipid enzymes, involved in the anabolic construction of phospholipids, are expressed in the epidermis. Additionally, the regulation of phospholipid synthesis in the epidermis has not been delineated. Further studies in this area are clearly needed, given the requirement for phospholipids for the formation of lamellar bodies, and as sources of free fatty acids, and for the low pH of the stratum corneum.

Finally, one example should accentuate the rather complex interplay between skin region, cells, lipids and metabolism. Lipids are secreted at the stratum granulosum-stratum corneum interface into an extracellular environment at neutral pH. Initial processing of the lipid bilayers into elongated, fused sheets is performed by enzyme(s) with a neutral pH optimum, such as the secretory phospholipase A_2. As the partially processed lipids migrate upward, the extracellular environment becomes progressively more acidic, activating β-glucocerebrosidase, and perhaps other enzymes with an acidic pH optimum, allowing the final processing step into mature lamellar multilayers (cited from Mauro et al. [33]).

All the mentioned studies prove that the stratum corneum is not a homogeneous matrix in itself; it instead forms a very heterogeneous structure and contains besides corneocytes and multilamellar lipid layers a variety of other biologically active substances. They are also interacting complexly, including the migration dynamics and gradients like pH and enzyme activities. The understanding of these mechanisms is just beginning to grow, despite the fact that skin (and especially the stratum corneum) is readily available for research, as it is situated at the outside of our body and samples are easy to obtain.

6.4 Phospholipid Skin Toxicity/Tolerance

It is widely documented that phospholipids (lecithin and hydrogenated lecithin) are well tolerated when administered to the skin [34]. Also, the US Cosmetic Ingredient Review (CIR) organization qualifies lecithin as safe as used [35]. In this safety assessment, the used concentrations of phospholipids in various commercially available products and cosmetic uses are described.

In order to confirm the high degree of skin compatibility of selected commercially available phospholipid products, the skin tolerability of various phospholipids was investigated by means of the HET-CAM (*Hen's egg test on chorioallantoic membrane*) test (data kindly provided by IADP (An-Institut für angewandte Dermatopharmazie, Halle (Saale) Germany), an in vitro model with chicken eggs to assess skin and ocular irritation potential. A 30% dispersion of the phospholipids in medium-chain triglycerides (MCT), which are well-tolerated in dermal and cosmetic products, was used [36]. The tested 30% dispersions of the phospholipids were ten times higher than previously tested [34] and appeared to be sufficiently transparent, allowing the more sensitive reaction-time method to assess the irritation potential. The eggs (n = 6 per test solution) were observed for 300 s, the changes of the morphology of the CAM were photographically documented and qualitatively assessed, also the occurrence of haemorrhages, lysis of veins and occurrence of blood coagulation (Table 6.3). Controls were 0.9% NaCl (non-irritating control) and 1–2% SDS (irritating control).

Table 6.3 Irritation potential of examples of phospholipid products in the HET-CAM test; (H) occurrence of haemorrhages, (L) lysis of veins (C) and occurrence of blood coagulation

Test product	H	L	C	Irritation potential
NaCl (0.9%)	–	–	–	No
SDS (1%)	Yes	–	–	Yes
SDS (2%)	Yes	–	–	Yes
MCT	–	–	–	No
Soybean lecithin 45[a]	–	–	–	No
Sunflower lecithin 50	–	–	–	No
Canola lecithin 50	–	–	–	No
Soybean lecithin (IP) 50	–	–	–	No
Soybean lecithin (IP) 75	–	–	–	No
Soybean lecithin 75	–	–	–	No
Egg phospholipids 80	–	–	–	No
Soybean phospholipids 80	–	–	–	No
Egg phosphatidylcholine 98				No
Soybean phosphatidylcholine (IP) 98	–	–	–	No
Soybean phosphatidylcholine 98	–	–	–	No

[a]The number indicates the content of phosphatidylcholine in the product

Eggs treated with 1–2% SDS showed changes of blood vessels like hyperaemia and bleeding [compare Fig. 6.9, *n* (before treatment) and *n** (after treatment)], as was expected. In spite of interference with the observations caused by film formation and color of the products, all tested phospholipid products showed, even tested with a 30% concentrated dispersion of egg phosphatidylcholine with 98% PC, the same lack of irritation potential as the non-irritating control with 0.9% NaCl and the used solvent MCT. Examples of the test results are provided in Fig. 6.9.

To further confirm the high degree of skin compatibility of selected commercially available phospholipid products, the following IQ ULTRA® occluded patch test, using 68 mm² polyethylene plastic moss chambers with incorporated filter paper soaked with lipid test solution or controls, was performed on volunteers (data kindly provided by Lipoid GmbH, Ludwigshafen, Germany). One patch test consisted of ten chambers that were fixed on the back of the test person with a hypoallergenic non-woven adhesive tape (see Fig. 6.10).

The filter papers were saturated with 0.02 ml of a 10% aqueous dispersion of the lecithin, phospholipid or emulsifier, respectively. The patches were in contact with the skin once for 48 h (48-hour occluded patch test). An empty patch was applied as "negative" control.

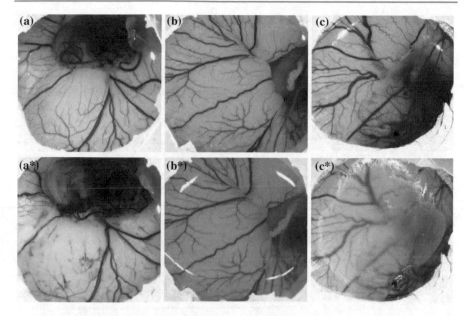

Fig. 6.9 Photographs of HET-CAMs before (n) and after application (n*) **a** SDS 1%, **b** NaCl 0.9%, **c** 30% aqueous dispersion of egg phosphatidylcholine with 98% PC

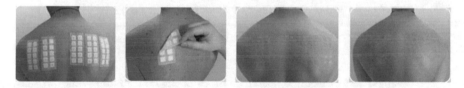

Fig. 6.10 Example of the performance of IQ ULTRA® patch test

The number of volunteers (female and male with healthy skin) was 22 or 24, respectively, and included only persons with normal skin, aged 18–70 years. The contact zone on the back of the volunteers was examined just before starting the study in order to apply the products on a skin surface free from macroscopic irritation marks, scars or any abnormalities which could interfere with the interpretation of the results. The area on which the patch was applied was cleaned with deionized water and dried with cellulose cotton wool tissue. The test skin areas were not in contact with any cosmetic products before and during the test. Withdrawal of the patches and the readings were made by a dermatologist.

The analysis of the skin reaction was performed for each volunteer, according to the following scale:

Erythema	
None	0
Slight erythema, hardly noticeable	1
Moderate and uniform redness	2
Significant and uniform redness	3
Dryness/desquamation	
No dryness	0
Dry with desquamation, smooth and stretched aspect, slight and	
Slight desquamation	1
Moderate desquamation	2
Severe desquamation with large scales	3
Oedema	
None	0
Slight	1
Noticeable	2
Significant	3
Blister	
None	0
Slight	1
Noticeable	2
Significant	3

The obtained results were compared to those of the control zone. The "Primary Irritation Index" was calculated according to the following formula:

$$\text{Primary Irritation Index} = \frac{\dfrac{\left(\sum \text{Marks after 48 Hours}\right) \text{Volunteer 1 to } n}{\text{Number of Readings}}}{\text{Number of Volunteers}}$$

The classification of the irritation potential was performed according to the following scale (Table 6.4).

Table 6.4 Classification of the irritation potential

Primary irritation index	Classification
0–0.08	Non-irritant
>0.08–0.16	Very slightly irritant
>0.16–0.56	Slightly irritant
>0.56–1.0	Moderate irritant
>1.0–1.6	Irritant
>1.6	Very irritant

Table 6.5 Results of the 48-hour occluded patch test performed with examples of commercially available phospholipid products

Product description	Irritation index	Classification
Soybean lecithin 50[a]	0	Non-irritant
Soybean phosphatidylcholine 85	0.04	Non-irritant
Soybean phosphatidylcholine 90	0.04	Non-irritant
Hydrogenated soybean lecithin 80	0	Non-irritant
Hydrogenated phosphatidylcholine 90	0	Non-irritant
Rapeseed lysolecithin 20[b]	0	Non-irritant

[a]The number indicates the content of phosphatidylcholine or [b]lysophosphatidylcholine in the product

In Table 6.5, the results of occluded patch test are provided. None of the investigated phospholipids show, under the used test conditions, any irritation potential.

Although the presented data and literature references suggest that phospholipids have no or a very low skin irritation potential, it should be realized that these findings have been made with specific in vitro models, not necessarily reflecting application on the human skin or in vivo settings with a limited number of volunteers (possibly not able to pick up skin reactions in sensitive individuals), using simple formulations ruling out synergistic effects with, e.g., (other) emulsifiers and/or actives. It is therefore recommended not to exceed the concentrations listed above and to test specific formulations separately.

The possibility that co-administration of phospholipids and (other) cosmetic actives, which generally do not permeate through the skin, will give rise to a systemic availability of the cosmetic active is negligible [4]. Even flexible liposomes do not permeate to a significant amount through the skin and cannot carry a (water soluble) encapsulated compound through the skin. In addition, as pointed out above, liposomes comprising unsaturated phospholipid tend to fall apart in the stratum corneum of the skin. Liposomes comprising saturated phospholipids do even have a lower tendency to penetrate into the skin. It can be assumed that other types of formulations like oil-in-water emulsions with phospholipids as emulsifier and a liquid oil phase will behave in the same way as flexible liposomes.

6.5 Use

Because of their chemical structure and physicochemical properties, the use of phospholipids in cosmetic formulations has several unique aspects. In the following, it will be apparent that phospholipids are not only technically useful surface-active ingredients, but they are also used as actives and used to enhance the skin interaction by means of deeper transport into the skin of other formulation components. The phospholipid which can be used for what purpose is dependent on the flexibility/rigidity of the phospholipids obtained at skin temperature. They can penetrate either deeper into the skin and enhance transport or they can support the barrier function of the skin. In the latter case, the phospholipids will keep the skin in healthy condition and can, therefore, be considered as cosmetic actives. These functional properties of phospholipids are the basis for the excellent skin tolerability of phospholipids [34, 37]. For these reasons, phospholipids are considered as multifunctional cosmetic ingredients and actives.

Phospholipids are particularly of interest for use in natural cosmetic products because certain qualities are certified as non-GMO materials and assessed as being genuinely natural by reputed organizations like Cosmetics to Optimize Safety (COSMOS).

In the following section, the multifunctional properties of phospholipids are discussed in detail. Firstly, the technical (functional) use of phospholipid in cosmetic formulations and secondly the use as cosmetic actives will be covered.

6.5.1 Surface Active

Due to their amphiphilic (hydrophilic as well as lipophilic) structure, phospholipids possess surface activity. They can be technically used in cosmetic formulations as an emulsifier, liposome/lamellar phase former, solubilizer or wetting agent.

6.5.1.1 Emulsifier
Diacylphospholipids are considered to be very mild detergents and excellent emulsifiers [38, 9]. After hydration of diacylphospholipid molecules which have a cylindrical shape, lamellar structures and/or liposomes are formed. For this reason, they do not emulsify oil with water spontaneously. Mechanical energy is needed to disrupt liposomal structures and simultaneously force the phospholipid molecules to position themselves at the oil–water interface. After applying mechanical energy (by means of stirring with, e.g., a rotor–stator mixer), very stable oil-in-water or water-in-oil emulsions/creams can be prepared. The diacylphospholipid emulsifiers can either be dissolved in the oil or dispersed in the water phase.

Monoacylphospholipids (lyso-phospholipids) form micelles upon dispersion in water. They are stronger detergents (cone-shaped) and require less mechanical energy for the preparation of oil-in-water emulsions because micelles are less stable compared to liposomes, and are in equilibrium with a much higher monomer concentration. Interestingly, products comprising diacyl- as well as monoacylphospholipids with about 60% lyso-PC form mixed micelles upon hydration and can be considered as intermediate mild emulsifiers, due to the presence of diacylphospholipids.

Published HLB values on phospholipid emulsifiers vary greatly and display a broad range, suggesting that they are suitable for the preparation of oil-in-water as well as water-in-oil emulsions using a large variety of oils differing in required HLB values [39, 40]. The variety and variability of reported HLB values and ranges of especially natural phospholipids may be explained by the heterogeneity in the composition of the tested phospholipid products and a sometimes inadequate poor description of their composition in specific publications, which makes a systematic comparison between published data difficult and the presented data unreliable. In addition, experimental methods for determination of HLB values are mostly short-term tests, which are not indicative of a long-term stability of the emulsions. It should be further realized that natural phospholipids if manufactured/fractionated reproducibly at high standards from, e.g., crude lecithin, do represent well-defined mixtures of phospholipids like PC, PE and PI. These phospholipid mixtures display a reproducible, origin (raw material)-specific, heterogeneous but reproducible fatty acid composition and are considered as well-defined mixtures of co-emulsifiers.

In cosmetic products, phospholipids are often used as co-emulsifier in combination with synthetic (non-natural) emulsifiers. Besides the fact that these products are not natural anymore, phospholipids can, as demonstrated in the following, also be used as sole emulsifiers in natural cosmetic products, eliminating the need for synthetic emulsifiers.

The following emulsification experiments were performed with a model formulation using various phospholipids as the only variable and sole emulsifier. In Table 6.6, the recipe of the model formulation (adjusted to pH 5.5) is shown.

A comparison of different emulsifying properties of phospholipids for the preparation of o/w emulsions with various oils (differing in required HLB value) is provided in Table 6.7.

The oil-in-water emulsions were prepared by using either MCT (medium-chain triglycerides, with required HLB 4), jojoba oil (with required HLB value 6), sunflower oil (with required HLB value 7) or squalane (with required HLB 11), or as oil phase, mixing components of Phase A (oil and phospholipid) and heating to 70 °C until a clear solution was obtained. Separately, Phase B (water, glycerine) was heated to 70 °C and added to Phase A under stirring. In case the phospholipid was

Table 6.6 Recipe of a model formulation to compare the emulsifying properties of phospholipid emulsifiers to prepare oil-in-water emulsions

Phase	INCI	Function	% (w/w)
A	Oil	Lipophilic phase	25.00
	Phospholipid[a]	Emulsifier	3.00
B	Aqua/water	Hydrophilic phase	64.00
	Glycerine	Moisturizer	6.00
C	Xanthan gum	Thickener	0.20
	Dehydroxanthan gum	Thickener	0.80
D	Benzyl alcohol, glycerine, benzoic acid, sorbic acid (Rokonsal™ BSB-N)	Antimicrobial preservative	1.00
	Sodium hydroxide	pH adjustment	q.s.
	Citric acid	pH adjustment	q.s.
Total			100.00

[a]Tested phospholipids are described in Table 6.7

Table 6.7 Influence of soybean phospholipids and oil phase HLB on the stability of model oil-in-water emulsions

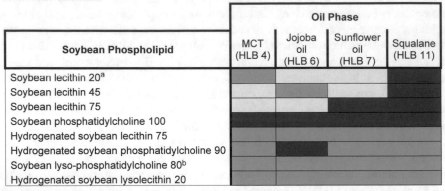

Soybean Phospholipid	Oil Phase			
	MCT (HLB 4)	Jojoba oil (HLB 6)	Sunflower oil (HLB 7)	Squalane (HLB 11)
Soybean lecithin 20[a]				
Soybean lecithin 45				
Soybean lecithin 75				
Soybean phosphatidylcholine 100				
Hydrogenated soybean lecithin 75				
Hydrogenated soybean phosphatidylcholine 90				
Soybean lyso-phosphatidylcholine 80[b]				
Hydrogenated soybean lysolecithin 20				

Green: stable for at least six weeks at 45 °C and five freeze–thaw cycles; Yellow: maximal stability three weeks at 45 °C, Red: no emulsion was formed (for further details see text). The numbers refer to the [a]PC content or [b]lyso-PC content; data kindly provided by Lipoid GmbH, Ludwigshafen am Rhein, Germany)

not soluble in the oil phase at this temperature (e.g. hydrogenated phosphatidyl-choline with >90% PC), it was dispersed in Phase B (water, glycerine), and Phase A was added to Phase B; then both phases were intensively emulsified using a rotor–stator mixer (5 min). Under stirring, the thickeners were added in portions to the emulsion and cooled down to 30 °C. Finally, the emulsion was homogenised again with a rotor–stator mixer for 1 min, the antimicrobial preservatives were added, and the pH was adjusted to pH 5.5 under stirring.

The stability of the resulting emulsions was tested after one, three and six weeks at 45 °C and subsequently in five freeze–thaw cycles and inspected for phase separation. The results are provided in Table 6.7.

From this table, it can be concluded that under the test conditions used, unsaturated diacyllecithins are less suitable as sole emulsifiers for oil-in-water emulsions compared to saturated and monoacyl-lipids. The poor results obtained with the pure PC fraction are probably due to the lack of stabilizing components like PE. It should be noted that these phospholipid fractions can still be suitable emulsifiers in other settings (e.g. combined with co-emulsifiers and using high-pressure homogenization).

Hydrogenated soybean phospholipids with 75 or 90% PC are suitable as sole emulsifiers for oils requiring HLB values form 4–11. The slightly better performance of the hydrogenated soybean phospholipids with 75% PC is interesting, as this intermediate grade lecithin may present an ideal mixture of phospholipid co-emulsifiers differing in the polar head group. In addition, this lipid provides a pleasant, non-greasy skin feel. It goes without saying that the phospholipids can also be used as co-emulsifiers.

As to be expected, the monoacylphospholipid products are (under the used test conditions) also excellent sole emulsifiers for oil-in-water emulsions over a broad range of HLB values. This result is not surprising since it is known that monoacylphospholipids due to their large polar head group surface area (phosphocholine) and the small area cross section of the fatty acid form a cone-like structure, fitting perfectly to the curvature of the oil–water interface in oil-in-water emulsions [41].

In general, Table 6.7 shows that the various phospholipid products differ enormously in emulsifying properties. It should be further realized that these findings are related to the chosen used test formulation. Other additional formulation components may convert a poor oil-in-water emulsion into an acceptable cosmetic formulation. For the design of oil-in-water emulsions, it is recommended to first start to explore a hydrogenated phospholipid fraction containing about 60–70% PC and not less than 10% PE as an emulsifier.

In summary, it has been shown that hydrogenated diacylphospholipids and (un)saturated monoacylphospholipids are excellent emulsifiers for oil-in-water emulsions. The hydrogenated phospholipids have the advantage that they give rise to a pleasant skin feel, whereas the use of monoacylphospholipids makes the manufacturing of such emulsions easier.

A similar approach was used to test the capability of phospholipids as sole emulsifiers for the preparation of water-in-oil emulsions. Pure soybean PC was not an adequate emulsifier for preparation of water-in-oil emulsions, whereas soybean fractions lower in PC and higher in PE yielded water-in-oil emulsions. Hydrogenated (i.e. fully saturated) PC and PC/PE fractions yielded oil-in-water emulsions. In the range of pure PE, phospholipids, e.g. unsaturated soybean PE with 98% PE, worked as an excellent emulsifier for water-in-oil emulsions, whereas the use of saturated compounds did not result in stable emulsions. This is in agreement with the geometry of the PE molecule which suites perfectly to the curvature of the oil–water interface in water-in-oil emulsions [41].

In summary, phospholipids, representing a broad range of chemical structures and polarities, show in general a very high emulsifying power for different oil phases and can be used as sole emulsifiers. Hydrogenated diacylphospholipids and (un)saturated monoacylphospholipids are very suitable to prepare oil-in-water emulsions for a broad range of oils. The diacylphospholipids are preferred since they are skin protective and have a pleasant non-greasy skin feel. Unsaturated phospholipid fractions lower in PC and higher in PE (which has a small polar head group and a larger lipophilic part) are suitable for preparation of water-in-oil emulsions with certain oil phases, whereas phospholipids with a large polar head group and a smaller lipophilic part, like lyso-phospholipids, are suitable for oil-in-water emulsions [42].

6.5.1.2   Liposome/Lamellar Phase Former

A unique feature of phospholipids compared to other emulsifiers is that they are able to form, upon hydration, closed vesicular structures called liposomes comprising a bilayer of phospholipids surrounding an aqueous vacuole. Although such vesicles can also be made with synthetic compounds like non-ionic surfactants of the alkyl or dialkyl polyglycerol ether class and cholesterol (Niosomes) [43], liposomes comprising phospholipids of natural origin are clearly preferred in natural cosmetic products.

Liposomes are intriguing because of their similarity with the basic structure of a natural cell membrane being a bilayer of phospholipid molecules which serves as a matrix for membrane proteins. This natural role makes clear that this structure must be biocompatible and also biodegradable. For use in cosmetic formulations, the following features are of importance:

The size and composition of the liposomes (as discussed above) may influence the degree of skin interaction (or kinetics of disintegration upon passage into the skin and distribution of phospholipids in the skin layers reached by the liposomes), without increasing the risk that the liposomes and their encapsulated content can become systemically bioavailable.

Several types of liposomes exist: small unilamellar vesicles (SUVs) with a particle size <100 nm, large unilamellar vesicles (LUVs) with a particle size 100–500 nm and multilamellar vesicles (MLVs) with particle size 500 nm to several microns. For cosmetic use, LUVs are most common, because the degree of encapsulation of water-soluble compounds is influenced by the morphology of the liposomes and size [44]. Large unilamellar vesicles have an encapsulated aqueous volume large enough to encapsulate water-soluble compounds efficiently. The size of these liposomes is also in the range enabling a good skin interaction (compare with properties of formulation like the Natipide® product, below) and they can be manufactured easily, especially when using the concept described below.

The lamellarity of liposomes, i.e. the number of membranes (bilayers) per liposomes, in the form of multilamellar liposomes is of limited importance because of the larger particle size which is not favourable for obtaining an adequate skin penetration. Lamellarity in the form of extensive lamellae which comprise saturated phospholipids is, however, of great importance. Extensive lamellae represent a skin like a barrier able to provide an excellent skin protective effect (an example of such a lamellar cream product is "Skin Lipid Matrix®").

The so-called phase transition temperature at which the fatty acids of the phospholipid molecules change (upon cooling) from a mobile state (liquid crystalline) to a rigid state (gel state) is very important for the use of phospholipid in cosmetic formulations. Below this temperature, phospholipids and liposomes have a more rigid character and do not tend to penetrate significantly into the skin but act more as a protective, skin compatible layer. Saturated phospholipids, which have a phase transition temperature around 40–60 °C, have such a rigid structure at the skin temperature of 32 °C. Unsaturated phospholipids have a phase transition temperature below 0 °C and are at skin temperature in the mobile state. The resulting liposomes comprising these lipids possess at skin temperature a flexible character, and they are able to penetrate deeper into the skin compared to rigid liposomes.

The lipid composition of the liposomes is therefore of interest to manipulate the cosmetic use with respect to a deeper skin penetration or application of liposomes as a skin protectant. At wish, the composition and flexibility of liposomes can be manipulated by adding, e.g., either other detergents [e.g. monoacylphospholipid (lysolecithin)] to the liposomes to make them more flexible, or to add saturated phospholipids to make them more rigid. The liposomes are of course, in addition, a source of phospholipids which may be considered as cosmetic actives (see below). In general, liposomes are, since the introduction of liposomes on the cosmetic market (CAPTURE® of Dior), regarded as the classical standard for an anti-ageing product.

6.5.1.3 Solubilizer

Phospholipids play a role in various cosmetic formulations as solubilizers. They are used as an emulsifier in an oil-in-water emulsion with an oil-droplet size of about 50 nm as part of micro-emulsions (an example of such product is PhytoSolve®). The oil phase may be an oil of cosmetic interest or/and serves as solubilizing phase, together with the phospholipid at the oil–water interface, to dissolve other (oil-soluble) lipophilic components of, e.g., plant extracts. Furthermore, phospholipids can be used in mixed micelles comprising mono- and diacylphospholipid or diacylphospholipids in combination with another detergent to solubilize lipophilic compounds. Also, liposomes have the possibility to solubilize compounds in the lipophilic part of the liposomal bilayer membrane. There are types of products available which are clear phospholipid concentrates in oily excipients that can be used as a solubilizing vehicle.

6.5.1.4 Wetting Agent

Due to their amphiphilic properties, phospholipids are excellent wetting agents to disperse lipophilic compounds, like color pigments in hydrophilic media [45, 46].

6.5.2 Actives

Phospholipids contribute to the efficacy of skin and hair care products. They form very thin mono- or oligolamellar layers on the skin which remain even after washing and have the following positive effects:

- better smoothness and suppleness of the skin
- protection against degreasing
- improved moisture balance of the skin
- improved combing quality and reduced electrostatic charge of the hair.

Saturated as well as unsaturated phospholipids can be used as actives. Referring to their multifunctionality, they can be used as surface-active agent as well as for skin transport enhancement of co-formulated compounds of cosmetic interest.

6.5.2.1 Moisturizer

As mentioned before, phospholipids possess hygroscopic properties and are able to bind water molecules. As a result, phospholipids which are able to penetrate into the skin act as a moisturizer and increase the degree of hydration of the skin. In addition, phospholipids act as a moisturizer as result of metabolism of the phospholipids in the skin. In the skin, PC is metabolized by phospholipase A (PLA) and phospholipase D (PLD) enzymes into choline, betaine and glycerophosphocholine (GPC), respectively. These metabolites are called osmoprotectants, also known as osmolytes and/or (most precisely) counteracting osmolytes, able to attract water molecules. These osmolytes are not able to penetrate into membranes and are able to generate an osmotic gradient over the cell membrane. These considerations explain the water binding and attracting properties and there resulting skin moisturizing effect of phospholipids.

The influence of the soybean PC concentration on skin humidity was studied after a single application to the skin of formulations containing 0% (control with 0.9% NaCl), or a 10% (w/w) liposomal dispersion containing phospholipids with varying PC content (10, 28 and 80%, respectively) to ten volunteers [47].

The obtained results (Fig. 6.11) show an acute and significant increase of the skin humidity for the formulation with a 10% (w/w) liposome dispersion containing phospholipids with 80% soybean PC. The formulation containing 10% (w/w) liposome dispersion with phospholipids containing 28% soybean PC showed a weaker effect. In comparison, the formulation with 10% (w/w) liposome dispersion with phospholipids containing 10% soybean PC demonstrated a reduction of the skin humidity. Since the only difference between these formulations is the content of soybean PC, the conclusion can be drawn that PC provides a moisturizing effect to the stratum corneum. This conclusion was underscored by a second multiple application study, which demonstrated an increase of skin humidity to steady-state levels.

6.5.2.2 Effect on Skin Roughness

The influence of the presence of soybean PC in topical formulations on the skin roughness was studied by comparing an aqueous soybean PC containing liposome

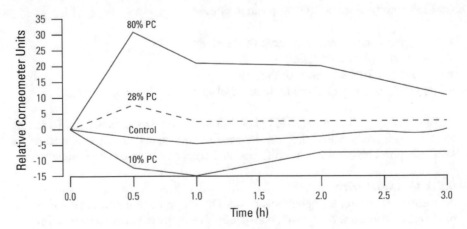

Fig. 6.11 Influence of soybean PC content in topical formulations on the degree of humidity of the human skin after a single application to ten healthy volunteers [48]

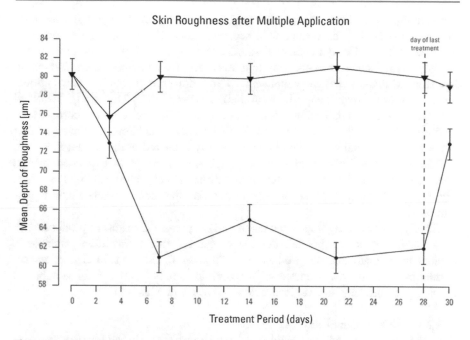

Fig. 6.12 Effect of a formulation with 20.6% soybean PC on the roughness of the human skin after repetitive application to healthy volunteers ($n = 20$). ●-● Soybean PC containing formulation; ▼-▼ o/w emulsion [48]

dispersion with 20.6% w/v soybean PC (with 93% PC) and ca. 16% ethanol with an oil-in-water emulsion after repetitive applications to 20 volunteers (Fig. 6.12) [48].

It was found that the skin roughness was significantly decreased in the group treated with formulation containing soybean PC.

6.5.2.3   Effect on Skin Barrier

Topical formulations comprising of saturated soybean PC possess a skin protective function. They restore and stabilize the skin barrier layers. Measurement of the transepidermal water loss (TEWL) showed that formulations with hydrogenated (saturated) soybean PC restore the natural TEWL level of irritated dry skin. Formulations with HSPC comprise lamellar structures, which are layered on top of each other very similar to the skin structures [49].

Beneficial effects by hydrogenated (saturated) soybean PC were demonstrated in a study [9] involving volunteers subjected to repeated arm washings with the skin irritant sodium lauryl sulphate (SLS). SLS challenge of human skin is generally recognized as the most convincing method for imitating dry skin conditions. The results demonstrated significant improvements in erythema associated with challenged skin samples treated with 1% aqueous hydrogenated (saturated) soybean PC dispersion samples following the washings compared to those without hydrogenated (saturated) soybean PC as determined by a chromameter. The results also showed significant increases in skin moisture contents for the hydrogenated (saturated) soybean PC dispersion. In a separate study involving the healthy skin, addition of 2 and 4% hydrogenated (saturated) soybean PC to conventional cream bases provided superior long-acting hydration activity and smoothness features to the skin of healthy female volunteers when compared to similar conventional creams without hydrogenated (saturated) soybean PC. The study involved twice-daily cream applications during a 28-day time period and corneometer readings taken at 28, 29 and 31 days. Both HSPC formulations displayed similar efficacy with optimal beneficial effect apparently achieved at the 2% level.

These beneficial effects of saturated soybean PC are also explained by the presence of large lamellae in the formulations which mimic skin layers (see Fig. 6.13) able to integrate seamlessly in the skin barrier to stimulate skin regeneration [50].

Finally, the use of hydrogenated (saturated) soybean PC results in products with a pleasant, very soft skin feel. For these reasons, this formulation principle is applied in dermatocosmetic products like Physiogel® AI for skin care of neurodermatitis patients. An example of a cosmetic lamellar cream base suitable to prepare such products is described below (Skin Lipid Matrix®).

6.5.2.4 Biochemical Actives

The positive impacts of phosphatidylcholine from soybean in the treatment of acne vulgaris, atopic dermatitis and other inflammations of the skin are described in numerous papers. Upon application to the skin, unsaturated diacylphospholipids like soybean PC act as a source of linoleic and linolenic acid (Table 6.1). The fraction of administered phospholipids containing linoleic acid reaching metabolic active skin cells may also strengthen the natural barrier function of the skin through

Fig. 6.13 Electron microscopic freeze-fracture picture of SLM 2026 a lamellar cream base comprising saturated phospholipids arranged in lamellar structures (photograph kindly provided by Lipoid GmbH)

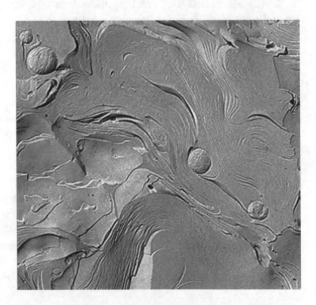

incorporation into skin ceramides. In addition, these phospholipids may play a role in the suppression of acne [51–53], neurodermatitis and psoriasis [49, 54]. The linolenic acid bound to phosphatidylcholine can be eventually converted to a certain extent to omega-3 fatty acids docosahexaenoic acid (DHA, 22:6n-3) and eicosapentaenoic acid (EPA, 20:5n-3).

In an in vitro cell study (data kindly provided by Lipoid GmbH, Ludwigshafen am Rhein, Germany), the effects of soybean phosphatidylcholine with 90% PC on the in vitro release of pro-MMP1 (MMP = matrix metalloprotease), on MMP1 activity and on the secretion of hyaluronic acid (HA) by Normal Human Dermal Fibroblasts (NHDF) fibroblasts, were investigated.

The analysis of the effects of PC (>90%) from soybean on the production of HA (hyaluronic acid) and on the production and activity of MMP1 (matrix metalloprotease-1) was performed on NHDF fibroblasts seeded in 24-well plates 24 h before the application of the phospholipid formulation and the reference molecules. The phospholipid was dispersed in 0.9% NaCl and applied during 72 h in the culture medium of NHDFs in serum-free DMEM medium. The reference molecules LPA (1-oleoyl-L-α-lysophosphatidic acid sodium salt) at 3 μM as control for HA release, TGF-β1 (tissue growth factor) at 20 ng/ml as control for MMP1 release and PMA (phorbol myristate acetate) at 60 ng/ml as control for MMP1 activity were applied in parallel (Table 6.8). The concentrations of released, HA, Pro-MMP1 and MMP1 were determined by compound-specific assay kits.

The PC product increased the release of extracellular hyaluronic acid and decreased MMP1 activity within cultured NHDFs fibroblasts. This underscores that soybean phosphatidylcholine with 90% PC has a moisturizing effect and a potential to increase the firmness of the dermis, which is associated with better maintenance of the collagen resulting in an anti-ageing effect.

Table 6.8 Quantification of HA release and pro-MMP1 release and MMP1 activity from NHDFs fibroblasts; for explanation of abbreviations, see text

Treatments	Concentrations	HA release	Pro-MMP1 release	MMP1 activity
LPA (compared to untreated control)	3 μM	↗↗↗	–	–
TGF-β1 (compared to untreated control)	20 ng/ml	–	↗↗↗	–
PMA (compared to untreated control)	60 ng/ml	–	–	↗↗
Soybean phosphatidylcholine 90 (compared to vehicle 0.9% NaCl)	0.0016%	↗↗	NS	↘

The statistical significance of the data, as compared to control conditions, is presented by ↘ (decrease) and ↗ (increase), highly significant differences are represented by ↗↗ (increase) and ↘↘ (decrease), and very highly significant differences are represented by ↗↗↗ (increase) and ↘↘↘ (decrease). NS—non-significant effects (data kindly provided by Lipoid GmbH)

In a further in vitro cell test on normal human epidermal keratinocytes (NHEKs), which were in a chemical induced inflammatory state, it could be proven that soybean phosphatidylcholine with 90% PC could suppress the secretion of TNF-α suggesting an anti-inflammatory potential (detailed data not shown).

Also, other phospholipids beside phosphatidylcholine may have specific benefits to be used in cosmetic formulations. For instance, it is claimed that phosphatidic acid stimulates hair growth [55].

In a patent application [56], it is claimed that phosphatidylinositol acts as vascular endothelial cell proliferation inhibitor and/or vascularization inhibitor. It is further claimed that PI could be advantageous as for cosmetic use by suppressing skin inflammation and related skin irritation.

Phosphatidylserine stimulates differentiation of keratinocytes which gives rise to an enhancement of the skin barrier [57].

6.5.3   Transport/Penetration Enhancement

The co-administration of cosmetic actives with phospholipids will influence the degree of skin interaction of these actives. Phospholipids represent a marvellous toolbox to design and predetermine a skin interaction profile of the formulated active. This profile can be influenced by the type of phospholipid used (saturated or unsaturated), type of formulation and particle size of the formulation. Also the selection of the type of formulation should be related to the solubility characteristics of the actives.

In order to stimulate transport/skin penetration of actives into the skin, liposomes or other formulations like oil-in-water emulsions comprising flexible unsaturated phospholipids are the first choice. It is assumed that the barrier properties of the membranes of the stratum corneum, which contain ceramides, are modified after incorporation of the flexible soybean PC molecules. The fluidity of the lipid barrier is thereby increased. The depth of modification is strongly dependent on the concentration of PC and which PC species is being used.

The skin penetration effect of unsaturated phospholipids (particularly soybean PC) has been the topic of several studies.

Verma et al. [58] showed that the very hydrophilic molecule carboxyfluorescein entrapped in flexible liposomes, comprising soybean phosphatidylcholine, penetrates through the stratum corneum into the viable epidermis to a small, but significant extent. Carboxyfluorescein only added outside of the liposomal suspension showed less permeation. A buffer solution of carboxyfluorescein failed to show any fluorescence in the viable epidermis.

In addition, it was shown that the depth of penetration of carboxyfluorescein-loaded liposomes was determined by the particle size [59]. Flexible liposomes with sizes of ca. 120 nm showed enhanced penetration of carboxyfluorescein (hydrophilic model drug) and of DiI (i.e. 1,1'-dioctadecyl-3,3,3',3'-tetramethylindocarbocyanine perchlorate, a lipophilic model drug) into the stratum corneum and epidermis. The penetration depth of the model drugs decreased gradually with larger liposomes; 810 nm sized liposomes could only to a smaller extent deliver the carboxyfluorescein restricted to the outer layers of the stratum corneum, whereas DiI due to its small size and lipophilic character permeates on its own to a slightly larger extent into the epidermis.

Lipophilic compounds may be associated with the liposomal bilayer, where the fatty acids, which are directed to each other, represent a solubilization domain in the membrane. In this way the liposome can carry and co-migrate a lipophilic compound [60].

In case of the lipophilic compound caffeine, it was shown that soybean lecithin liposomes enhanced the skin penetration of caffeine compared to the control formulation without phospholipids. It was further found that the better penetration of caffeine was caused by the presence of the phospholipids in the formulation and not necessarily by the encapsulation in liposomes [61].

Röding et al. [62, 63] assessed the degree of skin penetration of soybean phosphatidylcholine liposomes, ^{3}H-labelled (in fatty acid) liposomes after application to porcine skin non-occlusively at 1 mg phospholipid/cm^2 skin. After 30, 60 and 180 min, the phospholipid concentration in the stratum corneum and underlying skin layers was determined using liquid scintillation counting on samples obtained by 20-fold stripping and subsequent tissue sampling. The results provided in Table 6.9 show that more than 99% of the applied PC accumulates in the stratum corneum, suggesting that the liposomes strongly interact with the stratum corneum lipids.

The fluidization properties of the lipid matrix and resulting enhanced penetration of UV absorber bis-ethylhexyloxyphenol methoxyphenyl triazine caused by liposomal phospholipids may also explain the water resistance of skin treated with liposomal sunscreen formulations, maintaining the protective effect of the UV absorber [64]. In contrast to non-liposomal formulations, the liposomal formulation was more effective after water sweeping tests. Obviously, the sun protective effect is not diminished after contact with water, which is most probably caused by the better penetration of the UV absorber into the upper layers of the stratum corneum as result of the interaction with the skin.

Table 6.9 Penetration of ^{3}H-labelled liposomes into porcine skin three hours after single application at 1 mg phospholipid/cm^2 skin [62, 63]

Skin strips/cuts	Skin section	µg liposomal phospholipid/g tissue
20 strips	Stratum corneum	100,000
1 mm	Epidermis	500
2 mm	Dermis	20
3 mm	Subcutis	8
4 mm	Subcutaneous fat	8
5 mm	Subcutaneous fat	12

Also in case of TiO_2 sunscreen particles, beneficial skin penetration effects could be observed when using liposomes composed of soybean phosphatidylcholine compared to a mixed micellar formulation [65].

The skin whitening effect of linoleic acid could also be enhanced by the use of soybean phosphatidylcholine liposomes compared to a hydrogel formulations [66].

Corderch et al. [67] studied the percutaneous absorption of sodium fluorescein (NaFl) vehiculized in two different liposomes using either phosphatidylcholine or lipids mimicking the stratum corneum (ceramides, cholesterol, palmitic acid and cholesteryl sulphate), respectively. The effect of these vesicles on the stratum corneum lipid alkyl chain conformational order was evaluated at different depths of the stratum corneum by corneometer, tewameter and ATR-FTIR. The highest penetration of NaFl was observed with PC liposomes, which could be related to the increase in stratum corneum lipid disorder detected by ATR-FTIR.

The penetration of liposomes with encapsulated fluorescent dye carboxyfluorescein into the human abdomen skin studied by fluorescence microscopy showed that liposomes made from fluid-state PC (di-lauroyl-PC) compared to liposomes composed of gel state (rigid) DSPC (di-stearoyl-PC) are taken up by the skin more readily, permeate faster and penetrate beyond the stratum corneum [68]. These findings suggest that rigid saturated phospholipids, and most probably also the accompanying water, are taken up by the stratum corneum but not by the deeper layers of the skin. In addition, saturated phospholipids seem to perturb the lipid barrier to a lesser extent as unsaturated soybean phospholipids.

The differences of skin penetration pattern between liposomes comprising either unsaturated or saturated phospholipids were also studied with liposomes simultaneously labelled with the hydrophilic water-soluble fluorescent marker carboxyfluorescein and a lipophilic fluorescent marker N-rhodamine PE [69] employing confocal laser microscopy. Indeed, it was found that the liposomes comprising saturated phospholipids were more located at the surface of the stratum corneum, whereas liposome comprising unsaturated phospholipid resulted in a localization of both markers in deeper layers of the studied skin fragments underscoring their suitability as penetration enhancer (see Discussion section for photographs).

Perez-Cullell et al. demonstrated that the skin penetration of sodium fluorescein was higher from fluid liposomes (PC) than from rigid liposomes (hydrogenated phosphatidylcholine). They conclude that the liquid crystalline phospholipids interact with the lipids of the stratum corneum by means of disintegration of the vesicles during their passage through the lipid intercellular pathway in the stratum corneum [70].

Yokomizo and Sagitani [71] observed with ATN-FTIR that liposomes, loaded with prednisolone, comprising phospholipids that have unsaturated acyl chains induced higher and broader absorbance shifts in the C–H bond-stretching region of the stratum corneum while phospholipids that have saturated acyl chains induced lower and sharper absorbance shifts in the C–H bond-stretching region. A significant parallel between the amount of prednisolone penetrated and the lipid-chain fluidity of the stratum corneum was found. These results suggest that phospholipids may influence the percutaneous penetration of prednisolone by changing the lipid-chain fluidity of the stratum corneum.

Komatsu et al. examined the in vivo percutaneous penetration of butylparaben and the saturated dipalmitoylphosphatidylcholine (DPPC) from liposomal suspensions in guinea pigs by autoradiography. They reported that butylparaben penetrated through the skin, whereas DPPC was scarcely detected in the body, suggesting that the liposomes themselves remained on the skin surface [72].

The observation that skin penetration is more enhanced by formulations comprising unsaturated fatty acid containing phospholipids, compared to formulations comprising saturated fatty acid containing phospholipids is supported by the findings of Ibrahim and Li [73] who showed that also the skin penetration effects of unsaturated versus saturated fatty acids are equally dependent on their melting point.

A typical example of a product, comprising unsaturated soybean phospholipids, used to enhance the skin penetration of water-soluble compounds is described below in more detail.

As the pilosebaceous units (hair follicles) are a kind of extension of epidermis and dermis to the surface of the skin, several studies investigated the targeting of the hair follicles by phospholipid particles. Du Plessis et al. reported that the amount of liposomally entrapped IFN (γ-interferon) in the deeper skin strata were in the order of increasing number of follicles/hair in the skin species, suggesting that the transfollicular route is an essential pathway for liposomal topical therapeutics [74]. Lieb et al. showed that topical application of the liposomal based formulation resulted in a significantly higher accumulation of carboxyfluorescein in the pilosebaceous units than the application of any of the other non-liposomal formulations [75]. Niemiec et al. showed significantly enhanced topical delivery of hydrophilic protein alpha-interferon and hydrophobic peptide cyclosporin-A into pilosebaceous units by non-ionic liposomes [76]. Another study of Li et al. demonstrated [77] that phosphatidylcholine liposomes entrapping either the fluorescent dye calcein or the pigment melanin can deliver these molecules into the hair follicle and hair shafts of mice when applied topically. In a recent study [78], the influence of massage and occlusion on the follicular penetration depths of rigid and flexible liposomes loaded with a hydrophilic and lipophilic dye was investigated. The application of massage increased follicular penetration significantly. Occlusion resulted in an increased follicular penetration depth only for rigid liposomes.

Employing the hamster ear skin model in a mechanistic study, liposomes of PC, cholesterol and phosphatidylserine delivered the fluorescent hydrophilic dye, carboxyfluorescein, into the pilosebaceous units. They were more efficient than aqueous solutions even after incorporation of 10% ethanol or 0.05% sodium lauryl sulphate, or using propylene glycol as the donor vehicle.

Although the particle size and the nano-aspect play an important role for efficient delivery of phospholipids (and associated compounds/actives) to hair follicles to influence the condition of the hair, phospholipids can also be integrated in shampoos and hair conditioners, mostly as emulsifier for direct application to the hair [79]. The addition of phospholipids to hair cosmetics may have beneficial effects in conditions like dandruff, hairbrush, hair splitting, lacklustre hair and fat hair.

6.6 Examples of Phospholipid Formulations

In the following, typical examples of phospholipid-based products are summarized which highlight the various possibilities to demonstrate the use phospholipids in cosmetic products.

6.6.1 Lamellae Containing Formulations

For many years, conventional oil-in-water and water-in-oil emulsions have been the heart of skin care products. However, the classic emulsifiers needed in these products to maintain their stability can also interact with the lipid layers of the stratum corneum. This may cause wash out effects and skin irritations. A big step forward to minimize these disadvantages offers formulations based on lamellar structure (see Fig. 6.13) comprising saturated soybean phosphatidylcholine and which does not contain any classical emulsifier (brand name: Skin Lipid Matrix® (SLM)). The SLM is composed of skin-identical lipids that can regenerate and reconstruct the skin barrier.

The stratum corneum, the horny layer of the epidermis, is composed of two parts, the corneocytes and the lipid matrix which fills the space between them. A disruption of the lipid layer, or the simple lack of skin lipids, can create moisture deficiency and dry skin. The SLM acts as an excellent skin renovator by providing new skin lipids.

There are commercially available products based on this formulation principle with saturated/hydrogenated phosphatidylcholine, also from non-GMO soybean.

6.6.2 Phospholipid Containing Nano-emulsions

Highly concentrated solutions of polyols (e.g. glycerol, sorbitol) and phospholipids facilitate the dissolution of a large quantity of lipids and form transparent to translucent nano-emulsions of honey-like to gel-like consistency and with droplet sizes of less than 100 nm. A diverse range of numerous active lipophilic ingredients can be incorporated: vegetable oils, essential oils, mineral oils, vitamins, ceramides or oil-soluble UV filters and pharmaceutically active ingredients.

PhytoSolve® is a brand comprising such a sophisticated solubilization system based on phospholipids. This delivery system with only natural ingredients combines excellent solubilizing function with physiological safeness, being well-tolerated by the body and also characterized by ecologically unobjectionable behaviour [80] (Note: NanoSolve® was the original name of this patented technology; this was rebranded to PhytoSolve®).

The system works as a versatile medium that accelerates the effects of active ingredients and allows them to penetrate the skin upon dermal application.

The product can be used directly without further processing in cosmetics, topical formulations and may be easily incorporated into any kind of emulsion, gels, sunscreen sprays with oil-soluble UV filters.

6.6.3 Phospholipid Concentrates

It has been proven that phosphatidylcholine can improve the solubility and absorption of pharmaceutically active ingredients that are either insoluble or barely soluble in water or only soluble in physiologically unacceptable solvents. In pharmaceutical applications, phosphatidylcholine may reduce objectionable side effects of drug substances and thereby render them physiologically more tolerable. Besides their application in the pharmaceutical oral and parenteral fields, compounds with phosphatidylcholine in mixture with oil featuring higher concentrations of phosphatidylcholine are also suitable for dermal pharmaceutical and cosmetic applications [81]. The designations in the PHOSAL® brand names clearly indicate the medium present in a particular product, e.g. medium-chain triglycerides (MCT), sunflower oil (SB), safflower oil (SA) and propylene glycol (PG). All ingredients of the formulation with phosphatidylcholine are of food-grade quality and provide excellent carriers of natural phosphatidylcholine and lipophilic active ingredients, e.g. vitamins and essential oils. Liquid compounds can directly be incorporated into: emulsions, i.e. creams and lotions.

6.6.4 Liposomal Concentrates

A soybean PC-based product for cosmetic use was developed under the brand name Natipide® [82, 83]. This product, Natipide® II, is a breakthrough for the use of liposomes in cosmetic products because it avoids the need to apply high-pressure homogenizer procedures to make liposomes with sufficiently small size. Instead, this concentrate is a ready-to-use product, which can be used for large-scale production of liposomes by simply diluting the concentrate. It contains a yellowish-brown gel-like liposome concentrate containing a purified phospholipid fraction with high phosphatidylcholine content and ethanol (antimicrobial preservative) and is described as a vesicular gel consisting of densely packed unilamellar liposomes. Recently, a ethanol-free product with non-GMO phospholipids became available (Natipide® Eco).

Electron microscopy shows a gel of closely packed liposomes (Fig. 6.14). These liposomes have leaky membranes because of the presence of ethanol. Upon dilution with an aqueous medium containing a substance of cosmetic interest, the added

Fig. 6.14 Freeze-fracture
electron micrograph of
Natipide® II. Bar = 100 nm
(photograph kindly provided
by Lipoid GmbH)

substance equilibrates by means of diffusion over the membrane and is partially
encapsulated in the liposomes. This type of formulation represents therefore an
ideal skin penetration enhancing vehicle, being a supply of phospholipids which
makes the stratum corneum more permeable for the cosmetic actives encapsulated
in the liposomes and outside the liposomes.

6.7   Discussion

Phospholipids are used in cosmetic formulations, comprising, e.g., emulsions and
liposomes. Upon administration to the skin, the liposome or emulsion particles fall
gradually apart and release their content and simultaneously distribute the lipid in
the surrounding skin layers. This distribution process may be accompanied by a
metabolic degradation of the lipids. It is generally accepted that intact liposomes are
not able to permeate through the skin at any meaningful level. In spite of these skin
distribution properties, the particle size in the nano-range of the lipid particles may
determine the depth of penetration. A correlation between particle size and depth of
penetration has been found with liposomes but is less clear for (nano)-water-in-oil

emulsions. It is claimed that a particle size of ca. 100–200 nm is ideal for an optimal skin interaction.

Besides the particle size, the lipid composition/properties of the lipid particles determine to a great extent the degree of interaction with the skin. Diacylphosphatidylcholines are in the liquid crystalline state above the phase transition temperature and below this temperature in the gel state and rigid. Unsaturated phospholipids have a phase transition temperature below 0 °C and are at 32 °C in the flexible state. Saturated phospholipids have a transition temperature between ca. 40 and 60 °C and are in the gel (rigid) state at 32 °C. Lipids which are rigid at the skin temperature of 32 °C have a lower tendency to penetrate into the stratum corneum compared to lipid particles which comprise lipids which are more flexible at 32 °C.

Soybean PC, i.e. phosphatidylcholines with unsaturated fatty acids, forms flexible liposomes which are able to penetrate more deeply into the skin compared to rigid liposomes/lamellar phases comprising of saturated soybean PC which for these reasons reside predominantly in the upper layers of the skin. Soybean PCs are therefore suitable as penetration enhancers/transportation vehicles to carry co-encapsulated compounds more deeply into the stratum corneum. In addition, they make a contribution to maintain or restore a healthy condition of the skin by means of supply of polyunsaturated fatty acids (linoleic, linolenic and oleic acids). Saturated soybean PCs make another contribution to maintain or restore the healthy condition of the skin by stabilizing the barrier function of the skin in a physical way.

The differences of skin penetration pattern between liposomes comprising either unsaturated or saturated phospholipids were also studied with liposomes simultaneously labelled with a hydrophilic water-soluble fluorescent marker carboxyfluorescein and a lipophilic fluorescent marker N-rhodamine PE [69] and confocal laser microscopy (Fig. 6.15). The photographs impressively show that the liposomes comprising saturated phospholipids were more located at the surface of the stratum corneum, whereas liposome comprising unsaturated phospholipid resulted in a localization of both markers in deeper layers of the studied skin fragments underscoring their suitability as penetration enhancer.

The several ways phospholipid containing lipid particles may interact with the skin and the influence of flexibility/rigidity of the phospholipid molecule are schematically presented (Fig. 6.16).

The penetration depth of the encapsulated compound (or co-administered with the phospholipid formulation) will be substance specific and depends on the physiochemical properties and molecular weight. In Fig. 6.16, the various mechanisms of interaction with the stratum corneum of "free" cosmetic compounds without cosmetic carrier (A) and with cosmetic carrier comprising either saturated (B) or unsaturated phospholipids (C–E) are presented schematically. The carriers could be, e.g., liposomes/lamellar phase or emulsion particles. The compound could be hydrophilic or lipophilic but is in case of a carrier associated with this carrier.

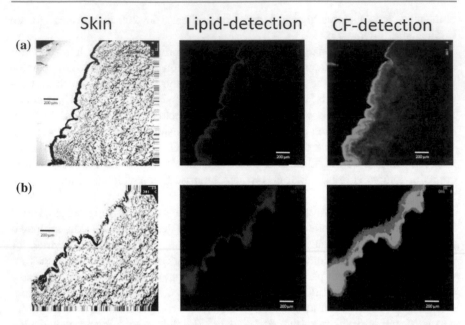

Skin   Lipid-detection   CF-detection

Fig. 6.15 Comparison of skin distribution pattern of liposomes containing the water-soluble marker carboxyfluorescein (CF detection) and the lipophilic marker N-rhodamine-PE (lipid detection), comprising either **a** unsaturated phospholipid or **b** saturated phospholipids (derived from Fahr et al. [69]). Bar = 200 μm

In case of the "free" compound, a relatively low degree of penetration into the stratum corneum compared to the scenario with cosmetic carrier can be expected (as depicted with the smaller red arrow) (A).

When the compound is co-administered with a carrier containing rigid, saturated phospholipids, the saturated phospholipid will be embedded in the upper layers of the stratum corneum as lamellar layers, thereby increasing/restoring the barrier function of the stratum corneum (as depicted by the diffuse yellow color of the lipids in the mortar). Depending on the physiochemical properties of the cosmetic compounds, the compound itself may be penetrating deeper and/or slowly released from the lamellae (B).

Co-administration with carriers containing flexible, unsaturated phospholipids results in various interaction possibilities with the stratum corneum. Which mechanism prevails (or the degree of relative contribution of the three discussed mechanisms) regarding penetration will be dependent on the properties of the cosmetic compound and carrier. Anyway, it can be expected, in general, that due to the flexible phospholipids, these phospholipids and the active cosmetic compound are able to penetrate deeper into the stratum corneum compared to the free compound and the scenario with saturated phospholipids (as depicted by the longer red arrow and the diffusion of the orange coloured unsaturated phospholipids and black dots (compounds) (C). Further, (D) represents the interaction option that the carrier

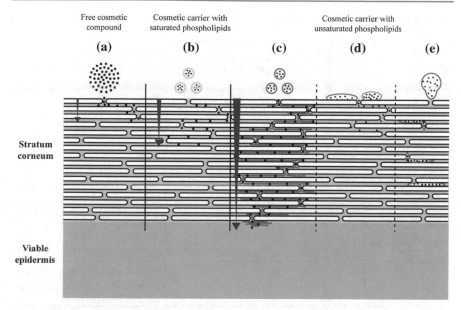

Fig. 6.16 Schematic presentation of the influence of the type of phospholipid (saturated or unsaturated) in emulsion- or liposome-based formulations on the interaction mode of encapsulated cosmetic compounds and used phospholipids with the stratum corneum. **a** Interaction of free cosmetic compound (black dots) with the stratum corneum, **b** interaction of a cosmetic carrier (emulsion particle or liposome) comprising a cosmetic compound (black dots) and saturated phospholipids (yellow color), **c** penetration enhancing process of cosmetic compound (black dots) by unsaturated phospholipids (orange color), **d** adsorption or fusion of cosmetic carrier comprising a cosmetic compound (black dots) and unsaturated phospholipids (orange) with the upper layer of the stratum corneum, **e** penetration of intact cosmetic carrier, comprising a cosmetic compound (black dots) and unsaturated phospholipid (orange). Red arrows represent the relative depth of penetration into the stratum corneum, of lipid and associated cosmetic compounds

fuses with or adsorbs on the upper layers of the stratum corneum and (E) represents the option that the compounds are co-migrating with intact vesicles (and unsaturated (orange) phospholipids) able to migrate into the stratum corneum.

In Fig. 6.16, the relative penetration depths of cosmetic compounds and phospholipids of the three administration options (free compound/compound with saturated phospholipids/compound with unsaturated phospholipids) are indicated by the lengths of the red arrows. These arrows represent a concentration gradient which ends in the lower layers of the stratum corneum/viable epidermis. As indicated above, the drawing is intended to explain interaction mechanisms and does not present the skin interaction and depth of presentation of any compound. In practice, compounds may penetrate deeper into the skin, especially when they have a low MW and are lipophilic. Also the phospholipids may penetrate deeper than indicated, but as explained above in extenso, to a minor extent. Regarding the skin interaction of unsaturated phospholipids, the degree of contributions of these

Table 6.10 Relevant characteristics of soybean PC and saturated soybean PC for cosmetic products

Parameter	Soybean PC	Saturated soybean PC
Fatty acid composition	Unsaturated fatty acids; linoleic, linolenic and oleic acid	Saturated fatty acids; stearic and palmitic acid
Phase transition temperature (°C)	Below 0	40–60
Structure at skin temperature	Flexible	Rigid
Structures upon hydration	Liposomes and lamellar structures dependent on process conditions	Liposomes and lamellar structures dependent on process conditions
Surface-active technical ingredient		
Emulsifier	Yes, for oil-in-water emulsions	Yes, for oil-in-water emulsions More stable than unsaturated soybean PC; pleasant skin feel, white/off-white color
Dispersing and solubilizing ability	Hydrophilic, amphiphilic and lipophilic compounds	Hydrophilic, amphiphilic and lipophilic compounds
Product examples	–	Saturated soybean phospholipid fractions
Cosmetic active		
Skin interaction	Not used for restoring skin barrier	Stabilizing the barrier function; conditioning the stratum corneum
Barrier compatibility	Yes, slightly enhancing TEWL[a]	Yes, stabilizing normal TEWL
Supply of linoleic and linolenic acid	Yes; beneficial effects of phospholipids containing omega-3 and omega-6 fatty acids on acne vulgaris, psoriasis, neurodermatitis. Stimulates in vitro excretion of hyaluronic acid	No
Product example	Soybean phosphatidylcholine	Formulations with saturated PC
Transport		
Transport properties	Penetration enhancement; conditioning the stratum corneum Penetrates into the stratum corneum and into the viable epidermis to a minor extent	Penetrates only into SC and not below
Product example	Liposomal formulations with soybean PC	–

[a]TEWL—transepidermal water loss

separate interaction modes (C)–(E), to the overall skin interaction is dependent on the cosmetic compound and formulation composition.

Depending on the oil solubility characteristic of the lipophilic actives, also nano-emulsions using rigid saturated or flexible unsaturated phospholipid as emulsifier may be used to carry especially actives with high solubility in the oil phase, or the oil phase as active in the skin.

The properties of unsaturated soybean PC and saturated soybean PC and related cosmetic products and relevant parameters for the design of cosmetic formulations using both classes of phosphatidylcholines are summarized in Table 6.10 (data from [48, 9, 84, 85]).

Table 6.10 addresses as well the multifunctional aspects of the use of the main classes of phospholipids (phosphatidylcholines). The table can also be used as guide to select the best possible phospholipid for obtaining a desired property of the cosmetic product

6.8   Conclusions

For decades, phospholipids have been used as ingredients in cosmetic products. Before the discovery of liposomes in the late sixties of the past century, the use of phospholipids as emulsifier in cosmetic formulations was prominent. The interest in phospholipids was then greatly enhanced by the introduction of liposomes in the cosmetic industry in 1987. Since then, extensive research, especially in the pharmaceutical area, has been performed on liposomes showing their skin penetration enhancing properties for various compounds. Simultaneously, the attributes of phospholipids as cosmetic and biochemical actives were established. In the last decade, the understanding of the influence of the phase transition temperature on the cosmetic use of phospholipids, either formulated in (mainly) liposomes or emulsions, became apparent.

The above-presented findings make also very clear that phospholipids are not synonymous with liposomes, but are valuable cosmetic ingredients in any other type of formulations as well. Phospholipids have several big advantages in nano-cosmetics. They are natural substances, present in the skin, addressing the cry for using sustainable products and refraining from synthetic compounds in cosmetic industry. By choosing the right components from available natural phospholipid products, they can replace missing or attenuated substances in the stressed skin (e.g. linoleic acid) via intrinsic metabolism pathways in the skin.

Nowadays, phospholipids are considered as unique natural compounds for formulators of cosmetic products. This is owed to their multifunctional properties as a surface-active technical ingredient, cosmetic active and use to enhance the skin interaction/penetration of co-formulated cosmetic actives, in combination with their high degree of skin tolerability.

Examples of natural phospholipids, which fulfil these properties as:

(1) surface-active ingredients are hydrogenated soybean lecithin with 75% PC and hydrogenated soybean phosphatidylcholine with 80% PC, which are excellent sole emulsifiers in oil-in-water emulsions,
(2) actives for improving barrier properties are lamellar creams with hydrogenated phosphatidylcholine, hydrogenated soybean phosphatidylcholine with 80% PC or 90% PC,
(3) biochemically active nutrients are soybean phosphatidylcholine with $\geq 90\%$ PC,
(4) skin transport enhancing modalities are liposomal concentrates, described above and soybean phosphatidylcholine with $\geq 90\%$ PC.

Besides soybean phospholipids, the use of sunflower phospholipid fractions and the respective unsaturated and saturated phosphatidylcholine is becoming increasingly popular.

In spite of the decade-long use of phospholipids, there are still research issues which would be of future interest. For instance, as pointed out above, the knowledge on skin metabolism (anabolism and catabolism) of endogenous and externally applied phospholipids is still very limited. Also, it would be of interest to compare the skin interaction of emulsions (nano- or macro-emulsions) comprising phospholipids with liposomes in much greater detail.

In nano-cosmetics, phospholipids play a role as an essential component in oil-in-water emulsions and liposomes in the size range typical for nanotechnology. Since they belong to the class of nano-materials which fall apart in the skin, they do not pose a risk for systemic absorption but represent material with an excellent skin interaction, which results in optimal cosmetic effects.

The availability of these natural compounds with controlled quality in various grades and modifications provides the formulator with a valuable toolbox for designing optimal cosmetic formulations and products.

References

1. U.S. Department of Health and Human Services, F., Center For Food Safety And Applied Nutrition. Guidance for Industry: Safety of Nanomaterials in Cosmetic Products; June 2014.
2. Dragicevic-Curic N, Scheglmann D, Albrecht V, Fahr A. Temoporfin-loaded invasomes: development, characterization and in vitro skin penetration studies. J Control Release. 2008;127:59–69.
3. Cevc G, Blume G, Schätzlein A. Transfersomes-mediated transepidermal delivery improves the regio-specificity and biological activity of corticosteroids in vivo. J Controlled Release. 1997;45:211–26.
4. Ashtikar M, Langeluddecke L, Fahr A, Deckert V. Tip-enhanced Raman scattering for tracking of invasomes in the stratum corneum. Biochim Biophys Acta. 2017;1861:2630–9.
5. Lasic DD. Novel applications of liposomes. Trends Biotechnol. 1998;16:307–21.
6. Kraft JC, Freeling JP, Wang Z, Ho RJ. Emerging research and clinical development trends of liposome and lipid nanoparticle drug delivery systems. J Pharm Sci. 2014;103:29–52.

7. Blaber, M. BCH 4053 Biochemistry I [Online]. Available: http://www.mikeblaber.org/oldwine/BCH4053/Lecture14/Lecture14.htm (2001). Accessed 20 Feb 2018.

8. Yee JY, Davis ROE. Accelerated method for determining moisture absorption. Ind Eng Chem Anal Ed. 1944;16:487–90.

9. Ghyczy M, Vacata V. Phosphatidylcholine and skin hydration. In: Leyden JJ, Rawlings AV, editors. Skin moisturization, vol. 15. NY: Marcel Dekker Inc., Basel Chapter; 2002. 303–321.

10. van Hoogevest P, Wendel A. The use of natural and synthetic phospholipids as pharmaceutical excipients. Eur J Lipid Sci Technol. 2014;116:1088–107.

11. USP_Monograph. Lecithin Monograph USP40-NF35; 2018.

12. Lekim D, Betzing H. Der Einbau von EPL-Substanz in Organe von gesunden und durch Galaktosamin geschädigten Ratten. Arzneim. Forsch. (Drug Res.). 1974;24:1217–21.

13. Elias PM, Friend DS. The permeability barrier in mammalian epidermis. The J Cell Biol. 1975;65:180–91.

14. Holbrook KA, Odland GF. Regional differences in the thickness (cell layers) of the human stratum corneum: an ultrastructural analysis. J Invest Dermatol. 1974;62:415–22.

15. Honeywell-Nguyen PL, de Graaff AM, Groenink HWW, Bouwstra JA. The in vivo and in vitro interactions of elastic and rigid vesicles with human skin. Biochim Biophys Acta. 2002;1573:130–40.

16. Itoh Y, Shimazu A, Sadzuka Y, Sonobe T, Itai S. Novel method for stratum corneum pore size determination using positron annihilation lifetime spectroscopy. Int J Pharm. 2008;358:91–5.

17. Ruddy S, Hadzija B. Iontophoretic permeability of polyethylene glycols through hairless rat skin: application of hydrodynamic theory for hindered transport through liquid-filled pores. Drug Des Discov. 1992;8:207–24.

18. Aguilella V, Kontturi K, Murtomäki L, Ramírez P. Estimation of the pore size and charge density in human cadaver skin. J Controlled Release. 1994;32:249–57.

19. Elias PM, Gruber R, Crumrine D, Menon G, Williams ML, Wakefield JS, Holleran WM, Uchida Y. Formation and functions of the corneocyte lipid envelope (CLE). Biochimica et Biophysica Acta (BBA)-Mol Cell Biol Lipids 2014;1841:314–318.

20. Bouwstra J, Honeywell-Nguyen P. Skin structure and mode of action of vesicles. Adv Drug Deliv Rev. 2002;54:S41–55.

21. Ohman H, Vahlquist A. In vivo studies concerning a pH gradient in human stratum corneum and upper epidermis. Acta Dermatovenereologica-Stockholm, 1994;74:375–375.

22. Korting H, Lukacs A, Vogt N, Urban J, Ehret W Ruckdeschel G Influence of the pH-value on the growth of Staphylococcus epidermidis, *Staphylococcus aureus* and *Propionibacterium acnes* in continuous culture. Zentralblatt fur Hygiene und Umweltmedizin = Int J Hyg Environ Med. 1992; 193:78–90.

23. Puhvel S, Reisner R, Sakamoto M. Analysis of lipid composition of isolated human sebaceous gland homogenates after incubation with cutaneous bacteria. Thin-layer chromatography. J Invest Dermatol. 1975;64:406–11.

24. Feingold KR, Elias PM. Role of lipids in the formation and maintenance of the cutaneous permeability barrier. Biochimica et Biophysica Acta (BBA)-Mol Cell Biol Lipids 2014;1841:280–294.

25. Feingold KR. The outer frontier: the importance of lipid metabolism in the skin. J Lipid Res. 2009;50:S417–22.

26. Mazereeuw-Hautier J, Redoules D, Tarroux R, Charveron M, Salles J, Simon M, Cerutti I, Assalit M, Gall Y, Bonafe J. Identification of pancreatic type I secreted phospholipase A2 in human epidermis and its determination by tape stripping. Br J Dermatol. 2000;142:424–31.

27. Mao-Qiang M, Jain M, Feingold KR, Elias PM. Secretory phospholipase A2 activity is required for permeability barrier homeostasis. J Invest Dermatol. 1996;106:57–63.

28. Cox P, Squier CA. Variations in lipids in different layers of porcine epidermis. J Invest Dermatol. 1986;87:741–4.

29. Lampe MA, Williams ML, Elias PM. Human epidermal lipids: characterization and modulations during differentiation. J Lipid Res. 1983;24:131–40.
30. Wertz PW, van den Bergh B. The physical, chemical and functional properties of lipids in the skin and other biological barriers. Chem Phys Lipids. 1998;91:85–96.
31. Lampe MA, Burlingame A, Whitney J, Williams ML, Brown BE, Roitman E, Elias PM. Human stratum corneum lipids: characterization and regional variations. J Lipid Res. 1983;24:120–30.
32. Bonte F, Saunois A, Pinguet P, Meybeck A. Existence of a lipid gradient in the upper stratum corneum and its possible biological significance. Arch Dermatol Res. 1997;289:78–82.
33. Mauro T, Grayson S, Gao WN, Man M-Q, Kriehuber E, Behne M, Feingold KR, Elias PM. Barrier recovery is impeded at neutral pH, independent of ionic effects: implications for extracellular lipid processing. Arch Dermatol Res. 1998;290:215–22.
34. Fiume Z. Final report on the safety assessment of lecithin and hydrogenated lecithin. Int J Toxicol. 2001;20(Suppl 1):21–45.
35. ORGANIZATION, U. C. I. R. C. Final Report Release Date: April 7, 2015. Safety Assessment of Lecithin and Other Phosphoglycerides as Used in Cosmetics, CIR Status.
36. Traul K, Driedger A, Ingle D, Nakhasi D. Review of the toxicologic properties of medium-chain triglycerides. Food Chem Toxicol. 2000;38:79–98.
37. Ghyczy M, Vacata V. Concepts for topical formulations adjusted to the structure of the skin. Chim Oggi. 1999;17:12–20.
38. Ghyczy M. Die Hautstruktur und adäquate Kosmetika. Beeinflussung der Permeabilitätsbarriere der Haut mittels Struktur der Kosmetika. Kosmetische Medizin. 1999;20:192–5.
39. Rebmann H. The composition of lecithin and its potential uses in cosmetics. Soap Parf Cosmet. 1977;50:361–70.
40. The Solae Company. Introduction to the use of lecithin, Technical Bulletin; 2003.
41. Knoth A, Scherze I, Fechner A. Emulgatoren zur Bildung von multiplen Emulsionen. Lecithin. In: Muschiolik G, Bunjes H, editors. Multiple Emulsionen: Herstellung und Eigenschaften. Hamburg: Behr's Verlag; 2007.
42. Heidecke CD, Van Hoogevest P, Seckler, D. Use of phospholipids as sole emulsifiers in water-in-oil emulsions. Poster. In: Fifth International Symposium on Phospholipids in Research; 2017; Heidelberg, Germany.
43. Schreier H, Bouwstra J. Liposomes and niosomes as topical drug carriers—dermal and transdermal drug delivery. J Controlled Release. 1994;30:1–15.
44. New RRC, editor. Liposomes: a practical approach. Oxford: IRL Press, Oxford University Press; 1990.
45. Fujita S, Suzuki K. Soyabean lysophospholipid as a surfactant: aqueous solution and wetting properties. Nippon Nogeikagaku Kaishi. 1990;64:1355–60.
46. Johansson D, Bergenståhl B. Lecithins in oil-continuous emulsions. Fat crystal wetting and interfacial tension. J Am Oil Chem Soc. 1995;72:205–11.
47. Artmann C, Röding J, Ghyczy M, Pratzel HG. Influence of various liposome preparations on skin humidity. Parfum Kosm. 1990;90:326.
48. Ghyczy M, Gareiss J, Kovats T. Liposomes from vegetable phosphatidylcholine: their production and effects on the skin. Cosmetics and toiletries. 1994;109:75–80.
49. Lautenschläger H. Starke Wirkung-Phospholipide in Kosmetika. Kosmetik Int. 2003;2:38–40.
50. Ghyczy M, Albrecht M, Vacata V. Phospholipids, Metabolites, and Skin Hydration. In: Lodén M, Maibach HI, editors. Dry skin and moisturizers: chemistry and function. Boca Raton: CRC Press Taylor & Francis Group.
51. Fabrizi G, Randazzo SD, Cardillo Λ, Tiberi L, Morganti P. Safety and efficacy of lamellar phosphatidylcholine emulsion to treat mild-to-moderate inflammatory acne. SOFW-J. 1999;125:12–4.
52. Ghyczy M, Nissen H, Biltz H. The treatment of acne vulgaris by phosphatidylcholine from soybeans, with a high content of linoleic acid. J Appl Cosmetol. 1996;14:137–46.

53. Kutz G. Galenische Charakterisierung ausgewählter Hautpflegeprodukte. PZ 142, 1997; 142:11–15.
54. Morck H. Liposomen bei Neurodermitis und Psoriasis. PZ. 1993;128:50–1.
55. Takahashi T, Kamimura A, Hamazono-Matsuoka T, Honda S. Phosphatidic acid has a potential to promote hair growth in vitro and in vivo, and activates mitogen-activated protein kinase/extracellular signal-regulated kinase kinase in hair epithelial cells. J Invest Dermatol. 2003;121:448–56.
56. JP2008143835. Vascular endothelial cell growth inhibitor and/or angiogenesis inhibitor. JP 2008143835; 2006.
57. Chung S-Y, Nam S-J, Choi W-K, Seo M-Y, Kim J-W, Lee S-H, Park C-S. Phosphatidylserine enhances skin barrier function through keratinocyte differentiation. J Soc Cosmet Sci Korea. 2006;32:17–22.
58. Verma DD, Verma S, Blume G, Fahr A. Liposomes increase skin penetration of entrapped and non-entrapped hydrophilic substances into human skin: a skin penetration and confocal laser scanning microscopy study. Eur J Pharm Biopharm. 2003;55:271–7.
59. Verma DD, Verma S, Blume G, Fahr A. Particle size of liposomes influences dermal delivery of substances into skin. Int J Pharm. 2003;258:141–51.
60. Chen M, Liu X, Fahr A. Skin penetration and deposition of Carboxyfluorescein and Temoporfin from different lipid vesicular systems: in vitro study with finite and infinite dosage application. Int J Pharm. 2011.
61. Kim C, Shim J, Han S, Chang I. The skin-permeation-enhancing effect of phosphatidylcholine: caffeine as a model active ingredient. J Cosmet Sci. 2002;53:363–74.
62. Röding J. Properties and characterization of pre-liposome systems. In: Braun-Falco O, Korting HC, Maibach HI, editors. Liposome dermatics. Heidelberg, Berlin: Springer; 1992.
63. Röding J, Artmann C. The fate of liposomes in animal skin. In: Braun-Falco O, Korting HC, Maibach HI, editors. Liposome dermatics. Heidelberg, Berlin: Springer; 1992.
64. Korting HC, Schollmann C. Resistance of liposomal sunscreen formulations against plain water as well as salt water exposure and perspiration. Skin Pharmacol Physiol. 2011;24:36–43.
65. Gottbrath S, Müller-Goymann C. Penetration and visualization of titanium dioxide microparticles in human stratum corneum—effect of different formulations on the penetration of titanium dioxide. SOFW-J. 2003;129:11–7.
66. Shigeta Y, Imanaka H, Ando H, Ryu A, Oku N, Baba N, Makino T. Skin whitening effect of linoleic acid is enhanced by liposomal formulations. Biol Pharm Bull. 2004;27:591–4.
67. Coderch L, de Pera M, Perez-Cullell N, Estelrich J, de la Maza A, Parra J. The effect of liposomes on skin barrier structure. Skin Pharm Physiol. 1999;12:235–46.
68. Van Kuijk-Meuwissen ME, Mougin L, Junginger HE, Bouwstra JA. Application of vesicles to rat skin in vivo: a confocal laser scanning microscopy study. J Controlled Release. 1998;56:189–96.
69. Fahr A, Schäfer U, Verma D, Blume G. Skin penetration enhancement of substances by a novel type of liposomes. SÖFW-J. 2000;126:48–53.
70. Perez-Cullell N, Coderch L, de la Maza A, Parra JL, Estelrich J. Influence of the fluidity of liposome compositions on percutaneous absorption. Drug Deliv. 2000;7:7–13.
71. Yokomizo Y, Sagitani H. Effects of phospholipids on the in vitro percutaneous penetration of prednisolone and analysis of mechanism by using attenuated total reflectance-Fourier transform infrared spectroscopy. J Pharm Sci. 1996;85:1220–6.
72. Komatsu H, Higaki K, Okamoto H, Miyagawa K, Hashida M, Sezaki H. Preservative activity and in vivo percutaneous penetration of butylparaben entrapped in liposomes. Chem Pharm Bull. 1986;34:3415–22.
73. Ibrahim SA, Li SK. Efficiency of fatty acids as chemical penetration enhancers: mechanisms and structure enhancement relationship. Pharm Res. 2010;27:115.
74. du Plessis J, Egbaria K, Ramachandran C, Weiner N. Topical delivery of liposomally encapsulated gamma-interferon. Antiviral Res. 1992;18:259–65.

75. Lieb LM, Ramachandran C, Egbaria K, Weiner N. Topical delivery enhancement with multilamellar liposomes into pilosebaceous units: I. In vitro evaluation using fluorescent techniques with the hamster ear model. J Invest Dermatol. 1992;99:108–13.
76. Niemiec SM, Ramachandran C, Weiner N. Influence of nonionic liposomal composition on topical delivery of peptide drugs into pilosebaceous units: an in vivo study using the hamster ear model. Pharm Res. 1995;12:1184–8.
77. Li L, Hoffman RM. Topical liposome delivery of molecules to hair follicles in mice. J Dermatol Sci. 1997;14:101–8.
78. Trauer, S., et al., Influence of massage and occlusion on the ex vivo skin penetration of rigid liposomes and invasomes. Eur. J. Pharm. Biopharm. 2014;86(2):301–6.
79. Orsinger K. Lecithin in der Haarkosmetik. SOFW J. 1983;109:495–9.
80. Wajda R, Zirkel J, Sauter K. NanoSolve: an advanced carrier system for cosmetic application. Cosmetic Sci Technol. 2009;1:11–4.
81. Hoepfner E-M, Fiedler HP. Fiedler Encyclopedia of excipients : for pharmaceuticals, cosmetics and related areas. Aulendorf, ECV Editio Cantor Verlag; 2007.
82. Morsy E. Applied liposomology natipide II. F2CO, 1992; 149–151.
83. Röding J. Natipide® II: new easy liposome system. SOFW J. 1990;116:509–15.
84. Lautenschläger H. Liposomes. In: Barel AO, Paye M, Maibach HI, editors. Handbook of cosmetics, science and technology. Boca Raton: CRC Press Taylor & Francis Group; 2006.
85. van Hoogevest P, Prusseit B, Wajda R. Phospholipids: natural functional Ingredients and actives for cosmetic products. SOFW-J. 2013;139:9–15.

Open Access This chapter is licensed under the terms of the Creative Commons Attribution 4.0 International License (http://creativecommons.org/licenses/by/4.0/), which permits use, sharing, adaptation, distribution and reproduction in any medium or format, as long as you give appropriate credit to the original author(s) and the source, provide a link to the Creative Commons license and indicate if changes were made.

The images or other third party material in this chapter are included in the chapter's Creative Commons license, unless indicated otherwise in a credit line to the material. If material is not included in the chapter's Creative Commons license and your intended use is not permitted by statutory regulation or exceeds the permitted use, you will need to obtain permission directly from the copyright holder.

SmartLipids—The Third Generation of Solid Submicron Lipid Particles for Dermal Delivery of Actives

Rainer H. Müller, Florence Olechowski, Daniel Köpke and Sung Min Pyo

Keywords

Solid lipid nanoparticles · Nanostructured lipid carriers · SmartLipids · Dermal bioavailability enhancement · Occlusive film formation · Transepidermal water loss · Anti-pollution effect · Nanotoxicological classification system

7.1 Introduction

Nanotechnology deals with particles or structures being in the nanodimension, which means from the *technical definition* below 1 µm. The size range is from a few nm to below 1000 nm ($\hat{=}1$ µm). There are two groups of particles: nanoparticles according to the *legal definition* being below 100 nm, and the particles between 100 and 1000 nm which are called submicron particles for better differentiation. For use in products, it is important, that all particles below approx. 1 µm in size have different physicochemical properties compared to respective bulk material (µm particles), which can be exploited in products.

The story of nanotechnology in dermal delivery started in the year 1986 when the company Dior introduced their product "Capture" to the cosmetic market. It is an anti-aging product, and as the name promises, it "captures the youth." Thus, in the first advertisements, it was announced as the "victory of science over time." The success of this product is evident by the fact that this brand is still on the market more than 30 years after its introduction, meanwhile in different variations.

R. H. Müller · D. Köpke · S. M. Pyo (✉)
Department of Pharmaceutical Technology, Freie Universität Berlin, Institute of Pharmacy, Kelchstr. 31, 12169 Berlin, Germany
e-mail: pyo.sungmin@fu-berlin.de

F. Olechowski
Berg + Schmidt GmbH & Co. KG, An der Alster 81, 20099 Hamburg, Germany

© Springer Nature Switzerland AG 2019
J. Cornier et al. (eds.), *Nanocosmetics*,
https://doi.org/10.1007/978-3-030-16573-4_7

Such a long-lasting success can definitely not be created by pure marketing promises, only when the product has a solid performance and clear advantages compared to traditional products on the market. It shows that nanotechnology can give additional features to dermal products. These features are manifold, as increased penetration of actives, better tolerability on the skin [1], and prolonged release with longer-lasting action.

However, the liposomes have some technical challenges, e.g., physical stability in emulsions or the limited loading of lipophilic actives inside the phospholipid bilayers. Thus, cosmetic but also the pharmaceutical industry is looking for complementary delivery systems to the liposomes since 1986, even with better performance. Examples are higher loading with lipophilic actives or further increased protection of incorporated, chemically labile actives to achieve sufficient shelf-lives of products.

Polymeric nanoparticles invented by P. P. Speiser and his PhD student Birrenbach in the middle of the 1970s [2] appeared as one potential complimentary system. The particles are not fluid as liposomes but in the solid state, i.e., they have a priori higher physical stability. In addition, the solid matrix can better protect labile incorporated actives and can control their release. However, the liposomes were no real market success neither in cosmetics nor in pharma for various reasons (e.g., no legal status of polymers, not existing or too costly large-scale production). Also many other nanostructured systems might "academically" showed good performance but could never be broadly established on the market as an alternative delivery system, e.g., niosomes [3], ethosomes [4], transfersomes [5, 6], pharmacosomes [7], herbosomes [8], colloidosomes [9], sphinosomes [10, 11], and cubosomes [12].

In 1991, the solid lipid nanoparticles (SLN) were developed as a delivery system combining the advantages of liposomes, nanoemulsions, and solid polymeric nanoparticles [13, 14]. The development was product orientated by considering industrial needs. In 1999, the second generation was developed, the so-called nanostructured lipid carriers (NLC) [15–19]. They appeared on the market in cosmetic products since 2005 (e.g., Dr. Rimpler GmbH Germany, Juvena Switzerland) and were perfected with the third-generation SmartLipids in 2014. This chapter provides an overview about general properties of particles made from sold lipids or lipid blends and focuses on the technical features of SmartLipids, the third-generation technology.

7.2 The Difference between the Three Solid Lipid Nanoparticle Generations: SLN, NLC, and SmartLipids (SL)

Literally, the "oldest" nanoparticles in dermal formulations are oil droplets in nanodimension. They were contained as a subfraction in traditional macroemulsion systems (Fig. 7.1, left), or when applying high dispersion forces, nanoemulsions were obtained (main population in the nm range (Fig. 7.1, right).

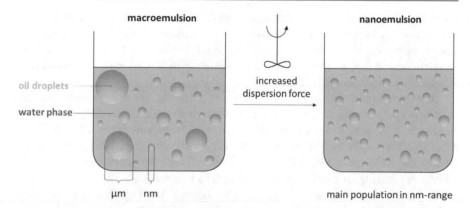

Fig. 7.1 Traditional macroemulsions already contained a few oil droplets in nanodimension as a subfraction (left). Nanoemulsions can be obtained by increasing the dispersion force on microemulsions. The main population of oil droplets was then reduced to nanodimension (right)

Liposomes are also a fluid system identical to the nanoemulsions. Both having the advantages of, e.g., being composed of well tolerated and regulatorily accepted excipients, and the ease of large-scale production by, e.g., high-pressure homogenization. Polymeric nanoparticles are in the solid state, which overcomes, e.g., stability limitations of fluid systems. In 1991, the advantages of these three systems were combined by replacing the liquid lipids (oils and phospholipids) by lipid in the solid state, generating the solid lipid nanoparticles (SLN) . They combine the advantages of liposomes (e.g., regulatorily accepted and well-tolerated excipients, case of large-scale production) with the advantages of polymeric nanoparticles (solid state of particle matrix providing chemical protection of incorporated actives and controlled release).

SLN technology is very simple and industry-friendly. Production takes place by high-pressure homogenization, identical to emulsions and liposomes but at elevated temperatures. The solid lipid is melted, the lipophilic active dissolved, this melt is dispersed in a hot surfactant solution and the obtained macroemulsion is passed through a high-pressure homogenizer to yield a hot nanoemulsion. Cooling leads to the solidification of lipids and the formation of SLN. This first generation was composed of typically one solid lipid only. The melting points of solid lipids were above body temperature. Thus, the particles remain in the solid state also after application to the skin.

The disadvantage was the limited loading capacity of the SLN, due to the lower solubility of actives in solid lipids than in liquid oils. Especially in particles with highly ordered particle matrices, drug expulsion took place, frequently described for particles from highly purity glyceride particles [20]. Often, the expulsion of incorporated active from the particle matrix into the water phase of aqueous particle suspensions took place during storage when lipids re-ordered to more densely and perfectly packed beta modification.

To increase the loading, the solid lipid was blended with oil, having a higher solubility for actives (in the ideal case, the solubilities are additive, and in special cases, "superadditive"). In addition, the oil should disturb the dense packing of solid lipid molecules by steric incompatibility. This second generation of particles was called nanostructured lipid carriers (NLC) , because of the trial to modulate the nanostructure of the particle matrix. The lipid blends were also solid at body temperature. In fact, the loading could be distinctly increased, e.g., for retinol from 2% in SLN to 5% in NLC [21].

However, when developing the NLC, it was not considered that oils can accelerate polymorphic transitions when admixed to solid lipids [22]. The gained loading during production by oil addition in the blend was often reduced during storage. This can be explained by the lowered viscosity of particle matrix by oil, giving the molecules more diffusional mobility. Thus, the acceleration of particle matrix re-ordering to more ordered structures is promoted.

In general, it was observed that particle loading was higher and stable when polydisperse lipids were used, meaning a mixture of mono-, di-, and triglycerides ideally with different fatty acids (e.g., Compritol 888 ATO instead of highly purified tristearin). As the next step, highly chaotic lipid mixtures were created for particle production, and the third generation was born. Considering this approach as a "*smart*" solution for improved "*lipid*" particle design, the particles were called "*SmartLipids*" [22–24]. The SmartLipids are characterized by their lipid mixtures consist of typically 5–10 blended solid lipids, or blends of solid and liquid lipids (Fig. 7.2). Obviously, such a "wild" lipid mixture with the presence of a limited amount of oils did not accelerate the transition as in NLC. Using a suitable lipid mixture, the crystalline structure remains unchanged during storage of the

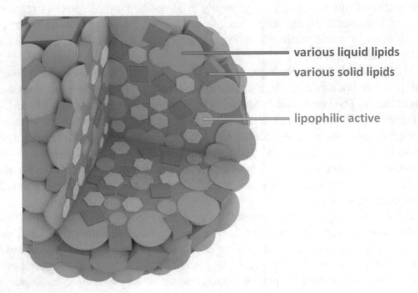

various liquid lipids

various solid lipids

lipophilic active

Fig. 7.2 Schematic representation of SmartLipids structure. Lipophilic actives (yellow pentagons) are incorporated in a solid lipid mixture being composed of typically 5–10 solid lipids (green squares) and liquid lipids (blue circles) (with permission from Stern-Wywiol Gruppe GmbH & Co. KG, 20099 Hamburg, Germany)

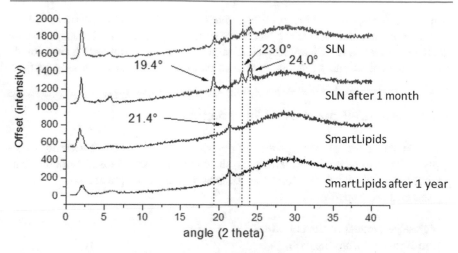

Fig. 7.3 X-ray diffractograms of SLN suspension after production (pink) and after 1 month of storage at room temperature (blue), compared to SmartLipids suspension after production (red) and 1 year of storage at room temperature (black) (modified after [22])

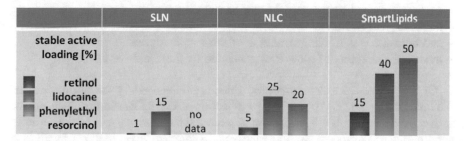

Fig. 7.4 Comparison of maximum loading capacities of retinol, lidocaine, and phenylethyl resorcinol in SLN, NLC, and SmartLipids (modified after [25])

suspension (Fig. 7.3) [22]. Therefore, distinctly higher loading can be achieved, e.g., retinol loading increased from 5% in NLC to 15% in the SmartLipids (Fig. 7.4) [22].

7.3   General Properties and Special Advantage of SmartLipids

There are general properties of lipid nanoparticles, mainly independent on their chemical composition, whereas certain special properties clearly depend on their composition. An example of an "independent" general property is the adhesion to surfaces, in this case, the skin. There is a general adhesiveness of nanoparticles to surfaces because by decreasing the size, adhesiveness increases. Apart from size, adhesiveness is additionally affected by surface properties, controllable by the type of used stabilizer, e.g., positively or negatively charged, non-ionic steric stabilizer. However, in case the surface properties are practically identical, because the same stabilizer was used, or at least very similar, adhesiveness is primarily a function of size and thereof resulting interaction area of particle surface and accepting surface.

7.3.1 General Properties of SmartLipids

Based on this, these "general" properties described for the first two generations also apply to the new-generation SmartLipids, since these properties result from the lipid particles themselves and not from the composition of their lipid matrix.

The general properties are:

- high adhesiveness to the skin, thus
- film formation onto the skin,
- repair of defects in the natural skin barrier and
- reinforcement of thin natural lipid film, and thus
- recreation of natural skin barrier, which has an
- anti-pollution effect,
- occlusion created by the particle film on the skin, which
- increases skin hydration
- and normalizes skin living condition of the keratinocytes,
- increased absorption of molecules generated by the occlusive effect.

Of course, there are also properties which are influenced by the composition of the lipid matrix, e.g., UV protection (= reflection of UV light) and firm inclusion of actives. Reflection of UV light is more pronounced the more crystalline the particles are, thus the more ordered the lipid matrix is. In contrast, incorporation of actives is firmer when the matrix is composed of a mixture of lipids in a more disordered state.

7.3.2 Special Advantages of SmartLipids

Special properties for SmartLipids were given by the chemical composition of their lipid particles. Examples are the high loading capacity of cosmetic actives or drugs and their firmness inclusion, the degree of chemical stabilization, and the release profile. The effect of particle composition on the release profile was shown by zur Mühlen and Mehnert [26], describing the three principle types of lipid nanoparticle structures: enriched core model, matrix model, and enriched shell model—applying basically to all lipid nanoparticles.

Special advantages of the SmartLipids compared to the previous generation are:

– increased loading capacity
– increased stability of the lipid particle matrix
– (showing no or less polymorphic transitions during storage), and
– high chemical stability of incorporated actives.

As discussed above, the loading of cosmetic retinol could be increased distinctly (Sect. 7.2, Fig. 7.5). Polymorphic transitions are limited or not existing (Sect. 7.2, Fig. 7.4). However, it needs to be pointed out that not every complex lipid matrix automatically shows this property. The underlying mechanisms are not yet fully understood to select a stable matrix in a predictive way. Screening experiments are therefore still necessary. Especially, the content of oils can be critical, when crossing the unknown borderline between the high lattice disorder and the critical minimum viscosity of the lipid matrix. Re-ordering will be promoted if viscosity is lowered to strongly by high oil content. Finally, improved chemical stability was found for the chemically labile molecule retinol when incorporated into SmartLipids. However, it should be kept in mind that an optimized NLC particle matrix can be more protective than a suboptimal SmartLipids matrix. In summary, SmartLipids possess in principle the outlined three technical advantages, but a suitable lipid composition cannot be predicted by computer simulation yet and therefore a manual screening is needed explaining the development costs for commercial formulations. However, the development is worth these investments since they can offer a superior market product with clear market differentiation.

LAB 40	Gaulin 5.5

small scale production	industrial scale production
batch size: 20-40ml	capacity: 150L/h

Fig. 7.5 Photographs of the Micron LAB 40, a laboratory-scale homogenizer with a batch size of 20–40 ml (left) and the Gaulin 5.5., a large-scale homogenizer with a capacity of 150 kg per hour (right)

7.4 Industrially Relevant Production Methods of Lipid Nanoparticles

7.4.1 General Considerations for Industrial Production

Basically, all three generations of lipid nanoparticles can be produced with the same methods, since a bulk lipid or lipid mixture needs to be reduced in size to the nanodimension in all cases. There are various production methods for lipid nanoparticles, e.g., microemulsion technique [27], solvent injection method [28], solvent emulsification–diffusion technique [29], multiple emulsion technique [30], membrane contractor technique [31, 32] and ultrasonic-solvent emulsification technique [33]. They are all suitable to produce particles on a small scale in the laboratory and many of them do not need expensive equipment, being convenient for the lab. In the mircroemulsion technique, for example, a hot microemulsion is formed containing the lipids, surfactants, and the active. The microemulsion is then poured into cold water. This breaks the microemulsion to nm-sized solid lipid particles. Therefore, only two beakers, a heating plate and a stirrer were necessary to produce these particles.

However, most of these methods are neither industry producible, nor suitable for cost-effective large-scale production. A nice example is the microemulsion technique. The dilution in water leads to particle suspensions with typically less than 1% particles (w/w). To use the particles in tablets, 99% of water needs to be removed again costly. To provide concentrates for admixing to dermal formulations, 10–20% particle suspensions are desirable. The user (e.g., company) prefers to dilute a concentrate by a factor 10–20 (e.g., 10 or 5 kg of concentrate in 100 kg of final product, corresponding to 1–2% particles in the final product). Thus, such highly diluted lipid particle suspensions with a particle content ≤1% are not suitable for the market.

Requirements for an industrially suitable production technology are:

1. The equipment used for production is preferably standard equipment in the industry and accepted by regulatory bodies.
2. The equipment should be able to be qualified and validated.
3. The production scale is sufficiently large, e.g., 100–1000 kg suspension per hour.
4. It should allow production with acceptable microbial quality, preferable aseptic production for special applications, e.g., sterile products for microneedling.

All these requirements are fulfilled by high-pressure homogenization (HPH). This technology is already well established in the food industry, cosmetics and pharma, and even for the production of intravenously administrated products. The lines are available in the industry and can be used lipid nanoparticles production. In the food industry, for example, homogenizers are used for milk production, with a production volume of more than 1 ton product per hour. Thus, the large scale does not present a problem. In the pharma industry, the lines are only possible when qualification and validation are possible. Depending on the capacity, the price for lines starts from 50.000–200.000 EUR/USD.

7.4.2 High-Pressure Homogenization (HPH)—Hot Process

The production process is identical to the production of emulsions. The only difference is the elevated temperature, which has to be above the melting point of the lipid/lipid mixture. A typical composition of a lipid suspension is 5–20% (lipid including the active), 1–2% stabilizer as well as water up to 100%. In the production process, the lipid or lipid mixture is melted, where the typical range for many lipids is between 50 and 70 °C. Then the active (cosmetic or drug) is dissolved in the hot melted lipid. Subsequently, this active-containing melt is dispersed by high-speed stirring in a hot surfactant solution with identical temperature to form a coarse pre-emulsion. This pre-emulsion is then passed at high pressure, being typically between 400 and 600 bar, through the homogenizer for 1–3 passages, called homogenization cycles.

Often laboratory-scale homogenizers are sometimes less effective than large-scale homogenizers, thus requiring 2–3 cycles. Large-scale homogenizers have often two homogenization valves in series. Thus, one passage is equal to two passages on a laboratory scale. For large-scale production, the pressure can be adjusted, that only one passage is needed for required product properties. This is more a cost-effective option than additional passages. Also, it is less stressful for heat-sensitive actives like retinol, leading to less or no degradation. For details, it is referred to [34].

For working in laboratory scale, the homogenizer should preferably have a low minimum volume of 20–40 ml (e.g., APV Micron LAB 40 (Fig. 7.5, left)). The production of Micron LAB was stopped, and most of the available homogenizers being currently on the market require a minimum volume of at least 100–200 ml (e.g., Panda Niro Soavi). This requires much higher amounts of materials in the screening process for finding a stable formulation, which is a clear disadvantage when high-priced actives are used (e.g., glabridin, 100 g costs approx. 1000 USD). An example of larger-scale homogenizer is the Gaulin 5.5 (APV Deutschland, Germany) with a capacity of 150 kg product per hour (Fig. 7.5, right). This is a capacity suitable for industrial production of SmartLipids concentrates. Half a ton particle suspension can be produced in about 3 h. There are many different homogenizers from various companies available. When selecting one, the flow through should be sufficiently high to minimize the heat exposure to the hot pre-emulsion.

7.4.3 High-Pressure Homogenization (HPH)—Cold Process

Instead of homogenizing the melted liquid lipid as an emulsion, it is also possible to homogenize the lipid mixture in the solid state as a suspension. In this so-called cold homogenization process, a suspension of solid lipid particles is passed through the homogenizer. It is a milling process comparable to the production of nanosuspension made from drugs [35] or cosmetic actives (SmartCrystals [36, 37]).

In this process, the active is dissolved in the melted bulk lipid, and subsequently, this mixture is cooled to recrystallize. In the next step, milling of this solid mixture is performed, and the so obtained µm-sized particles are dispersed in a stabilizer solution. This macrosuspension is then passed through the homogenizer. The forces are high enough to break the solid materials into nanodimension. Obviously, more energy is needed to break solid materials than disperse liquid oils. Thus, instead of 1–3 cycles rather 5–10 cycles are required to achieve a similar size to the hot process with melted lipid.

For this reason, the cold process should only be applied, when it provides clear advantages compared to the hot process in case of:

(a) highly thermolabile actives,
(b) generation of a certain particle matrix structure.

In case of highly thermolabile actives with very fast decomposition, heat exposure during hot homogenization can be avoided. Heat exposure is therefore limited to the dissolution process of actives in the melted lipid. Another reason for the cold process is the generation of a certain structure of the lipid matrix. Recrystallization in a bulk material and in a nanodimension can lead to different structures, e.g., lipid modifications. In case, the structure of the bulk crystallization is required to achieve a certain release profile, cold homogenization can be applied. However, it should be kept in mind that this process is costlier, industrially preferred cost-wise is the hot homogenization.

7.5 How Do the Particles Interact with the Skin?

The interaction of the lipid particles with the skin is the reason for various general properties which lipid particles evolve on or in the skin. This basic interaction is generally independent on the lipid composition. In general, the lipid particles are dispersed in the water phase of a dermal formulation, for example, a gel or an oil-in-water system, e.g., o/w lotion or cream. When the formulation is spread on the skin, the particles are evenly dispersed in the outer water phase of the formulation (Fig. 7.6, upper). By and by the particles start to adsorb onto the skin surface (Fig. 7.6, lower). They form a film on the skin, which repairs damaged natural lipid

directly after application

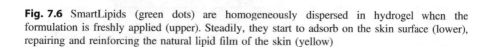

hydrogel with
SmartLipids
evenly dispersed

damaged natural lipid film

skin

adsorption of SmartLipids on the skin

hydrogel with
SmartLipids
adsorbed to the skin

repaired natural lipid film

skin

Fig. 7.6 SmartLipids (green dots) are homogeneously dispersed in hydrogel when the formulation is freshly applied (upper). Steadily, they start to adsorb on the skin surface (lower), repairing and reinforcing the natural lipid film of the skin (yellow)

films. Also, they reinforce a weakened lipid film and strengthen it. The tendency to adhere is higher for nanosized materials (lipid nanoparticles) than for microsized particles (e.g., oil droplets). Thus, also in creams, preferential adsorption of the lipid particles will take place. In general, the adsorptive forces are higher with a smaller size (lipid submicron particles >100 nm to <1000 nm, lipid nanoparticles "legally" <100 nm). Thus, the size of the submicron particles should be preferentially as close as possible to the size limit of 100 nm of legal nanoparticles, that means about 200–300 nm. In this case, they exhibit their high adsorptive properties without being legally nanoparticles and therefore no addition of "(nano)" to INCI nomenclature is required. The formed film generates the general properties of lipid nanoparticles listed in Sect. 3.1, where special properties of SmartLipids (Sect. 3.2) are generated by their matrix structure being different from SLN and NLC.

7.6   Properties on Skin—Active-Free Particles

7.6.1   Restoration of Natural Skin Barrier—Controlled Occlusion

As outlined above, the particles adsorb onto the skin restoring a damaged natural lipid film acting as a barrier for the skin. The density of the film can be controlled by the concentration of the lipid nanoparticles incorporated in the formulation (Fig. 7.7). Based on the size of the particles, it can be calculated how many particles theoretically are required for generating a monolayer in hexagonal packaging. The degree of occlusion can thus be controlled via the particle concentration in the product—occlusive enough but not too occlusive to allow "breathing" of the skin (controlled contact with air). In case lipid particles are admixed to oil droplets containing cream, a mixed film on the skin will be formed, requiring less lipid nanoparticles. Based on this, as a rule of thumb 0.5–2%, particles are recommended in commercial products, being equivalent to 2–10% commercial SmartLipids concentrate to be added to the dermal formulation (Fig. 7.8).

The film formation of lipid nanoparticles on the skin can experimentally be proven by corneometer measurements. The corneometer quantifies the moisture in the skin by measuring the dielectric constant D of the skin [39]. D is 80 for water, about 5 for many lipids without water, the higher the reading is, the more moisture is in the skin. The corneometer works as a parallel-plate capacitor and measures the capacitance, which is a function of the dielectric constant of the material between the plates. In the example, the reading having arbitrary units is about 37 when

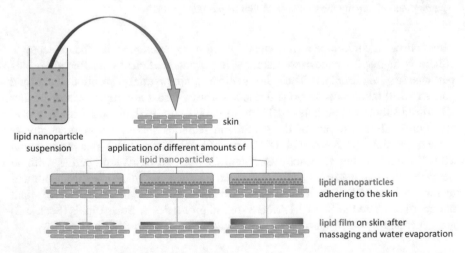

Fig. 7.7 Application of lipid nanoparticle suspensions (gel, cream, or lotion) to the skin leads to the adsorption and formation of a film. Its density depends on the concentration of lipid nanoparticles applied and increases with increasing particle concentration (from left to right). Dense film formation is promoted after water evaporation due to capillary forces (with permission from [38])

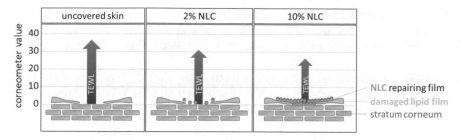

Fig. 7.8 Proof of film formation by lipid nanoparticles by corneometer measurements: The corneometer readings (y-axis) decrease with an increasing film thickness from left to right (with permission from [38])

placing the corneometer directly on the skin (Fig. 7.8, left). When a film on the skin is applied acting as an insulator, the reading decreases. The thicker the film is, the more pronounced is the decrease. A corneometer decrease was found after application of lipid nanoparticles (Fig. 7.8, middle and right) proving the film formation. The film density increases with increasing particle concentration from 2 to 10%, as proven by the decrease in corneometer reading from about 30–24. The degree of occlusion can thus be easily controlled via the lipid particle concentration.

7.6.2 Anti-pollution Effect

Anti-pollution cosmetics are a growing trend in Asia, and in the last 2 years, interest is also increasing strongly in Europe. When looking at the fairs such as in-cosmetics in Europe, more and more companies advertise anti-pollution products. Pollution, especially smog is a big problem in the large cities of Asia. Pollution promotes also aging; meanwhile, the term "anti-aging" is being partially replaced by "anti-polluaging." Cosmetic industry reacts with placing products on the market such as Urban Skin Protect (Nivea, Germany), Hydro Effect Serum (lavera, Germany), and City Miracle Cream (Lancôme, France). Environmental problems are fine dust, smog, smoke, exhaust gases (e.g., NO_2), etc., which cannot only damage the lungs but also interact with the skin. Dirty air accelerates skin aging. Not only the particles themselves are a problem, but they can also have organic compounds and heavy metals adsorbed onto their surfaces. This can lead to skin irritation, inflammation, and acceleration of aging processes. Also oxidative stress increases because vitamins such as vitamin E are reduced in the skin. Further, soot can increase the pigmentation of the skin. In a long-term study by Jean Krutmann (Leibniz-Institut für Umweltmedizinische Forschung (IUF), Düsseldorf), an increase of 20% of pigment spots was found in women living in areas with polluted air [40].

Considering the concurrent causes for the skin damage by pollution, SmartLipids are an ideal anti-pollution system. They can:

1. restore an effective natural protective barrier,
2. effectively deliver anti-pollution actives such as vitamins and anti-oxidants,
3. recreate white skin complexion via improved penetration of whitening agents.

Anti-pollution formulations with SmartLipids should thus be active-loaded particles, having a synergistic mixture of actives with anti-pollution effect. A separate treatment can be performed locally with whitening SmartLipids.

7.7 Regulatory Aspects

7.7.1 Certification of Natural Cosmetic Ingredients

Today's consumers pay considerable attention to the quality of a product which is considered to be better when the ingredients of the product or the product itself are of natural origin. As a result in the cosmetic field, certification bodies appeared almost simultaneously in different countries, establishing their own standards and definition of "natural" with dissimilar threshold values. Consumers were therefore often lost in the so-called "jungle of logos."

In order to provide a better overview for consumers the five organizations ECOCERT, BDIH, ICEA, cosmebio, and The Soil Association joined forces and developed a harmonized standard at European level, now providing a uniform definition of "natural." This standard is known as COSMOS, being an alliteration of COSmetic Organic Standard. Still, more standards exist, e.g., Black swan, Austrian codex, Natrue, etc. However, COSMOS is the best-known and well-excepted standard from the consumers and, therefore, is preferred by the manufactures.

COSMOS conform raw materials are not only of natural origin, their whole production process has also to be safe for people and the environment. Even though the certification is voluntary, manufacturers participate, expecting a marketing advantage.

SmartLipids consist of lipids, surfactants, and preservatives besides the actives and water. COSMOS conform surfactants are available to a very limited extent possessing being a challenge for the production of SmartLipids meeting the COSMOS standards. Since no reliable prediction can be made for surfactants, an intensive screening for their type and concentration is recommended. Also, the production parameters can be adapted to obtain physically stable SmartLipids suspensions correspond to COSMOS standards.

7.7.2 Submicron Versus Nanoparticles

Following the unit system, nanoparticles possess a size in the nanometer range, i.e., from 1 nm to 1000 nm. However, following the EU cosmetic regulation [41],

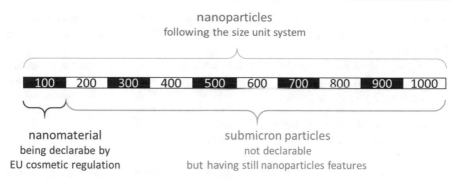

Fig. 7.9 Illustration of the different size definitions of nanoparticles (blue), nanomaterials (red), and submicron particles (green) on a size scale with nm as a unit

nanomaterials are particles with sizes below 100 nm. If these nanomaterials were used in the final product, they must be declared in the ingredient list by adding (nano) behind their INCI name.

Consumer's concern about nanomaterials increases continuously, whether legitimately or not. Thus, the market reacts by foregoing nanodeclaration required particles. Nevertheless, they know well about the beneficial properties of nanomaterials in dermal formulations and are, therefore, highly interested in integrating them into their products.

In this case, submicron particles can be used. They provide almost equally pronounced nanoproperties as nanomaterials without being nano declarable since they are particles being in the size range of 100–1000 nm (Fig. 7.9). SmartLipids with their typical size range of 200–400 nm belong to the category of submicron particles and therefore meet both desires.

7.8 Tolerability

All the excipients used in SmartLipids are regulatorily accepted and skin-friendly. Solid lipids and oils can be selected, so that they have additional skin-caring features to their particle matrix material function, e.g., argan and kukui oil. The lipids are natural or naturally derived, considering the nowadays expectations of the consumers. And stabilizers with very high skin tolerability can be used, e.g., polyglucosides.

However, increasing concern of the consumers is nanotechnology. Nano is accepted in pharma and still in cosmetics, but less in food. Cosmetic product manufacturers take this into account by increasingly incorporating delivery systems which are no nano according to the law and regulations, the so-called submicron particles. After the unit system, nanoparticle's size range is between 1–1000 nm. However, nanoparticles after EU cosmetic regulation [41] have particle sizes between 1–100 nm. Particles having a dimension being nano after the unit system but no nano after the legal definition is called submicron particles (size between 100–1000 nm).

To judge the tolerability and potential toxicity, the nanotoxicological classification system (NCS) was established [42]. The particles below 1000 nm are classified according two categories:

1. size: above 100 nm and below 100 nm,
2. persistency: biodegradable or non-biodegradable.

In a simplified way: Particles below 100 nm can potentially enter each cell via the process of endocytosis. Each cell can take up particles by endocytosis, thus these particles have a toxicological risk. Particles larger than approx. 100 nm can only be taken up by special cells in the body, the phagocytes being localized in the liver and spleen. Thus, when submicron particles are applied to the skin, there is no risk of particle uptake.

Even when particles are taken up by a cell, they will disappear again when they are biodegradable in contrast to non-biodegradable particles which will stay forever in the cell potentially causing damage. Thus, biodegradable particles applied are generally toxicologically not problematic.

Based on this, the particles were divided into four classes I–IV (Fig. 7.10) and labeled in accordance with the traffic signal colors green, yellow, and red. The SmartLipids are green particles (Class I) which are well tolerated especially when applied not systemically but only on the skin surface.

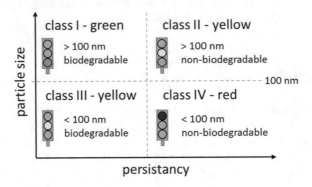

Fig. 7.10 Nanotoxicological Classification System (NCS) divides the particles into four classes according to their size and persistence. Traffic light scheme: green—Class I with no or low potential risk, yellow—Class II and III with medium and red—Class IV with high potential risk (modified after [42]). The toxicological risk is nearly non-existing or distinctly reduced for Class I and II when particles were only to the skin surface, and not administered internally (e.g., food, injection)

7.9   Incorporation of Concentrates into Products

Incorporation into final products is very straight forward. The particles are typically localized in the water phase of dermal formulations. Suitable formulations are, thus, gels and o/w systems, but not ointments (e.g., petrolatum, etc.).

In case of gels, the formulation is produced with reduced water content, and after production, a SmartLipids concentrate is added by gentle stirring, preferentially at room temperature to avoid melting of the lipid. In case air bubbles are incorporated during the incorporation process of the concentrate, the particle suspension can also be added before adding the gelling agent and subsequent gelation process. In this case, it needs to be ensured, that the gelling agent does not interact with the particles, may cause aggregation. Creams and lotions are prepared with reduced water content identical to gels. During the cooling phase, the SmartLipids concentrate can be added as the last production step when the temperature is below 30 °C.

7.10   Summary and Outlook

The SmartLipids are an alternative to liposomes for the dermal delivery of lipophilic actives in cosmetics but also in pharma. Compared to liposomes, they possess a higher loading capacity since the lipophilic particle fraction for dissolving actives is higher (100% lipid matrix in SmartLipids compared to the lipid bilayers in water-filled liposomes). The physical stability in dermal formulations is very high due to the solid state of the particle matrix, identical to polymeric particles. They consist of regulatorily accepted excipients and lipids with skin-caring properties can be used.

Looking into the future, an important prerequisite for entering the market in products is fulfilled: the availability of commercial concentrates which can be purchased by companies for their production. There is also the possibility to order tailor-made particles from Berg + Schmidt (Hamburg, Germany). These submicron particles are legally no nanoparticles, but they possess properties like nanoparticles for improved dermal delivery (e.g., prolonged residence on the skin). Their ability to protect chemically labile actives like retinol opens new perspectives for stable products with improved performance.

First products with lipid nanoparticles (second-generation NLC) are already on the market since 2006. SmartLipids as the improved third-generation offer even better features for more effective dermal products.

References

1. Paris Match no. 3531, 19 au 25 Janvier 2017 2 b, France.
2. Birrenbach G, Speiser PP. Polymerized micelles and their use as adjuvants in immunology. J Pharm Sci. 1976;65(12):1763–6.

3. Cerqueira-Coutinho C, dos Santos EP, & Mansur CRE. Niosomes as nano-delivery systems in the pharmaceutical field. Critical Reviews™ in Therapeutic Drug Carrier Systems, 2016:33 (2).

4. Touitou E, Dayan N, Bergelson L, Godin B, Eliaz M. Ethosomes—novel vesicular carriers for enhanced delivery: characterization and skin penetration properties. J Controlled Release. 2000;65(3):403–18.

5. Cevc G, Blume G, Schätzlein A. Transfersomes-mediated transepidermal delivery improves the regio-specificity and biological activity of corticosteroids in vivo1. J Controlled Release. 1997;45(3):211–26.

6. Rajan R, Jose S, Mukund VB, Vasudevan DT. Transferosomes-A vesicular transdermal delivery system for enhanced drug permeation. J Adv Pharm Technol Res. 2011;2(3):138.

7. Semalty A, Semalty M, Rawat BS, Singh D, Rawat MSM. Pharmacosomes: the lipid-based new drug delivery system. Expert Opin Drug Deliv. 2009;6(6):599–612.

8. Ajay, S., Vikas, P., Rajesh, S., Punit, B., & Suchit, J. (2012). Herbosomes: A Current Concept of Herbal Drug Technology, An Overview. Journal of Medical Pharmaceutical and Allied Sciences, 1.

9. Dinsmore AD, Hsu MF, Nikolaides MG, Marquez M, Bausch AR, Weitz DA. Colloidosomes: selectively permeable capsules composed of colloidal particles. Science. 2002;298 (5595):1006–9.

10. Webb MS, Bally MB, Mayer LD. (1996). U.S. Patent No. 5,543,152. Washington, DC: U.S. Patent and Trademark Office.

11. Saraf S, Gupta D, Kaur CD, Saraf S, Res IJCS. Sphingosomes a novel approach to vesicular drug delivery. Int J Cur Sci Res. 2011;1(2):63–8.

12. Karami Z, Hamidi M. Cubosomes: remarkable drug delivery potential. Drug Discovery Today. 2016;21(5):789–801.

13. Gasco, M. R. (1993). U.S. Patent No. 5,250,236. Washington, DC: U.S. Patent and Trademark Office.

14. Müller RH, Mäder K, Gohla S. Solid lipid nanoparticles (SLN) for controlled drug delivery–a review of the state of the art. Eur J Pharm Biopharm. 2000;50(1):161–77.

15. Müller RH, Jenning V, Mäder K, Lippacher A. (2000). European Patent. EP1176949A2.

16. Müller RH, Jenning V, Mäder K, Lippacher A. (2014). U.S. Patent No. 8,663,692. Washington, DC: U.S. Patent and Trademark Office.

17. Pardeike J, Hommoss A, Müller RH. Lipid nanoparticles (SLN, NLC) in cosmetic and pharmaceutical dermal products. Int J Pharm. 2009;366(1–2):170–84.

18. Doktorovova S, Souto EB. Nanostructured lipid carrier-based hydrogel formulations for drug delivery: a comprehensive review. Expert Opin Drug Deliv. 2009;6(2):165–76.

19. Garcês A, Amaral MH, Lobo JS, Silva AC. (2017). Formulations based on solid lipid nanoparticles (SLN) and nanostructured lipid carriers (NLC) for cutaneous use: A review. European J Pharm Sci.

20. Bunjes H, Westesen K, Koch MH. Crystallization tendency and polymorphic transitions in triglyceride nanoparticles. Int J Pharm. 1996;129(1–2):159–73.

21. Jenning V, Gohla SH. Encapsulation of retinoids in solid lipid nanoparticles (SLN). J Microencapsul. 2001;18(2):149–58.

22. Ruick R SmartLipids—die neue Generation der Lipidnanopartikel nach SLN und NLC. Doctoral dissertation, 2016 Freie Universität Berlin.

23. Müller RH, Ruick R, Keck CM. (2014). smartLipids®—the next generation of lipid nanoparticles by optimized design of particle matrix. DPhG-Jahrestagung, Frankfurt, PT.27.

24. Müller RH, Ruick R, Keck CM. smartLipids®—the new generation of lipid nanoparticles after SLN and NLC. San Diego: AAPS Annual Meeting; 2014. p. T3134.

25. Pyo SM, Müller RH, Keck CM. (2017). Encapsulation by nanostructured lipid carriers. In Nanoencapsulation Technologies for the Food and Nutraceutical Industries, 114–37.

26. zur Mühlen A, Schwarz C, Mehnert W. (1998). Solid lipid nanoparticles (SLN) for controlled bioactive delivery–bioactive release and release mechanism. Europ J Pharm Biopharm 45 (2):149–55.
27. Gasco MR. (1993). U.S. Patent No. 5,250,236. Washington, DC: U.S. Patent and Trademark Office.
28. Schubert MA, Müller-Goymann CC. Solvent injection as a new approach for manufacturing lipid nanoparticles–evaluation of the method and process parameters. Eur J Pharm Biopharm. 2003;55(1):125–31.
29. Trotta M, Debernardi F, Caputo O. Preparation of solid lipid nanoparticles by a solvent emulsification–diffusion technique. Int J Pharm. 2003;257(1–2):153–60.
30. Garcia-Fuentes M, Torres D, Alonso MJ. Design of lipid nanoparticles for the oral delivery of hydrophilic macromolecules. Colloids Surf B. 2003;27(2–3):159–68.
31. Charcosset C, Fessi H. Preparation of nanoparticles with a membrane contactor. J Membr Sci. 2005;266(1–2):115–20.
32. Ahmed El-Harati A, Charcosset C, Fessi H. Influence of the formulation for solid lipid nanoparticles prepared with a membrane contactor. Pharm Dev Technol. 2006;11(2):153–7.
33. Luo Y, Chen D, Ren L, Zhao X, Qin J. Solid lipid nanoparticles for enhancing vinpocetine's oral bioavailability. J Controlled Release. 2006;114(1):53–9.
34. Müller RH, Dingler A, Schneppe T, Gohla S. Large scale production of solid lipid nanoparticles (SLN™) and nanosuspensions (DissoCubes™). Handbook of pharmaceutical controlled release technology; 2000:359–76.
35. Keck CM, Müller RH. Drug nanocrystals of poorly soluble drugs produced by high pressure homogenisation. Eur J Pharm Biopharm. 2006;62(1):3–16.
36. Romero GB, Chen R, Keck CM, Müller RH. Industrial concentrates of dermal hesperidin smartCrystals®–production, characterization & long-term stability. Int J Pharm. 2015;482(1–2):54–60.
37. Hatahet T, Morille M, Hommoss A, Dorandeu C, Müller RH, Bégu S. Dermal quercetin smartCrystals®: Formulation development, antioxidant activity and cellular safety. Eur J Pharm Biopharm. 2016;102:51–63.
38. Müller RH, Sinambela P, Keck CM. NLC–the invisible dermal patch for moisturizing & skin protection. EuroCosmetics. 2013;6:20–3.
39. Courage W. (1994). Hardware and measuring principle: corneometer. In: Elsner P, Berardesca E, Maibach HI, editors. Bioengineering of the skin: water and the Stratum corneum. CRC Press, 17, 1–176.
40. Vierkötter A, Schikowski T, Ranft U, Sugiri D, Matsui M, Krämer U, Krutmann J. Airborne particle exposure and extrinsic skin aging. J Invest Dermatol. 2010;130(12):2719–26.
41. Buzek J, Ask B. Regulation (EC) No 1223/2009 of the European Parliament and of the Council of 30 November 2009 on cosmetic products. Official J Europ Union. 2009:59.
42. Keck CM, Müller RH. Nanotoxicological classification system (NCS)–a guide for the risk-benefit assessment of nanoparticulate drug delivery systems. Eur J Pharm Biopharm. 2013;84(3):445–8.

Nanocrystals for Dermal Application

8

Olga Pelikh, Steffen F. Hartmann, Abraham M. Abraham
and Cornelia M. Keck

Abstract

Dermal application of actives aims at delivering the active to the desired place of action, typically to the deeper layers of the skin. Passive diffusion is the main driving force of absorption into the skin, and a main prerequisite for effective passive diffusion is a sufficient amount of dissolved active within the formulation, because only dissolved molecules can be taken up. Many cosmetic and cosmeceutical actives possess poor solubility and can therefore not be delivered to the skin by classical formulation approaches. One of the modern and most powerful strategies to overcome poor solubility is the use of nanocrystals, which are addressed in this chapter.

Keywords

Nanocrystals · Nanosuspensions · Poor solublity · Noyes-Whitney equation · Kelvin equation · Dissolution rate · Solubility

8.1 Introduction

Dermal application of actives aims at delivering the active to the desired place of action, typically to the deeper layers of the skin. Passive diffusion is the main driving force of absorption into the skin, and a main prerequisite for effective passive diffusion is a sufficient amount of dissolved active within the formulation, because only dissolved molecules can be taken up [1–4]. Many cosmetic and

O. Pelikh · S. F. Hartmann · A. M. Abraham · C. M. Keck (✉)
Department of Pharmaceutics and Biopharmaceutics, Philipps-Universität Marburg,
Robert-Koch-Str. 4, 35037 Marburg, Germany
e-mail: cornelia.keck@pharmazie.uni-marburg.de

© Springer Nature Switzerland AG 2019
J. Cornier et al. (eds.), *Nanocosmetics*,
https://doi.org/10.1007/978-3-030-16573-4_8

cosmeceutical actives possess a sufficient high solubility, however—especially natural antioxidants—possess poor solubility and can therefore not be delivered to the skin by classical formulation approaches. Much research has been done to overcome poor solubility and to improve the bioactivity of poorly soluble actives. Chemical modification of such actives is one possibility. This means the active is chemically modified, e.g. by adding water-soluble substituents to the molecule. The procedure has several drawbacks. It modifies the chemical structure which might also lead to modified biological activities and moreover, it modifies a natural compound into a chemical, non-natural compound, which cannot be used for natural products anymore. Another possibility to improve the solubility of poorly soluble actives is the use of innovative carrier systems. Examples are microemulsions, liposomes, micelles, cyclodextrins or polymeric micro- and nanoparticles [5–8]. One of the most powerful strategies to overcome poor solubility is the use of nanocrystals, which are addressed in this chapter.

8.2 Properties of Nanocrystals

Nanocrystals are solid particles with sizes below 1 μm. They are composed of 100% active (Fig. 8.1) and possess special properties when compared to larger-sized bulk material [9]. Special properties include:

• Increased solubility of poorly soluble actives
• Increased dissolution velocity
• Increased adhesiveness
• Increased concentration gradient
• Improved bioavailability.

8.2.1 Increased Solubility of Poorly Soluble Actives

Nanocrystals are the most suitable formulation principle if poorly soluble actives need to be formulated. The reason for the improved properties of the crystals is their small size. Due to the small size, the curvature of the particles is higher when

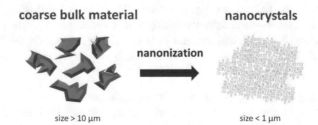

Fig. 8.1 Properties of nanocrystals: Nanocrystals are composed of 100% active, possess a size below 1 μm and are obtained from large sized bulk material, typically by wet milling techniques

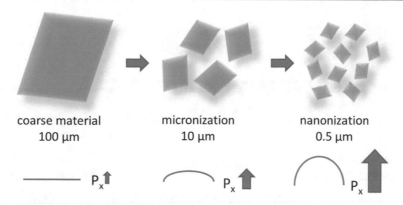

Fig. 8.2 Properties of nanocrystals: Principle of increase in kinetic solubility with decreasing size

compared to the larger-sized material (Fig. 8.2). The curvature is related to the tension, i.e. the energy within the solid material. The higher the tension of the surface, the higher will be the tendency to "release" this energy, which results in a higher dissolution pressure and thus an increased kinetic solubility. The "curvature effect" is also well known for liquids. Small-sized droplets, which possess a large curvature, will evaporate faster than larger-sized droplets, because the vapour pressure, i.e. the tendency to evaporate, increases with decreasing size. The physical explanation is the Kelvin equation (formula 1), where Δp is the pressure, R the radius of curvature and γ the surface tension [9–15].

Formula 1: Kelvin equation.

$$\Delta p = \frac{2\gamma}{R} \tag{1}$$

8.2.2 Increased Dissolution Velocity

A decrease in size increases the surface area of the active, and thus more solvent molecules get in contact with the surface of the material, which allows for a faster dissolution of the active (Fig. 8.3). A simple example of this "size effect" is known from daily life. Tea can be sweetened with rock candy or with granulated sugar. The rock candy consists of larger particles, whereas the granulated sugar possesses a smaller size and thus a larger surface. If both types of sugar are used for sweetening the same cup of tea, the smaller-sized sugar will dissolve faster and thus will lead to a faster sweetening of the tea. Smaller-sized sugar, i.e. powdered sugar, will sweeten the tea even faster. Consequently, nanosized sugar, i.e. with a size being about 1000 times smaller than granulated sugar, would dissolve within seconds. This principle is used when nanocrystals are employed (Fig. 8.3). The physical explanation is the Noyes–Whitney equation (Formula 2), where dc/dt is the

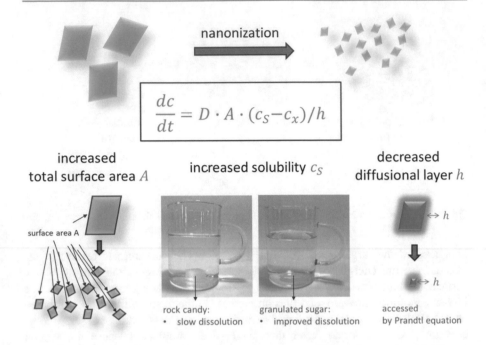

Fig. 8.3 Properties of nanocrystals: Principle of increase in dissolution velocity with decreasing size

dissolution velocity, A the total surface area, D the diffusion coefficient, $c_s - c_x$ the concentration gradient and h the thickness of the diffusional layer. A reduction in size increases both, the total surface area A and the solubility c_s. Furthermore, the diffusional layer around the particle decreases with decreasing size (Prandtl equation) [9–15].

Formula 2: Noyes–Whitney equation.

$$\frac{dc}{dt} = D \cdot A \cdot (c_S - c_x)/h \tag{2}$$

8.2.3 Increased Adhesiveness

If large-sized material is comminuted into small-sized particles, the number of particles increases and the total surface area is increased. Hence, the mass : surface ratio decreases, which in turn leads to an increased adhesiveness on surfaces. This can be explained by the fact that less weight is carried per attaching point. The "gecko effect" is also well known in daily life, i.e. powdered sugar sticks much better to bakery than granulated sugar (Fig. 8.4). The gecko effect is named after the animal, which is known to be able to walk upside down, even on extremely smooth

Fig. 8.4 Images of German bakery ("Berliner"—covered with granulated and powdered sugar, respectively) to prove the influence of particle size on the adhesiveness. With decreasing size the adhesiveness on surfaces increases

surfaces. The reason for this ability is the unique structure of the feet, which consists of millions of nanofibres, the so-called spatula. Hence, the weight of the gecko is carried by many spatulas, making the mass to be carried for each spatula extremely small.

8.2.4 Increased Concentration Gradient

A concentration gradient is the difference between the concentrations of dissolved active at two different regions. It can be expressed as $c_a - c_b$, where c_a is the concentration of dissolved active in the region a and c_b the concentration of dissolved active in the region b. If c_a is the region in which the solid material is dispersed, c_a corresponds to the solubility c_s of the active. As nanocrystals possess an increased solubility, c_s ($= c_a$) is higher than for bulk material, which will also lead to an increase in concentration gradient (Fig. 8.5).

8.2.5 Increased Passive Diffusion Through Biological Barriers

Passive diffusion is the most common procedure for the uptake of actives upon dermal application. It means that the active is penetrating without the help of active transporters into the stratum corneum or deeper skin layers. The driving force of passive diffusion is the concentration gradient; i.e. the higher the concentration gradient, the more effective is the penetration of the active. As nanocrystals possess a higher concentration gradient, when compared to larger-sized material, passive diffusion can be increased with the use of small-sized crystals (Fig. 8.5).

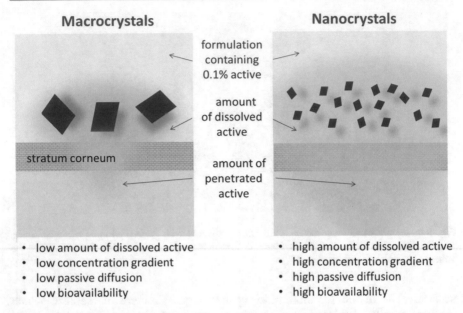

Macrocrystals   **Nanocrystals**

- low amount of dissolved active
- low concentration gradient
- low passive diffusion
- low bioavailability

- high amount of dissolved active
- high concentration gradient
- high passive diffusion
- high bioavailability

Fig. 8.5 Properties of nanocrystals: Principle of increase in bioavailability with decreasing size. The increased solubility of smaller-sized particles leads to an increased concentration gradient between formulation and skin. This leads to an improved passive diffusion, resulting in an increased bioavailability of the active in the skin

8.2.6   Improved Bioavailability

Bioavailability (BA) is a measure of the absorption efficacy. It is 100% if all actives are absorbed. Only dissolved active, i.e. molecularly dispersed active, can be absorbed by the body. Hence, sufficient solubility is the main prerequisite for effective delivery of actives. The penetration efficacy of dermally applied actives depends on various parameters and can generally be described by Fick's law (Formula 3), where dQ/dt is the amount of active penetrating out of the formulation, D the diffusion coefficient in the stratum corneum, V_k the distribution coefficient between stratum corneum and vehicle, A the area of skin on which the formulation is applied, d the thickness of stratum corneum and c_v the concentration of dissolved active in the vehicle. The latter explains why small-sized particles with higher solubility and increased concentration gradient lead to improved penetration (Fig. 8.5). The principle is not limited to dermal application, but can also be employed for oral, pulmonal, ocular, nasal and other parenteral application routes [16–20].

Formula 3: Fick's law.

$$\frac{dQ}{dt} = \frac{D \cdot V_k \cdot A}{d} \cdot c_v \tag{3}$$

8.3 Application of Nanocrystals

Nanocrystals increase the solubility of poorly soluble actives and with this the bioavailability of poorly soluble actives [20–23]. The increased adhesiveness of the nanocrystals prolongs the retention time at the application site, thus further promoting the uptake of the active. Hence, nanocrystals are the best choice if poorly soluble actives need to be formulated. Nanocrystals for pharmaceutical use were invented in the early 1990s and the superior properties, i.e. to increase the solubility and the bioavailability of poorly soluble actives, were proven many times since then. Nanocrystals helped to overcome the big issue of poor solubility of new chemical entities, which could not be formulated before this. Many highly promising drug entities could not be further developed, and new treatment innovations could not reach the patients. With the invention of the nanocrystals, many pharmaceutical products employing the nanocrystal technology entered the market since then and new therapies could be invented. Due to this big success and contribution in the pharmaceutical field in 2010, the nanocrystals were even awarded as the most successful drug delivery technology innovation (Rapamune®, Emend®) [24, 25]. The nanocrystal technology can be used not only in the pharmaceutical field, but everywhere, where poor solubility becomes an issue. Examples for the use of nanocrystals are poorly soluble food additives, e.g. vitamins or supplements, for use in food and nutraceuticals or poorly soluble agricultural products, e.g. pesticides or herbicides [9]. Nanosized herbicides decrease the amount of herbicides needed, because the retention time on leaves is longer and the penetration is more effective. In addition, organic solvents to dissolve the poorly soluble compounds are not necessary anymore, thus protecting the environment and the health beneficial properties of the plants.

8.4 Production of Nanocrystals

Nanocrystals are a universal formulation principle for all poorly soluble actives and can be produced by either top-down or bottom-up technologies [9, 26–28]. Bottom-up production techniques involve the dissolution of the raw material in organic solvents and a subsequent precipitation by transferring this solution into a medium in which the active does not dissolve [29, 30]. Top-down processes use various wet milling strategies to reduce the size of large-sized bulk material. Most often applied techniques are bead milling [31], high-pressure homogenization [32] or combinations of both techniques [32]. New approaches use high-speed stirring and subsequent high-pressure homogenization [33, 34]. Especially in the cosmetic field, bottom-up technologies are not frequently used. The reason is the need to use organic solvents, which must be removed after the precipitation. As the removal of organic solvents is costly, more effective strategies are preferred. To date, the most commonly applied process for the production of cosmetic nanocrystals is the combination of bead milling and high-pressure homogenization, as it combines the advantages of both technologies. In most cases, bead milling of coarse material leads to smaller sizes than high-pressure homogenization. However, as it is a low-energy process, the time which is required for the nanonization is long. Depending on the material and the required size of the end product, milling can take as long as 48 h or even longer [35, 36]. High-pressure homogenization is a high-energy process, and thus nanonization is much faster. However, typically the size and size distribution achieved are larger and broader [37]. In addition, the high-energy input during the homogenization is mainly transferred into heat, which results in a strong increase in temperature within the formulation. This can lead to agglomeration of the nanocrystals and, thus, must be avoided [38]. Effective cooling systems are therefore required to ensure a constantly low product temperature during production to circumvent agglomeration of the product. Combining both methods is beneficial as milling time can be reduced and the homogenization pressure can be reduced. Thus, no cooling is required during the high-pressure homogenization step. Nanocrystals that are produced in this way are known as smartCrystals® [12, 20, 39, 40]. The process can be run in laboratory scale and can be easily scaled up to industrial scale. Equipments for different batch sizes are available on the market and are proven to fulfill all regulatory requirements to be used for the production of cosmetic products.

Independent of the process used, all wet milling techniques involve the use of liquids in which the active is dispersed. Hence, nanocrystals are obtained from coarse suspensions which are transferred into nanosuspensions during the milling process. For cosmetic use, these nanosuspensions are not transferred into dry powders, but used as liquids for further processing. Due to the small size, the particles do not sediment (Stokes law) and the particles remain homogenously distributed within the dispersion medium during storage (Fig. 8.6). This enables a precise and fast dispensing. If formulated correctly, nanocrystals remain physically stable for at least 12 months.

Fig. 8.6 Macrosupensions (left) versus nanosuspensions (right) from different materials. In contrast to large-sized particles, nanocrystals—due to their small size—remain homogenously distributed within the water phase and no sediment is formed. The colour of the suspension depends on the colour of the active used

8.4.1 Production of Cosmetic Formulations with Nanocrystals

Cosmetics containing nanocrystals can be produced by simply admixing the required amount of nanosuspension to the final product, e.g. cream, gel or ointment. The addition of the nanosuspensions to the formulation at elevated temperatures must be avoided, as the solubility is strongly influenced by the temperature. Hence, heat will increase the solubility of the active and thus promote Ostwald ripening, i.e. particle growth during storage, which would lead to a loss of the special properties of the nanocrystals. Due to the high efficacy of the nanocrystals, typically only low volumes of nanocrystals are required in the final formulations. Typical concentrations range from about 0.02 to 0.2% (w/w) active in the final cosmetic product.

Nanosuspensions contain water and thus are subject to microbial contamination. To ensure a high microbial quality, preservatives need to be added to the formulation. Recent studies proved that preservatives that are highly hydrophilic are most suitable to preserve nanosuspensions. Lipophilic preservatives with poor solubility in water tend to adsorb on the surface of the nanocrystals. This can lead to a reduction of the zeta potential and to agglomeration of the crystals [41–43]. It is therefore advisable to use only hydrophilic preservatives, not only in the nanosuspension as an intermediate product, but also in the vehicle in which the nanocrystals are incorporated. In some cases, lipophilic perfumes were also found to cause agglomeration of the nanocrystals. In general, all additives which possess charges, e.g. electrolytes, can interfere with the physical stability of the nanocrystals [44]. Thus, if possible, only non-charged additives should be used for the formulation.

To sum up, cosmetic formulations containing nanocrystals can be simply produced by admixing nanosuspensions. However, the compatibility of the ingredients of the vehicle with the nanocrystals must be ensured and tested, to avoid agglomeration of the nanocrystals and a loss of the special properties of the nanocrystals.

8.5 Nanocrystals for Dermal Application

Nanocrystals for dermal application were invented in the beginning of the twenty-first century by using rutin nanocrystals. Results confirmed that the dermal penetration and bioactivity of the poorly soluble natural compound rutin can be greatly enhanced when formulating it as nanocrystals. Rutin is a flavonoid that is produced in plants as a secondary metabolite. Rutin possesses many health beneficial properties, e.g. anti-oxidative and anti-inflammatory properties. Moreover, it is known to decrease the capillary permeability and to protect blood vessels, which, for example, can be used to treat rosacea. However, its most prominent cosmetic use is due to its excellent UV protecting properties [45].

Prior to the invention of the nanocrystals, rutin as such could not be formulated in dermal products, due to its poor solubility and thus poor dermal bioavailability. Chemical modification of rutin was done to improve the water solubility of the active by adding a sugar entity to the original molecule. The resulting rutin glucoside was sufficiently soluble in water and could be used in dermal, cosmetic formulations. The UV protecting effect was about 30% higher, when compared to untreated skin if the active was used in a concentration of about 5%. In contrast, nanocrystals of the unmodified rutin, with a concentration of dissolved active, being as low as 0.01%, increased the UV protection to about 160%. This means, the nanocrystals—with a 500-fold lower concentration—doubled the biological activity. In fact, they possessed a 1000-fold higher bioactivity, when compared to the chemically modified rutin glycoside [46, 47]. Besides this invention, data could prove that nanocrystals in general are highly beneficial for improved dermal delivery of poorly soluble actives.

In the light of this success, various nanocrystal-based formulations were developed. Compounds used include different flavonoids, e.g. hesperetin [27, 48], hesperidin [39], apigenin or quercetin [49]. All these compounds possess poor solubility and thus poor dermal bioavailability. Nevertheless, their anti-oxidative and thus anti-ageing potential is high and can be unlocked by simple nanonization of these compounds. There are many other cosmetic or cosmeceutical compounds that possess poor solubility. Examples are curcumin, glabridin or resveratrol. Transforming these compounds into nanocrystals increases their health beneficial properties, i.e. antioxidant capacity (Fig. 8.7), solubility and thus dermal penetration efficacy [40, 50–52].

Recent studies even suggest that nanonization of whole plant materials is an effective method (a) to unlock the health beneficial potential of plants and (b) to reduce organic waste, because especially peels or parts of plants that cannot be used in food industry contain high amounts of secondary plant metabolites [54–57]. These compounds possess, for example, high anti-oxidative properties that are highly interesting for use in cosmetic formulations. One example is pomegranate. Pomegranate is already used in food, food supplements and cosmetic products. In

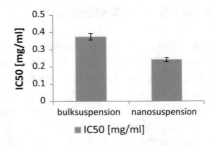

Fig. 8.7 Influence of particle size on antioxidant capacity of natural compounds (modified after [53]), expressed as IC50 values, which indicate the amount needed to scavenge 50% of a radical; i.e. the smaller the IC50 value, the higher the antioxidant capacity of the formulation

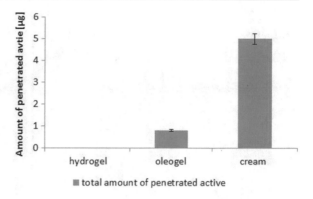

Fig. 8.8 Influence of vehicle on the penetration efficacy of poorly soluble actives from nanocrystals (modified after [68])

most cases, especially in food industry, pomegranate juice is used. Only in some cases, e.g. in nutraceuticals, parts of the peel are used as dry and ground powders. However, transferring these peels into "submicron peel" increased the antioxidant capacity up to 20-fold [58–65]. In fact, transferring plants or parts of plants into small-sized particles seems to be a highly promising and innovative strategy to unlock or improve the bioactivity of natural actives, to save costs, material and to reduce organic waste. In this way, one can not only improve productivity but also contribute to a lower carbon dioxide footprint in future. Due to the natural origin, such formulations can also be used in natural cosmetic products.

Nanocrystals are also in focus of recent scientific studies. Current research is done to identify most suitable vehicles that further promote the passive penetration of poorly soluble actives. Preliminary data suggest that creams and emulsions, containing both water and oil, are vehicles of choice, whereas hydrogels and moisturizers that increase skin hydration seem to hamper the penetration efficacy [40, 66, 67] (Fig. 8.8).

8.5.1 Nanocrystals and Hair Follicle Targeting

A relatively new field, especially in the cosmetic field, is the targeting of actives to the hair follicles [69–71]. The total surface area of all the hair follicle epithelia is believed to be similarly large as the surface area of the stratum corneum [72]. Therefore, the hair follicle as application site is highly interesting. Despite the large surface area, it bears further benefits. For example, it can act as a reservoir for the active and promote a penetration of the active into deeper skin layers, because active is penetrating through the epithelium of the hair follicle, thus bypassing the skin barrier, i.e. the stratum corneum (Fig. 8.9). Even though research in hair follicle targeting is relatively new, there is already evidence that actives can only be transported into the hair follicles if they are incorporated into or at least bound to particles [73–80]. Actives in solution will not penetrate the hair follicle (Fig. 8.9).

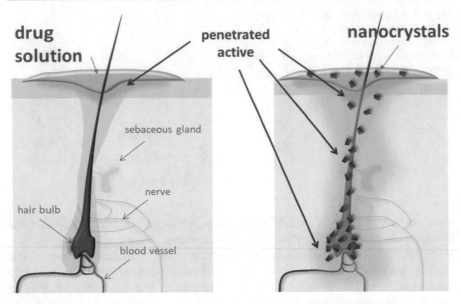

Fig. 8.9 Principle of hair follicle targeting and improved bioavailability of actives by nanocrystals

The influence of the particle size on the hair follicle targeting efficacy was also investigated and revealed that particle sizes of about 600 nm are most efficient to target the hair follicle [81].

As nanocrystals can be produced in different sizes by varying the production parameters, recent studies investigate the influence of the particle size of nanocrystals on the hair follicle targeting efficacy. First results confirm that smaller nanocrystals increase passive diffusion through the stratum corneum, whereas larger-sized nanocrystals lead to improved uptake of the nanocrystals into the hair follicles, thus forming a depot of active in the follicle, from which the active can be released over a longer period. Taken into account that nanocrystals are composed of 100% active, nanocrystals can be regarded to be the most effective delivery system for the delivery of actives to the hair follicle.

8.6 Regulatory Aspects

Nanocrystals for cosmetic use are sold as intermediate products and thus require all documents needed for cosmetic ingredients, e.g. INCI nomenclature, REACH documentation (in the European Community) and safety tests. According to the European cosmetic regulation, nanomaterial is defined as material being "insoluble or biopersistent and intentionally manufactured material with one or more external dimensions, or an internal structure, on the scale from 1 to 100 nm" [82].

As nanocrystals are soluble and not biopersistent and possess sizes well above 100 nm, they do not fulfill these definitions and thus are not regarded as nano-material. Nanocrystals possess no "nano"-hazards [83], and thus, unlike other nanomaterial-containing cosmetic products, cosmetic products containing nanocrystals do not require a special "nano"-notification to the Cosmetic Products Notification Portal (EC) and no "nano"-declaration at the INCI list. All other parts of the world are currently setting up their own "nano-regulations". It is expected that the EU regulation will serve as an example.

8.7 Conclusion

Nanocrystals are the formulation principle of choice for the formulation of poorly soluble actives. They consist of 100% active and possess sizes below 1 μm. Due to the small size, the solubility is increased, which leads to an improved penetration of active after dermal application and thus to an improved bioactivity of the active. Nanocrystals are also the formulation principle of choice if hair follicle targeting is desired. In this case, larger-sized nanocrystals need to be applied. The production of nanocrystals complies with cosmetic GMP in both small and large scales, and the process is easy to scale up. Due to their high solubility and size >100 nm, nanocrystals are not regarded as nanomaterials according to the European cosmetic regulation. Many new cosmetic actives, especially plant actives, are poorly soluble, and thus nanocrystals are believed to attract even more attention in the near future. First products on the market are already available and proved the superiority of the formulation principle for the formulation of innovative and effective cosmetic products.

References

1. Scheuplein RJ. Analysis of permeability data for the case of parallel diffusion pathways. Biophys J. 1966;6:1–17.
2. Scheuplein RJ, Blank IH. Permeability of the skin. Physiol Rev. 1971;51:702–47.
3. Godin B, Touitou E. Transdermal skin delivery: predictions for humans from in vivo, ex vivo and animal models. Adv Drug Deliv. 2007;59:1152–61.
4. Norlén L. Skin barrier formation, The membrane folding model. J Invest Dermatol. 2001;117 (4):823–9.
5. Kreilgaard M. Influence of microemulsions on cutaneous drug delivery. Adv Drug Deliv. 2002;54:77–98.
6. Pawar KR, Babu RJ. Lipid materials for topical and transdermal delivery of nanoemulsions. Crit Rev Ther Drug Carrier Syst. 2014;31:429–58.
7. Junyaprasert VB, Teeranachaideekul V, Souto EB, et al. Q10-loaded NLC versus nanoemulsions: stability, rheology and in vitro skin permeation. Int J Pharm. 2009;377:207–14.
8. Roberts MS, Mohammed Y, Pastore MN, et al. Topical and cutaneous delivery using nanosystems. J Control Release. 2017;247:86–105.

9. Müller RH, Gohla S, Keck CM. State of the art of nanocrystals—special features, production, nanotoxicology aspects and intracellular delivery. Eur J Pharm Biopharm. 2011;78(1):1–9.
10. Keck CM, Müller RH. Nanodiamanten - Erhöhte Bioaktivitat. Labor & More. 2008;01 (08):64–5.
11. Keck CM, Müller RH. Drug nanocrystals of poorly soluble drugs produced by high pressure homogenisation. Eur J Pharm Biopharm. 2006;62(1):3–16.
12. Keck CM, Al Shaal L, Müller RH. SmartCrystals®—review of the second generation of drug nanocrystals. Handbook of materials for nanomedicine. In: Torchilin VP, Amiji MM, Editors. Pan stanford series on biomedical nanotechnology, London: Pan Stanford Publishing; 2010. p. 555–80.
13. Müller RH, Keck CM. Challenges and solutions for the delivery of biotech drugs—a review of drug nanocrystal technology and lipid nanoparticles. J Biotechnol. 2004;113(1–3):151–70.
14. Mauludin R, Müller RH, Keck CM. Kinetic solubility and dissolution velocity of rutin nanocrystals. Eur J Pharm Sci. 2009;36(4–5):502–10.
15. Müller RH, Keck CM. Twenty years of drug nanocrystals. Where are we, and where do we go? Eur J Pharm Biopharm. 2012;80(1):1–3.
16. Mauludin R, Müller RH, Keck CM. Development of an oral rutin nanocrystal formulation. Int J Pharm. 2009;370(1–2):202–9.
17. Rabinow BE. Nanosuspensions in drug delivery. Nat Rev Drug Discov. 2004;3(9):785–96.
18. Merisko-Liversidge E, Liversidge GG, Cooper ER. Nanosizing: a formulation approach for poorly-water-soluble compounds. Eur J Pharm Sci. 2003;18(2):113–20.
19. Stegemann S, Leveiller F, Franchi D, et al. When poor solubility becomes an issue. From early stage to proof of concept. Eur J Pharm Sci. 2007;31(5):249–61.
20. Müller RH, Keck CM. Second generation of drug nanocrystals for delivery of poorly soluble drugs: SmartCrystal®-technology. Eur J Pharm Sci. 2008;34(1):20–1.
21. Liversidge GG, Cundy KC. Particle size reduction for improvement of oral bioavailability of hydrophobic drugs: I. Absolute oral bioavailability of nanocrystalline danazol in beagle dogs. Int J Pharm. 1995;125(1):91–7.
22. Merisko-Liversidge E, Liversidge GG. Nanosizing for oral and parenteral drug delivery. A perspective on formulating poorly-water soluble compounds using wet media milling technology. Adv Drug Deliv Rev. 2011;63(6):427–40.
23. Zhai X, Lademann J, Keck CM, et al. Nanocrystals of medium soluble actives—novel concept for improved dermal delivery and production strategy. Int J Pharm. 2014;470(1–2):141–50.
24. Liversidge GG, Cundy KC, Bishop JF, et al. Surface modified drug nanoparticles. US Patent 5,145,684; (1992).
25. Michael JM, Thomas ER, Atkins J. Microencapsulated 3-piperidinyl-substituted 1,2-benzisoxazoles and 1,2-benzisothiazoles. US Patent 5,965,168; 1992.
26. Müller RH, Akkar A. Drug nanocrystals of poorly soluble drugs. Encyclopedia of Nanoscience and Nanotechnology (Nalwa HS, Editor), American Scientific Publishers; (2004). p. 627–38.
27. Mishra PR, Al Shaal L, Müller RH, et al. Production and characterization of Hesperetin nanosuspensions for dermal delivery. Int J Pharm. 2009;371(1–2):182–9.
28. Gassmann P, List M, Schweitzer A, Sucker H. Hydrosols—Alternatives for the parenteral application of poorly water soluble drugs. Eur J Pharm Biopharm. 1994;40:64–72.
29. List M, Sucker H. Pharmaceutical colloidal hydrosols for injection. GB Patent 2,200,048; 1988.
30. Auweter H, Bohn H, Heger R, et al. Precipitated water-insoluble colorants in colloid disperse form. US Patent 6,494,924; 2002.
31. Merisko-Liversidge E, Sarpotdar P, Bruno J, et al. Formulation and antitumor activity evaluation of nanocrystalline suspensions of poorly soluble anticancer drugs. Pharm Res. 1996;13(2):272–8.

32. Petersen RD. Nanocrystals for use in topical formulations and method of production thereof; 2006. PCT/EP2007/009943.
33. Scholz P, Arntjen A, Müller RH, et al. ARTcrystal® process for industrial nanocrystal production—optimization of the ART MICCRA® pre-milling step. Int J Pharm. 2014;465 (1–2):388–95.
34. Scholz P, Keck CM. Flavonoid nanocrystals produced by ARTcrystal®-technology. Int J Pharm. 2015;482(1–2):27–37.
35. Salazar J, Müller RH, Möschwitzer JP. Performance comparison of two novel combinative particle-size-reduction technologies. J Pharm Sci. 2013;102(5):1636–49.
36. Scholz P, Keck CM. Ibuprofen nanocrystals produced by ArtCrystal-technology. Pharm Ind. 2016;9(16):1340–54.
37. Salazar J, Ghanem A, Müller RH, et al. Nanocrystals: comparison of the size reduction effectiveness of a novel combinative method with conventional top-down approaches. Eur J Pharm Biopharm. 2012;81(1):82–90.
38. Keck CM. Cyclosporine nanosuspensions: optimised size characterisation & oral formulations, PhD-thesis, Freie Universität Berlin; 2006.
39. Kobierski S, Keck CM. Production of Hesperidin dermal nanocrystals by novel smartCrystal® combination technology, 10th European Workshop on Particulate Systems, Berlin/Germany, 30th–31th May; 2008.
40. Romero GB, Chen R, Keck CM, et al. Industrial concentrates of dermal hesperidin smartCrystals®–production, characterization & long-term stability. Int J Pharm. 2015;482 (1–2):54–60.
41. Al Shaal L, Müller RH, Keck CM. Preserving hesperetin nanosuspensions for dermal application. Die Pharmazie. 2010;65(2):86–92.
42. Kobierski S, Ofori-Kwakye K, Müller RH, et al. Resveratrol nanosuspensions. Interaction of preservatives with nanocrystal production. Die Pharmazie. 2011;66(12):942–7.
43. Obeidat WM, Schwabe K, Müller RH, et al. Preservation of nanostructured lipid carriers (NLC). Eur J Pharm Biopharm. 2010;76(1):56–67.
44. Rachmawati H, Rahma A, Al Shaal L, et al. Destabilization mechanism of ionic surfactant on curcumin nanocrystal against electrolytes. Sci Pharm. 2016;84(4):685–93.
45. Kessler M, Ubeaud GJL. Anti- and pro-oxidant activity of rutin and quercetin derivatives. J Pharm Pharmacol. 2003;55:131–42.
46. Stahr P, Keck CM. Tailor-made nanocrystals for optimised dermal drug delivery, 5th Galenus Workshop, Berlin/Germany, 16th–18th November; 2016.
47. Braun A, Stahr P, Schäfer K-H, Keck CM. SmartCrystals® for neuroprotection, Menopause, Andropause, Anti-Aging-Kongress, Vienna/Austria, 8th–10th December; 2016.
48. Al Shaal L, Mishra PR, Müller RH, et al. Nanosuspensions of hesperetin. Preparation and characterization. Die Pharmazie. 2014;69(3):173–82.
49. Hatahet T, Morille M, Hommoss A, et al. Dermal quercetin smartCrystals®. Formulation development, antioxidant activity and cellular safety. Eur J Pharm Biopharm. 2016;102:51–63.
50. Vidlářová L, Romero GB, Hanuš J, et al. Nanocrystals for dermal penetration enhancement— effect of concentration and underlying mechanisms using curcumin as model. Eur J Pharm Biopharm. 2016;104:216–25.
51. Shegokar R, Müller RH. Nanocrystals. Industrially feasible multifunctional formulation technology for poorly soluble actives. Int J Pharm. 2010;399(1–2):129–39.
52. Lohan SB, Bauersachs S, Ahlberg S, et al. Ultra-small lipid nanoparticles promote the penetration of coenzyme Q10 in skin cells and counteract oxidative stress. Eur J Pharm Biopharm. 2015;89:201–7.
53. Abraham A. (in prep.), PlantCrystals for improved delivery of antioxidants, PhD-thesis, Philipps-Universität Marburg.
54. Griffin S, Tittikpina NK, Al-marby A, et al. Turning waste into value: nanosized natural plant materials of Solanum incanum L. and Pterocarpus erinaceus Poir with promising antimicrobial activities. Pharmaceutics. 2016;8(2):11.

55. Griffin S, Sarfraz M, Hartmann SF, et al. Resuspendable powders of lyophilized chalcogen particles with activity against microorganisms. Antioxidants (Basel). 2018;7(2):23.
56. Griffin S, Sarfraz M, Farida V, et al. No time to waste organic waste. Nanosizing converts remains of food processing into refined materials. J Environ Manage. 2018;210:114–21.
57. Griffin S, Masood MI, Nasim MJ, et al. Natural nanoparticles. A particular matter inspired by nature. Antioxidants (Basel). 2017;7(1):3.
58. Sinambela P, Egorov E, Löffler BM, et al. Anti-Aging rutin smartCrystals® for reduction of brown and red skin spots—an in vivo study, Menopause, Andropause, Anti-Aging-Kongress, Vienna/Austria, 6th–8th December; 2012.
59. Keck CM, Müller RH, Gohla S. Nanokristalle - innovatives Formulierungsprinzip für schwerlösliche Anti-Aging Wirkstoffe, Menopause, Andropause, Anti-Aging-Kongress, Vienna/Austria, 6th–8th December; 2007.
60. Sinambela P, Egorov E, Löffler BM, et al. Combination of rutin smartCrystals® and peptide-loaded liposomes for wrinkle reduction—an in vivo study, Menopause, Andropause, Anti-Aging-Kongress, Vienna/Austria, 6th–8th December; 2012.
61. Gerst M, Rostamizadeh K, Arntjen A, et al. ARTCrystal®-technology for improved dermal penetration of rutin nanocrystals, nanocrystals for improved antioxidant capacity of flavonoids, NutriOx 2014: Nutrition and Ageing, Metz/France, 1th–3th October; 2014.
62. Keck CM, Pyo SM, Jin N, et al. Rutin smartCrystals®—most effective anti-oxidant activity & skin penetration, Menopause, Andropause, Anti-Aging-Kongress, Vienna/Austria, 11th–13th December; 2014.
63. Jin N, Staufenbiel S, Keck CM, et al. SmartCrystals®—enhancement of drug penetration without penetration enhancer, Menopause, Andropause, Anti-Aging-Kongress, Vienna/Austria, 11th–13th December; 2014.
64. Keck CM, Monsuur F, Höfer HH, et al. SmartPearls®—new dermal injection-like delivery system without use of a needle, Menopause, Andropause, Anti-Aging-Kongress, Vienna/Austria, 11th–13th December; 2014.
65. Knauer J, Pyo SM, Keck CM, et al. Antioxidant activity of rutin—watersoluble derivatives vs. rutin smartCrystals®, Menopause, Andropause, Anti-Aging-Kongress, Vienna/Austria, 10th–12th December; 2015.
66. Pelikh O, Stahr P, Dietrich H, et al. Anti-aging actives for dermal application—size matters, Menopause, Andropause, Anti-Aging-Congress, Vienna/Austria, 6th–9th December; 2017.
67. Pelikh O, Stahr P, Dietrich H, et al. Anti-aging actives for dermal application—the vehicle is the key for efficacy, Menopause, Andropause, Anti-Aging-Congress, Vienna/Austria, 6th–9th December; 2017.
68. Pelikh O, Stahr P, Gerst M, et al. Nanocrystals for improved dermal drug delivery. Eur J Pharm Biopharm; 128:170–8.
69. Vogt A, Mandt N, Lademann J, et al. Follicular targeting—a promising tool in selective dermatotherapy. J Investig Dermatol Symp Proceed. 2005;10(3):252–5.
70. Toll R, Jacobi U, Richter H, et al. Penetration profile of microspheres in follicular targeting of terminal hair follicles. J Invest Dermatol. 2004;123(1):168–76.
71. Vogt A, Hadam S, Deckert I, et al. Hair follicle targeting, penetration enhancement and Langerhans cell activation make cyanoacrylate skin surface stripping a promising delivery technique for transcutaneous immunization with large molecules and particle-based vaccines. Exp Dermatol. 2015;24(1):73–5.
72. Lademann J, Richter H, Schaefer UF, et al. Hair follicles—a long-term reservoir for drug delivery. Skin Pharmacol Physiol. 2006;19(4):232–6.
73. Blume-Peytavi U, Massoudy L, Patzelt A, et al. Follicular and percutaneous penetration pathways of topically applied minoxidil foam. Eur J Pharm Biopharm. 2010;76(3):450–3.
74. Knorr F, Lademann J, Patzelt A, et al. Follicular transport route—research progress and future perspectives. Eur J Pharm Biopharm. 2009;71(2):173–80.
75. Patzelt A, Knorr F, Blume-Peytavi U, et al. Hair follicles, their disorders and their opportunities. Drug Discov Today: Dis Mech. 2008;5(2):173–81.

76. Blume-Peytavi U, Vogt A. Human hair follicle. Reservoir function and selective targeting. Br J Dermatol. 2011;165(2):13–7.
77. Raber AS, Mittal A, Schäfer J, et al. Quantification of nanoparticle uptake into hair follicles in pig ear and human forearm. J Control Release. 2014;179:25–32.
78. Wosicka H, Cal K. Targeting to the hair follicles. Current status and potential. J Dermatol Sci. 2010;57(2):83–9.
79. Otberg N, Patzelt A, Rasulev U, et al. The role of hair follicles in the percutaneous absorption of caffeine. Br J Clin Pharmacol. 2008;65(4):488–92.
80. Lauterbach A, Müller-Goymann CC. Comparison of rheological properties, follicular penetration, drug release, and permeation behavior of a novel topical drug delivery system and a conventional cream. Eur J Pharm Biopharm. 2014;88(3):614–24.
81. Lademann J, Knorr F, Richter H, et al. Hair follicles—an efficient storage and penetration pathway for topically applied substances. Summary of recent results obtained at the Center of Experimental and Applied Cutaneous Physiology, Charité - Universitätsmedizin Berlin, Germany. Skin Pharmacol Physiol. 2008;21(3):150–5.
82. European Parliament. Regulation (EC) No 1223/2009 of the European parliament and of the council of 30 November 2009 on cosmetic products. Off J Eur Union: L 342/59; 2009.
83. Keck CM, Müller RH. Nanotoxicological classification system (NCS)—a guide for the risk-benefit assessment of nanoparticulate drug delivery systems. Eur J Pharm Biopharm. 2013;84(3):445–8.

Part III
Characterization and Production of Nanoparticles and Nanocosmetics

Characterization of Nanoparticles for Cosmetic Applications

9

Steffen F. Hartmann, Ralph W. Eckert, Daniel Knoth
and Cornelia M. Keck

Abstract

Nanoparticles are tiny and cannot be seen by the naked eye. They possess different properties than macro-sized material and most of the well-established characterization methods for larger sized materials cannot be applied for nanomaterials. Hence, different techniques need to be used for a meaningful characterization of the nanosized material. This chapter will focus on the most important characterization methods that need to be applied to characterize and develop nanocarrier for cosmetic applications.

Keywords

Dynamic light scattering · Laser diffraction · Microscopy · Zeta potential · Physical stability · Size · Size distribution

9.1 Introduction

Particles as such possess various properties that can be analysed. Therefore, prior to any characterization, it is of uppermost importance to define what properties of the particles are of interest. Hence, the question that should be answered by the analysis of the particles or the sample needs to be defined exactly. In the next step, a characterization method that is able to answer this question needs to be selected. The method should be robust, lead to reproducible results, when performed correctly and should enable a clear discrimination between samples that possess the

S. F. Hartmann · R. W. Eckert · D. Knoth · C. M. Keck (✉)
Department of Pharmaceutics and Biopharmaceutics, Philipps-Universität Marburg,
Robert-Koch-Str. 4, 35037 Marburg, Germany
e-mail: cornelia.keck@pharmazie.uni-marburg.de

© Springer Nature Switzerland AG 2019
J. Cornier et al. (eds.), *Nanocosmetics*,
https://doi.org/10.1007/978-3-030-16573-4_9

required properties and those samples which do not possess the required properties. Very often, it is not possible to determine properties of particles directly. Hence, indirect methods need to be applied. This is especially true for the characterization of nanoparticles. Nanoparticles are tiny and cannot be seen by the naked eye. They possess different properties than macro-sized material and most of the well-established characterization methods for larger sized materials cannot be applied for such materials. Hence, with the increasing interest in nanoparticles also sophisticated characterization methods for nanosized material needed to be developed. Most frequently used characterization methods for nanoparticles involve microscopic techniques, methods for size characterization and methods that allow for a specific characterization of material properties, e.g. surface charge, hardness, crystallinity, melting behaviour, hydrophobicity, etc. [1–7].

The use of nanosized materials in cosmetic applications is relatively novel. Hence, the variety of nanosized materials used in cosmetic products is quite small, especially when compared to other fields, e.g. pharmaceutics, medical devices, physics, automotive, sports or in space industry. However, the nanoparticles used in cosmetic products so far, clearly demonstrate the superior advantages of such materials over larger sized material. But what characteristics of nanoparticles are important for their use in cosmetic products? Clearly, the most important information is the size of the material, because only small-sized material will possess the beneficial properties. Hence, to ensure the special "nano-properties", the size and changes in size during storage need to be monitored during product development and during storage for each nanocarrier-based system. Additional characterization methods might be needed to gain more specific information on the particular carrier system. Additional characterization methods may include high-resolution imaging, e.g. electron microscopy, atomic force microscopy, confocal laser scanning microscopy, Raman microscopy and methods to determine the crystalline state of the material, e.g. differential scanning calorimetry or X-ray diffraction. The list of available methods being available today for the characterization of nanoparticles can be "endlessly" extended [1–7]. However, this chapter will focus on the most general methods that need to be applied to characterize and develop a nanosized system in the cosmetic field. Due to the high importance of monitoring size, changes in size and physical stability, this chapter will focus on methods which allow for a precise characterization of nanosized cosmetic ingredients in this regard. Special focus will be put on the methods that are suggested by the ISO 13320 and 22412 [8, 9].

9.2 Size Characterization

9.2.1 Light Microscopy

Nanoparticles possess a size well below 1 μm. This means a nanoparticle cannot be seen by classical light microscopy, because the resolution is too low and allows only for the detection of particles >500 nm [10]. However, this can be greatly used

(a)   **(b)**   **(c)**   **(d)**

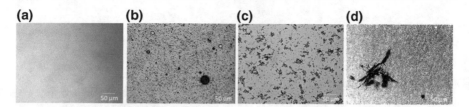

Fig. 9.1 Light microscopy is a simple approach to distinguish between nanosized samples and samples which contain larger sized particles. The image demonstrates this for nanocrystals: **a**—small nanosized sample, **b**—sample with larger particles, **c**—sample with agglomerates, **d**—sample with large crystals that grew during storage due to Ostwald ripening (shape and size of large crystals differs from the size and shape of the original bulk material)

to discriminate between samples that contain larger particles and samples which only contain small-sized nanoparticles (Fig. 9.1). It also bears the advantage that—in most cases—the origin of the larger particles can be determined. For example, larger sized particles that remained in the product due to insufficient milling methods of the bulk material (Fig. 9.1b), agglomerates that occurred due to the use of an inappropriate stabilizer (Fig. 9.1c) or particle growth over time due to Ostwald ripening (Fig. 9.1d) can be easily distinguished from each other. The advantage of light microscopy is the easy to perform analysis, the relatively low cost equipment and the possibility to analyse the original sample. Therefore, light microscopy should always be used for the characterization of nanoparticles.

High-resolution imaging enables the observation of the nanoparticles as such. However, these techniques are quite expensive, require sound knowledge and experience and—moreover—a sophisticated sample preparation prior to the analysis. This can lead to changes of the nanoparticles (e.g. agglomeration) and might lead to artefacts and misinterpretation of the results obtained [11–14]. According to the regulatory requirements [8, 9] independent on the microscopic method used, microscopy is not able to obtain a statistical relevant mean size and size distribution, which is due to the low number of particles that are observed within the analysis (known as breakfast cereal problem). Therefore, to obtain statistically relevant size results, other methods are required.

9.2.2   Dynamic Light Scattering

Dynamic light scattering (DLS), also known as photon correlation spectroscopy (PCS), is an indirect method for the determination of size and size distribution of nanosized samples. It determines the velocity (diffusions constant) of the particles in a liquid medium, which is then used to calculate the particle size (hydrodynamic diameter = z-average) [15].

9.2.2.1   The Measuring Principle of Dynamic Light Scattering
The measuring principle is based on two facts. First, the diffusion constant is inversely proportional to the size of nanoparticles (smaller particles move faster

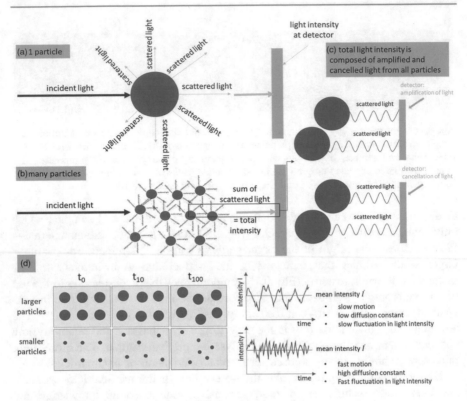

Fig. 9.2 Principle of size analysis via DLS: explanations cf. text

than larger particles) and second, when light hits a particle the light is scattered around the particle, leading to a so-called scattering pattern around the particle (Fig. 9.2a). When a sample which contains many nanoparticles with different sizes is illuminated with light, all particles will scatter light and the sum of the scattered light (light intensity I) can be detected (Fig. 9.2b). The total light intensity detected is a combination of amplified and cancelled light, which is scattered from the different particles (Fig. 9.2c). Due to the movement of the particles, fluctuations in the light intensity will occur over time and depending on the size of the particles, fluctuations will be fast or slow (Fig. 9.2d).

From the detected fluctuations in light intensity, the calculation of the diffusion constant D can be done by transferring the detected fluctuations in light intensity over time into an autocorrelation function $G(\tau)$ (Fig. 9.3). $G(\tau)$ is simply calculated by using formula 9.1, where $I(t)$ is the detected light intensity at time t, $G(t + \tau)$ the detected light intensity at the next timepoint of the measurement, τ the time (delay time) between each measurement and $\bar{I}$ the mean intensity of the light fluctuation from all particles. The calculation of $G(\tau)$ results in an exponential curve (Fig. 9.3), [15].

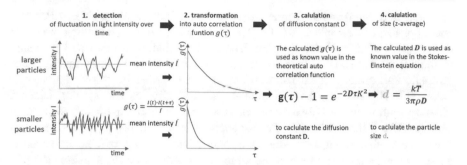

Fig. 9.3 Principle of size analysis via DLS: transformation of analysis result into the particle size explanations cf. text. Modified after [16] and [15]

The curve possesses a fast decay if small particles are analysed and a slower decay in case larger particles are analysed. In the next step, the measured and calculated function $G(\tau)$ is used for the calculation of the diffusion constant with the help of formula 9.2 (Fig. 9.3), where $G(\tau)$ is the measured correlation function, D the diffusion constant, τ the time between each measurement (delay time) and K the absolute value of the scattered light vector, which can be calculated by using formula 9.3, where n is the refractive index of the dispersant medium, λ the wavelength of the incident laser light and θ the angle of scattered light, i.e. the detection angle. In the last step, the calculated diffusion constant D is used to calculate the particle size by using the Stokes-Einstein equation (formula 9.4), where d is the diameter, k the Boltzmann constant, T the absolute temperature, and η the dynamic viscosity of the dispersion medium. The size obtained is the so-called z-average. It is an intensity-weighted hydrodynamic diameter and corresponds to a size of a sphere that would possess a similar diffusion velocity of the particles in the sample being analysed [15].

$$g(\tau) = \frac{I(t) \cdot I(t+\tau)}{\bar{I}} \tag{9.1}$$

$$g(\tau) - 1 = \mathrm{e}^{-2D\tau K^2} \tag{9.2}$$

$$K = \frac{4\pi n}{\lambda} \sin\left(\frac{\theta}{2}\right) \tag{9.3}$$

$$d = \frac{kT}{3\pi\rho D} \tag{9.4}$$

$$PdI = \frac{2c}{b^2} \tag{9.5}$$

In addition to size, also the polydispersity index (*PdI*) can be calculated. The *PdI* is a measure for the broadness of the size distribution and can possess values in the range between 0 and 1. A monodisperse formulation would possess a *PdI* of 0 and a polydisperse sample a *PdI* of 1. The *PdI* is analysed by determining the width of the distribution c of the logarithmized autocorrelation function and by using formula 9.5, where b is the slope of the logarithmized autocorrelation function. For cosmetic and cosmeceutical nanocarriers, the *PdI* should be below 0.2, ideally below 0.1. Broader distributions indicate a physically instable system, because phase separation and/or Ostwald ripening is more pronounced in broadly distributed samples [15]. If the *PdI* is >0.3, the size analysis is not reliable anymore. Hence, in such cases, other sizing techniques are required to obtain a reliable information about size and size distribution of the sample. However, for a rough estimation of a "good" or "bad" sample, this information will be sufficient.

9.2.2.2 Practical Hints for Size Analysis with Dynamic Light Scattering

Today's instruments for size analysis via dynamic light scattering are easy to handle and the measurements are fast and easy to perform. However, misleading results can be obtained due to various parameters. First of all, it is highly important that samples do not change during the whole time of the measurement. Samples which settle or float during analysis create additional movements which can interfere with the measured diffusion speed of the particles. Samples which agglomerate, dissolve or air bubbles and dust can also disturb the size analysis. Thus, to enable meaningful and reproducible results, all these parameters must be excluded [16, 17]. Another major prerequisite for correct and reproducible size analysis is the temperature and the viscosity, because these values are needed for the size calculation with the Stokes-Einstein equation (formula 9.4). Only small inaccuracies can already lead to tremendous differences in the size results obtained [16–18]. It is therefore advisable to adjust the temperature of the sample and the dispersion medium prior to the measurement and to analyse the viscosity of the dispersion medium, exactly at this temperature [16, 17].

9.2.3 Laser Diffraction

The measuring range of DLS is typically in the range between 1 nm and 6 µm. Hence, possible larger particles cannot be detected via this method. Therefore, other methods that detect larger particles are highly important. Light microscopy is an easy to perform method. However, it does not provide statistically relevant data within a short time. Therefore, the ISO 13321 suggests laser diffraction (LD) as valuable, additional tool for size characterization of nanoparticles. Modern LD instruments possess a broad measuring range, typically in the range between 20 nm and 2000 µm. Hence, small-sized particles can be detected in parallel to larger sized particles within only one measurement. This "all-in-one" size analysis is possible by the combination of different size analysis techniques which are combined in only one instrument. Larger sized particles, i.e. >400 nm, are analysed by classical laser diffraction and the smaller sized particles are analysed by determining the scattering intensity, which is related to the size of the particles. The results obtained from the two different mechanisms are combined together into one size result.

Size results obtained by laser diffraction are typically expressed as volume-weighted median diameters. Typically, the diameters $d(v)0.10$, $d(v)0.50$, $d(v)0.90$, $d(v)0.95$ and $d(v)0.99$ are used to express the median size and the size distribution of the particles. The values represent the volumes of the particles that are equal or below the given size, i.e. the $d(v)$ 0.50 represents the upper size for 50% of the volume of the particles within a sample. The broader the difference between $d(v)$ 0.1 and $d(v)$ 0.99—the broader is the size distribution and typically the poorer is the quality of the sample. Size results are also expressed as size distribution curves, which might enable the discrimination between monomodal or bi- or trimodal distributions.

9.2.3.1 The Measuring Principle of Modern Laser Diffractometry

Early laser diffractometers could analyse particles with sizes >400 nm by using a simple instrumental set-up, consisting of a light source, optical lenses, a cuvette for the sample and a detector for the detection of scattered light (Fig. 9.4). As mentioned above, when particles are illuminated with light, particles will scatter light, leading to a size depended scattering pattern. Scattered light is the sum of diffracted light, refracted light, reflected light and light that was absorbed and re-radiated (Fig. 9.5a). Laser diffraction (LD) aims at detecting diffracted light. The principle of LD is the detection of the diffraction pattern, because the angle of diffraction (first minimum of intensity distribution of diffracted light, Fig. 9.5b) increases with decreasing particle size and the intensity of diffracted light decreases with decreasing size [15]. These phenomena are described in the Airy-formula (formula 9.6), where d is the diameter of the particle, θ the angle of scattered (diffracted) light and λ the wavelength of the incident light.

$$d = \frac{1.22\lambda}{\theta} \tag{9.6}$$

However, in "real-world" samples, not only one size but different sizes exist in the sample. Therefore, the diffraction pattern detected will correspond to an overlay of the different diffraction patterns from the different particles. To calculate the size

Fig. 9.4 Set-up of a classical laser diffractometer. Modified after [15]

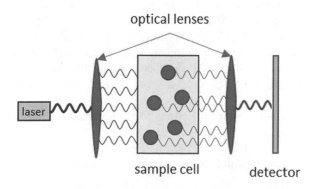

optical lenses

laser

sample cell

detector

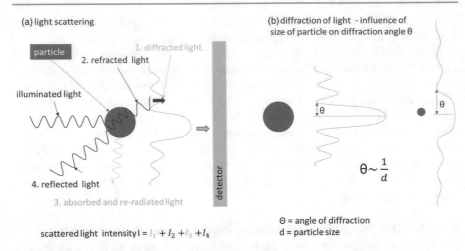

Fig. 9.5 Basic principles of light scattering and laser diffraction. **a**—scattered light is the sum of reflected, refracted, diffracted and re-radiated light, **b**—laser diffraction uses the phenomenon that the angle of diffraction is correlated to the size of the particle

distribution of such samples, the software uses an algorithm which is based on the Mie-theory to simulate various particle distributions and the corresponding diffraction patterns that would occur. The pattern which fits best to the detected pattern is used and the corresponding size distribution is presented.

Particles <400 nm cannot be detected by classical LD analysis, because the diffraction angles are too large and the intensity of diffracted light is too small when compared to larger particles. Hence, small particles become "invisible" at the detector. To increase the measuring range of LD instruments in the late 1980s, modern LD instruments with an additional technique were introduced. New instruments detect not only the diffracted light, but also detect the scattered light at different angles and with the use of different wavelengths. This is reasonable, because the intensity of scattered light depends on the size, the incident wavelength and the angle of detection (Fig. 9.6).

The first additional technique on the market was the so-called PIDS technology [15]. PIDS used three different wavelengths and six different angles of detection for the detection of the scattering patterns. The changes in light intensity can be correlated to the size of the particles, i.e. the stronger the changes, the smaller are the particles (Fig. 9.6). This information is used and combined with the original LD data to obtain a "modern" LD result. Additional techniques are meanwhile employed in almost all modern LD instruments. The principle is always similar, however, the set-up might vary for each manufacturer.

9.2.3.2 Practical Hints for Size Analysis via Laser Diffraction

Similar to size analysis via DLS, also size measurements with modern LD instruments are easy and fast to perform. However, due to the complexity, especially with the combination of the additional techniques, special caution must be taken when

Principle of additional techniques:

1. sample is illuminated with different wave lengths ➡ 2. the intensity of scattered light is measured at different angles θ

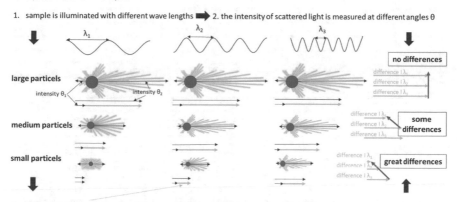

3. differences in light intensity I at different angles θ and wavelengths λ are used to estimate the size

Fig. 9.6 Basic scheme of principle of additional techniques for size analysis of submicron-sized particles with modern LD instruments (explanations c.f. text.)

analysing complex nanosized samples via LD. Also LD requires the use of stable samples, which do not change during analysis [16]. Special caution must be taken when selecting the measuring procedure, i.e. the stirring speed during analysis. High-speed stirring can destroy agglomerates and too low stirring can lead to sedimentation of large particles, which then escape from the sample cell and cannot be detected anymore [10, 11]. The best option to gain meaningful information of a particular sample is the repeated analysis of the sample by using different measuring set-ups [19].

Another major prerequisite for accurate size analysis is the use of correct optical parameters, which are needed for the size calculation by the software. The software uses the Mie-theory for size analysis, which is a complex formula describing the scattering pattern of a particle of a given size in dependency of wavelength and the angle of scatter. As scattering is the sum of all phenomena that occur when light hits a particle, also refracted light (c.f. Figure 9.5a) is involved in the correct solution of the Mie-theory. The influence of the optical parameters on the size results is tremendous and is demonstrated for a nanosuspension below (Fig. 9.7).

Unfortunately, for many solid actives that are used in cosmetics, the optical properties are unknown and need to be determined for a meaningful size analysis with LD. This is possible, for example by the determination of the Becke line or by determing dn/dc, i.e. the changes in real refractive index with increasing concentration [16–18]. These methods are not highly accurate, however, the accuracy, e.g. two digits after the comma, is sufficient for LD analysis [18].

Recent results also proved that the combination of classical LD and additional technique can lead to an overestimation of small-sized particles, which means that larger particles might be overlooked [16–19]. If this happens, LD fails its purpose, because as described above, it is especially used to detect possible larger size particles within a small-sized system. The problem can be simply overcome by

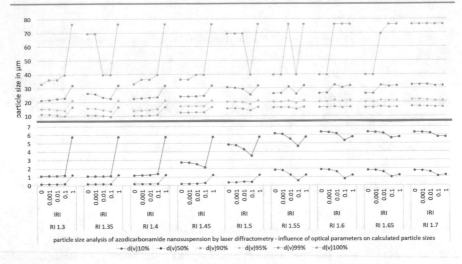

Fig. 9.7 LD measurements can be misleading if incorrect optical properties are used for size analysis. This is exemplarily shown for an LD analysis of an azodicarbonamide nanosuspension, which was analysed once and the size was calculated using different optical properties. Modified after [18]

size analysis with classical LD alone, i.e. analysis of samples without additional technique and by using light microscopy to confirm the data obtained by LD analysis [20].

9.3 Determination of Physical Stability

9.3.1 Physical Stability

A sufficient small size is the main prerequisite for the performance of the nanosized material in cosmetic products. Therefore, the size needs to be maintained during the whole time of the shelf life of the product. This involves the nanocarriers as intermediate product prior to the incorporation into the final product, but also in the final product. Measuring the size and changes in size during storage is, therefore, a major tool in product formulation and development. Physical stability for the nanocarriers is typically assessed over a period of 24 months. Different storage conditions (e.g. room temperature and storage at 4 and 40 °C) should be used to enable a more detailed evaluation of the physical stability of the sample. The packaging can have a great influence on the stability of the particles [21]. The best packaging is glass of pharmaceutical quality, i.e. type I glass containers (borosilicate glass) and storage without exposure to light and oxygen. However, such packaging is not always suitable for cosmetic products or intermediate products. Therefore, to investigate the "real" physical stability of the sample, already during formulation development, samples should be stored in the real packaging of the final product. Size characterization during storage should be performed at adequate time points, i.e. day of production, day 1, 7, 14 and 28 after production and after 3, 6, 12 and 24 months of storage. In addition, stress, e.g. accelerated shaking or heat and cooling cycles, can be performed.

9.3.2   Zeta Potential Analysis

Obtaining changes in size during storage is an easy method to determine the physical stability of nanosized samples. However, the drawback is the time needed to obtain the final results. Therefore, this section describes a method that can be used to predict the physical stability of nanoparticles, directly at the day of production. The method is known as zeta potential analysis. The zeta potential is defined as the potential, i.e. the charge of a particle at the plane of shear and is related to the surface charge of the particle [22]. Zeta potential analysis is, therefore, an indirect measure of the surface charge of particles, which cannot be measured directly. The reason for this is the adsorption of ions on the surface of particles, which cannot be removed completely (Fig. 9.8).

Physical stability of a nanodisperse system is given when the particles remain homogeneously distributed in the dispersion medium, i.e. no sedimentation, flotation or agglomeration over time and no changes in size occur. Agglomeration occurs if the free energy of the system is high, thus particles tend to agglomerate to decrease their surface area which is in contact with the dispersion medium. Physical stabilization can be done by two mechanisms. The first mechanism is the introduction of a steric hindrance, which prevents that particles can get into direct contact with each other. The principle is comparable with bumper cars in amusement parks (Fig. 9.9). The stabilization mechanism is known as steric stabilization

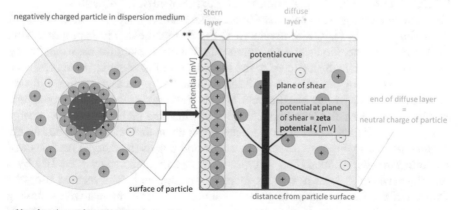

** surface charge of particle = Nernst potential:
    cannot be measured directly due to unremovable Stern layer – zeta potential ζ is measured instead

Fig. 9.8 Theoretical scheme of zeta potential, which is the potential of a particle at the plane of shear. Modified after [22]

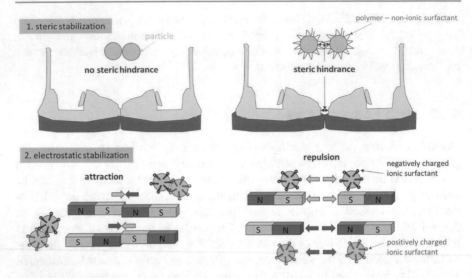

Fig. 9.9 Basic principles of stabilization of particles. Upper: steric stabilization. Lower: electrostatic stabilization

and can be typically achieved with polymers, i.e. non-ionic surfactants. The second mechanism is electrostatic stabilization, where likely charges are introduced onto the surface of the particles. If the charges are high enough, the particles repulse each other, similar to magnets with similar charge (Fig. 9.9). Electrostatic stabilization is achieved with the use of ionic surfactants. In addition, small molecules with high affinity to surfaces, e.g. citrate or phosphate, can be used to increase the charge of nanoparticles. Such molecules are known as anti-flocculants [22].

9.3.2.1 The Measuring Principle of Zeta Potential Analysis

Zeta potential analysis can help to estimate if the stabilizer and the concentration of the stabilizer used for the formulation of the nanoparticles are appropriate and, moreover, it can help to determine ingredients that can destabilize the nanosystems and should, therefore, be avoided for the formulation of nanocarrier-based cosmetic products. The zeta potential is analysed indirectly by measuring the diffusion constant of particles in an electric field, which is referred to be the electrophoretic mobility (EM). Particles with high charge move fast and particles with low charge move slowly. The direction of the particles depends on their charge, i.e. negatively charged particles will move towards the anode and positively charged particles will move towards the cathode (Fig. 9.10a). Modern instruments analyse the EM by a technique which is known as laser Doppler anemometry. The method uses the Doppler effect, i.e. changes in frequency that occur if waves interact with a moving object. The phenomenon is often recognized when an ambulance is passing by. If the ambulance is approaching towards the passengers, the pitch of the horn becomes higher, and if the ambulance has passed the passengers, the pitch gets lower (Fig. 9.10). The same happens if a sample containing nanoparticles is moving

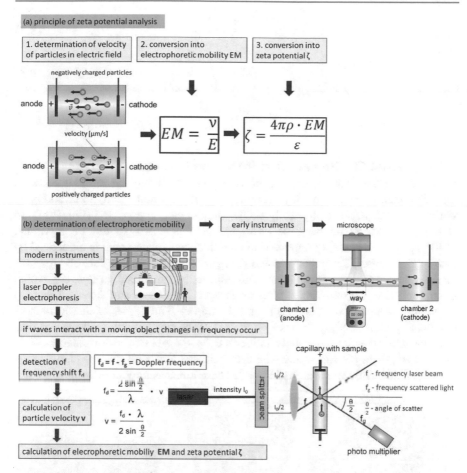

Fig. 9.10 Basic principle of zeta potential analysis and determination of frequency shift and electrophoretic mobility (explanations c.f. text)

in an electric field and gets illuminated with light (Fig. 9.10b). The change in frequency f_d of the light upon interaction with the moving particle is correlated to the velocity v of the particle (formula 9.7), where θ is the angle and λ the wavelength of the incident light. The electrophoretic mobility (EM) can be calculated by dividing the velocity v by the electric field strength E applied during the measurements (formula 9.8). Finally, the zeta potential is calculated, typically by using the Helmholtz-Smoluchowski equation (which is a simplified version of the Henry equation) where ζ is the zeta potential, ρ the viscosity of the dispersion medium, EM the electrophoretic mobility and ε the dielectric constant of the dispersion medium (formula 9.9), [22].

$$f_d = \frac{2 \sin \theta/2}{\lambda} \cdot v \qquad (9.7)$$

$$\mathrm{EM} = \frac{v}{E} \qquad (9.8)$$

$$\zeta = \frac{4\pi\rho \cdot \mathrm{EM}}{\varepsilon} \qquad (9.9)$$

9.3.2.2 What Can Be Expected from Zeta Potential Analysis?

In case steric stabilization is used, the zeta potential should be low, i.e. close to zero. This means the polymer layer around the particle is sufficiently thick and good steric stabilization, i.e. a physically stable sample, can be expected (Fig. 9.11). In case electrostatic stabilization is used, the zeta potential should be high, i.e. $>|30|\ mV$ [22]. It is highly important to note that the zeta potential is influenced by many parameters. This is because the charges of a particle are derived from the molecules from which it is composed off. Especially, functional groups, e.g. amino groups or hydroxy groups, act as electron acceptors or donors and thus contribute to the surface charge of the particles (Fig. 9.12 upper). The degree of dissociation of functional groups strongly depends on the pH of the surrounding medium and thus also the zeta potential is strongly influenced by the pH of the formulation. This means changes in pH can change the zeta potential and with this the stabilization efficacy of the stabilizer (Fig. 9.12 lower). Therefore, it is important to perform zeta

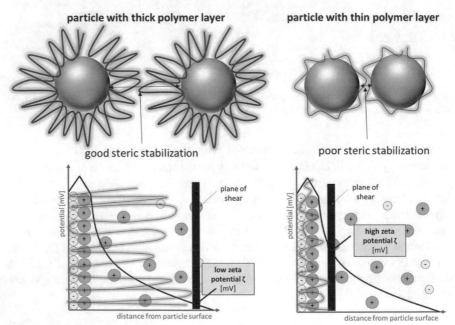

Fig. 9.11 Influence of thickness of non-ionic surfactant layer on physical stability and zeta

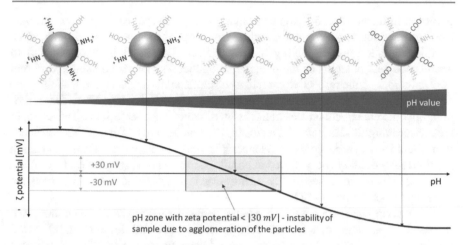

Fig. 9.12 Influence of pH on particle charge, zeta potential and physical stability

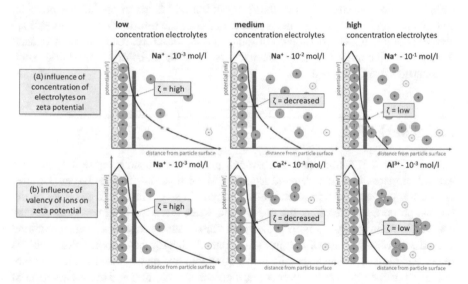

Fig. 9.13 Influence of concentration of electrolytes and ion valency on zeta potential and physical stability. Modified after [22]

potential measurements always at a constant pH and to investigate the stability of a formulation at the pH of the intended final formulation.

The zeta potential is also influenced by electrolytes in the dispersion medium (Fig. 9.13). Low concentrations of electrolytes yield a slow decay of the potential curve and with increasing concentration the decay is fast. As the zeta potential is the potential at the plane of shear, it decreases with increasing concentrations of electrolytes, in case electrostatic stabilization is used for the stabilization of the particles, leading to a destabilization of the sample and strong agglomeration of the particles. In formulation development, many excipients can carry charges and thus can influence the zeta potential and the stability of the particles. If in doubt if an excipient might cause instability of a nanosized system, it is recommended to analyse the zeta potential of the particles in aqueous solutions containing different concentrations of the respective excipient. Excipients that often cause instability are electrolytes, i.e. sodium chloride, preservatives or perfumes. Also, the reproducibility of zeta potential measurements is influenced by electrolytes. Therefore, to ensure meaningful zeta potential analysis, the conductivity of the dispersion medium should always be constant. Hence, pH and conductivity of the dispersion medium should always be analysed in parallel and should be standardized. Analysis in water is performed best, when using water which is adjusted to a constant conductivity of 50 µS/cm [22].

When taking into account all these facts, zeta potential analysis—even though, it is based on a rather complicated theory—is a great tool for a straightforward development of innovative nanocarrier-based cosmetic products [23–30]. It yields useful information about suitable stabilizers and the required concentrations of stabilizers, optimal pH values of the formulation and helps to predict the physical stability. It can identify excipients which destabilize the formulation without tedious stability studies which are time-consuming and is, therefore, recommended in each formulation laboratory where nanocarriers and formulations containing such nanocarriers are developed.

9.4 Conclusions

Nanoparticles and nanocarriers in cosmetic products are used because they possess special properties when compared to larger sized material. Hence, larger particles would not possess these properties and changes in size would lead to a loss of the special properties. Therefore, the major task, when characterizing nanoparticles for use in cosmetic products, is to prove that these "nano-properties" are maintained over the whole shelf life of the product. Characterization of nanoparticles involves many techniques. Besides specific techniques that are mainly used for the characterization of specific material properties, size characterization and the observation of possible changes in size during storage are the most important tools for the formulation of effective nanocarrier-based cosmetic products. Size characterization involves microscopic techniques and indirect analysis methods. The most simple method is the use of light microscopy which enables a fast detection of possible larger particles or agglomerates and is, therefore, a reliable method to select instable samples, without the need of complex measurement procedures or cost-intensive equipment. If a more precise analysis is required, dynamic light scattering should be used to analyse the mean particle size of the sample. However, due to the limited measuring range of DLS, additional methods to DLS are required to detect possible larger particles within a small-sized population. Laser diffraction is the most suitable method for this. However, as each method can lead to artefacts and misleading results therefore, each analysis result should be confirmed by an additional, independent method. Size characterization of nanoparticles is, therefore, performed best by combining microscopic imaging with DLS and LD measurements. The physical stability of nanosized samples should be determined by observing possible changes in size during storage at different temperatures and in the original packaging. In addition, zeta potential analysis should be used to gain detailed information on suitable excipients, e.g. surfactants, surfactant concentrations, other excipients (perfumes, thickener, preservatives, electrolytes) and suitable pH values for the dispersion medium for the formulation of small-sized and physically stable "nanocosmetics".

References

1. Particle Allen T. Particle size measurement. 4th ed. Berlin: Springer; 1990.
2. Anderson W, et al. A comparative study of submicron particle sizing platforms: accuracy, precision and resolution analysis of polydisperse particle size distributions. J Colloid Interface Sci. 2013;405:322–30.
3. Calabretta M, et al. Analytical ultracentrifugation for characterizing nanocrystals and their bioconjugates. Nano Lett. 2005;5(5):963–7.
4. Carney RP et al. Determination of nanoparticle size distribution together with density or molecular weight by 2D analytical ultracentrifugation. Nat Commun. 2011;2:335.
5. Kumar A, Dixit CK. Methods for characterization of nanoparticles. Advances in nanomedicine for the delivery of therapeutic nucleic acids. Cambridge: Woodhead Publishing; 2017. p. 43–58.
6. Müller RH. Colloidal carriers for controlled drug delivery and targeting: modification, characterization, and in Vivo distribution. Boca Raton: CRC Press; 1991.
7. Cho EJ, et al. Nanoparticle characterization: state of the art, challenges, and emerging technologies. Mol Pharm. 2013;10(6):2093–110.
8. ISO 13320. ISO 13320: Particle size analysis—Laser diffraction methods. International Organization for Standardization; 2009.
9. ISO 22412. ISO 22412: Particle size analysis—Dynamic light scattering (DLS). International Organization for Standardization; 2017.
10. Stephens DJ, Allan VJ. Light microscopy techniques for live cell imaging. Science. 2003;300 (5616):82–6.
11. Burrows ND, Penn RL. Cryogenic transmission electron microscopy: aqueous suspensions of nanoscale objects. Microsc Microanal Official J Microsc Soc Am Microbeam Anal Soc Microsc Soc Can. 2013;19(6):1542–53.
12. Chen S, et al. Avoiding artefacts during electron microscopy of silver nanomaterials exposed to biological environments. J Microsc. 2016;261(2):157–66.
13. Tiede K et al. Detection and characterization of engineered nanoparticles in food and the environment. Food Addit Contam Part A Chem Anal Control Exposure Risk Assess. 2008; 25 (7):795–821.
14. Woehl TJ et al. Experimental procedures to mitigate electron beam induced artifacts during in situ fluid imaging of nanomaterials. Ultramicroscopy. 2013;127:53–63.
15. Müller RH, Schuhmann R. Teilchengrößenmessung in der Laborpraxis. Wissenschaftliche Verlagsgesellschaft Stuttgart; 1996.
16. Keck CM. Particle size analysis of nanocrystals: improved analysis method. Int J Pharm. 2010;390(1):3–12.
17. Keck CM, Müller RH. Size analysis of submicron particles by laser diffractometry–90% of the published measurements are false. Int J Pharm. 2008;355(1–2):150–63.

18. Keck CM. Cyclosporine nanosuspensions—Optimised size characterisation & oral formulations. PhD thesis. Freie Universität Berlin; 2006.
19. Kübart Acar S, Keck CM. Laser diffractometry of nanoparticles: frequent pitfalls & overlooked opportunities. J Pharm Technol Drug Res. 2013;2:17.
20. Keck CM. Partikelgrößenanalytik für Nanopartikel: Ein Kinderspiel oder doch eine verflixte Kiste? TechnoPharm. 2012;4:279–87.
21. Acar Kübart S. Menthol-beladene Lipidnanopartikel für Consumer-Care: Entwicklung & optimierte Charakterisierung. PhD thesis. Freie Universität Berlin; 2017.
22. Müller RH. Zetapotential und Partikelladung in der Laborpraxis. Wissenschaftliche Verlagsgesellschaft Stuttgart; 1996.
23. Al Shaal L, Müller RH, Keck CM. Preserving hesperetin nanosuspensions for dermal application. Pharmazie. 2010;65(2):86–92.
24. Zhai X, et al. Nanocrystals of medium soluble actives–novel concept for improved dermal delivery and production strategy. Eur J Pharm Biopharm. 2014;470(1–2):141–50.
25. Mauludin R, Müller RH, Keck CM. Development of an oral rutin nanocrystal formulation. Int J Pharm. 2009;370(1–2):202–9.
26. Mishra PR, et al. Production and characterization of Hesperetin nanosuspensions for dermal delivery. Int J Pharm. 2009;371(1–2):182–9.
27. Müller RH, Gohla S, Keck CM. State of the art of nanocrystals–special features, production, nanotoxicology aspects and intracellular delivery. Eur J Pharm Biopharm Official J Arbeitsgemeinschaft fur Pharma Verfahrenstechnik. 2011;78(1):1–9.
28. Romero GB, et al. Industrial concentrates of dermal hesperidin smartCrystals®–production, characterization & long-term stability. Int J Pharm. 2015;482(1–2):54–60.
29. Schwarz JC, et al. Ultra-small NLC for improved dermal delivery of coenyzme Q10. Int J Pharm. 2013;447(1–2):213–7.
30. Zhai X, et al. Dermal nanocrystals from medium soluble actives—physical stability and stability affecting parameters. Eur J Pharm Biopharm. 2014;88(1):85–91.

Characterization of Nanoparticles in Dermal Formulations

10

D. Knoth, R. W. Eckert, S. F. Hartmann and C. M. Keck

Abstract

The characterization of nanoparticles in dermal formulations is challenging and there are several different aims to characterize nanoparticles in their respective dermal formulations. The major aim will be to observe the stability of the nanoparticles, i.e., to monitor if size, crystalline state, distribution of the particles within the vehicle and/ or incorporation of the actives are maintained during the shelf life of the product. Another reason to characterize nanoparticles in dermal products might be derived from regulatory bodies, which might be interested to determine if a dermal product contains nanoparticles or not. Due to the diverse nature of the different nanocarriers being available today and due to the diverse characterization aims no standard procedure is available for this. Hence, for each type of formulations an individual analysis protocol needs to be established. The aim of this chapter is to provide a brief overview of the existing methods for the characterization of nanoparticles in dermal formulations and to provide basic information on how to characterize nanoparticles in dermal formulations on a day to day basis.

Keyword

Nanocarriers · Size · Crystallinity · Stability · Agglomeration · Vehicle · Microscopy · Spectroscopy · Separation

D. Knoth · R. W. Eckert · S. F. Hartmann · C. M. Keck (✉)
Department of Pharmaceutics and Biopharmaceutics, Philipps-Universität Marburg,
Robert-Koch-Str. 4, 35037 Marburg, Germany
e-mail: cornelia.keck@pharmazie.uni-marburg.de

© Springer Nature Switzerland AG 2019
J. Cornier et al. (eds.), *Nanocosmetics*,
https://doi.org/10.1007/978-3-030-16573-4_10

10.1 Introduction

The characterization of nanoparticles in dermal formulations is more challenging than the characterization of the nanoparticles in original—unformulated—state. This is because particles are embedded in a matrix; this means the vehicle of the dermal product. In cosmetics, many types of vehicles are available. Examples include classical vehicles, e.g., emulsions, creams or gels, modern vehicles, e.g., gel-creams, foams or complex structures, e.g., phospholipid-based vehicles containing lamellar structures or "suspo emulsions", where both particles and droplets are formulated in one vehicle. Depending on the composition and the colloidal structure of the vehicle nanoparticles can be localized in different phases of the vehicle. The nanoparticles can be homogeneously dispersed in either the water or the oil phase of the vehicle. They can intercalate at the interphase of the oil and the water phase or at other interphases in the vehicle, e.g. on droplets, air, particles etc. Another possibility is agglomeration of the particles in the vehicle, which would lead to a loss of the "nanoproperties" (Fig. 10.1).

The distribution of the particles in the vehicle depends not only on the vehicle but also on the properties of the nanoparticles itself. Important properties that contribute to the distribution in the vehicle include particle size, charge of the

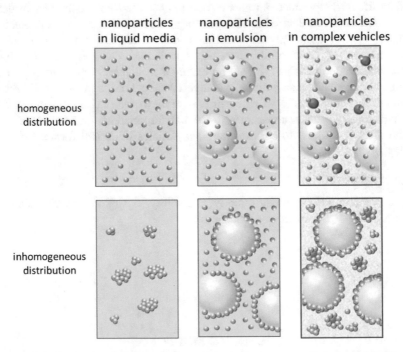

Fig. 10.1 Scheme of nanoparticles dispersed in liquid, emulsion, or complex vehicles. Upper—no interactions and no instability—"nano properties" of the nanoparticles are maintained; lower—Instabilities due to agglomeration and/or interaction with other compounds (structures) of the vehicles—nanoparticles lose their special properties

particles, surface hydrophobicity, and mechanism of stabilization. Larger particles possess smaller kinetic energy and thus are less prone to agglomeration. Non-charged particles are less prone to interactions with charged excipients in the vehicle, whereas charged nanoparticles easily interact with oppositely charged molecules. Hence, if the vehicle contains oppositely charged molecules, it is most likely that the nanoparticles interact with these molecules by forming complexes or agglomerates. This leads to a loss of the "nano properties" on one hand and most often to a loss or modification of the properties of the other excipients on the other hand. As many excipients in cosmetics possess charges—examples are ionic surfactants, preservatives, perfumes or other active ingredients, e.g., proteins—non-charged nanoparticles are typically less prone to interactions and agglomeration upon the addition to the vehicle.

Nanoparticles with hydrophilic surface properties tend to locate in the water phase, whereas nanoparticles with hydrophobic surface tend to localize in the oil phase. If hydrophilic nanoparticles are incorporated into a hydrophobic vehicle, nanoparticles will try to reduce their total surface area to reduce their thermodynamic energy, which will result in agglomeration of the nanoparticles. The same will occur if hydrophobic nanoparticles are added to a hydrophilic vehicle. Agglomeration of the particles can be circumvented by changing the type of vehicle or the surface hydrophobicity of the particles. The latter can be achieved with the use of stabilizers. For this, non-ionic stabilizers—providing steric stabilization—or ionic stabilizers—providing electrostatic stabilization can be exploited. Also, during storage of the dermal products, nanoparticles can change their properties. Depending on the properties of the nanoparticles, nanoparticles can agglomerate, grow in size, modify their crystalline state, or dissolve (Fig. 10.2). In addition, if the nanoparticles were used as a carrier system for the incorporation of actives, drug expulsion can occur over time (Fig. 10.3). In any case, the modification will modify the performance of the dermal products and thus must be avoided to ensure a safe and high-performing dermal, cosmetic product.

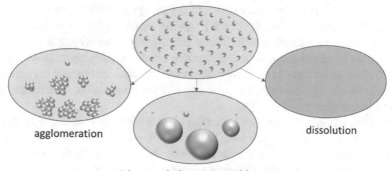

agglomeration

dissolution

particle growth due to Ostwald ripening

Fig. 10.2 Overview of instabilities of nanoparticles in dermal formulations during storage

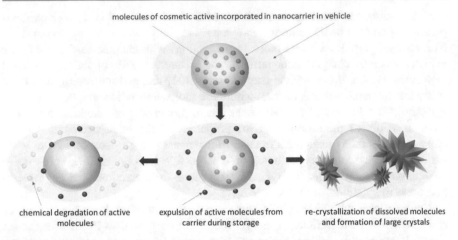

molecules of cosmetic active incorporated in nanocarrier in vehicle

chemical degradation of active molecules

expulsion of active molecules from carrier during storage

re-crystallization of dissolved molecules and formation of large crystals

Fig. 10.3 Overview of instabilities of actives loaded into nanocarriers in dermal formulations during storage

Due to the above-mentioned problems that can occur during production and storage of nanoparticle-containing cosmetic products, there are several different aims to characterize nanoparticles in their respective dermal formulations. The major aim will be to observe the stability of the nanoparticles, i.e., to monitor if size, crystalline state, homogeneous distribution, and incorporation of the actives are maintained during the shelf life of the product. Another major reason to characterize nanoparticles in dermal products might be derived from regulatory bodies, which might be interested to determine if a dermal product contains nanoparticles or not.

Various techniques are available to characterize nanoparticles. However, prior to the selection of a characterization technique, it is of uppermost importance to determine the purpose of the analysis, which might be different in each case. Only if the purpose of characterization is known, a method being suitable for this purpose can be selected. In addition, it is highly important to realize that most techniques require a preparation of the sample prior to the analysis. Sample preparation means that the original sample, i.e., the dermal product, needs to be dispersed in another liquid or needs to be dried, heated, frozen etc. During these preparational steps the original sample can change. Especially, nanoparticles are highly sensitive to heat, cold, dilution or changes in environment due to drying. This means very often the nanoparticles change their properties during sample preparation, which can lead to artifacts in the subsequent analysis.

Examples of artifacts that can occur during sample preparation are agglomeration or dissolution of the particles. Therefore, to avoid misinterpretation of the characterization results, it is best to select methods that allow for the analysis of the original sample without any modification prior to the measurements. However, very often, this is not possible. In these cases, special care must be taken with (i) the preparation of the sample and (ii) the interpretation of the data. Further, to ensure a correct interpretation of the data, suitable controls and additional independent techniques for the stratification of the results should be used.

For the selection of a suitable characterization method, more parameters need to be considered. Preferably, the method should be easy and fast to perform and the characterization instrument should be affordable—not only the acquisition—but also in the maintenance and the consumables. Today, many characterization methods are available for the characterization of nanoparticles and many more are about to come in the future. Unfortunately, the more sophisticated the technique, typically, the more complex is the measurement procedure and the more expensive is the method. This means most of the techniques being available today for the characterization of the nanoparticles are not suitable for daily analysis but only for basic research. The aim of this chapter is to provide a rough overview of the existing methods for the characterization of nanoparticles in dermal formulations and to provide basic information on how to characterize nanoparticles in dermal formulations on a day to day basis in both—product development and production of dermal products.

10.2   Overview of Analytical Methods

The variety of the techniques being available today for the characterization of nanocarrier-based formulations is high (Fig. 10.4) [1, 2]. The section is giving a brief overview of the most relevant techniques used today for the characterization of nanoparticles in dermal products.

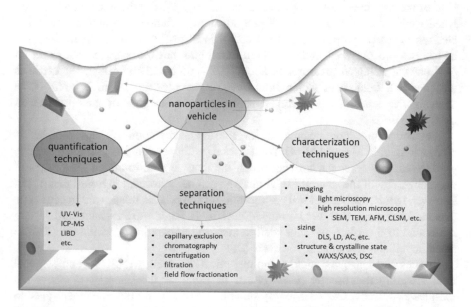

Fig. 10.4 Overview of analytical methods being available for the characterization of nanoparticles in dermal formulations, modified after [1]

10.2.1 Microscopic and Microscopy-Related Techniques

An old saying in the scientific world is "seeing is believing". This means you can only trust things you have seen with your own eyes. As nanoparticles are too small to be seen by naked eye, microscopic imaging and microscopy-related techniques are highly important for the characterization of nanoparticles. Due to the diffraction limit of visible light, light microscopy is not able to visualize particles with sizes < 500 nm. Therefore, to gain detailed information on size, shape and structure of nanoparticles, high-resolution microscopic techniques are required.

The most well-known microscopic techniques for the characterization of nanoparticles involve transmission electron microscopy (TEM) and scanning electron microscopy (SEM) [3]. TEM is a microscopic technique in which electrons are transmitted through an ultrathin sample (thickness < 100 nm). Similar to light microscopy, where the image is formed due to the interaction of the light with the sample as the beam is transmitted through the specimen, in TEM, the image is formed due to the interaction of the electron beam with the specimen. In SEM, the surface of the sample is scanned with a focused beam of electrons, which interact with the atoms of the surface of the sample. The obtained signals are combined with the position of the laser beam to obtain an image of the topography of the surface of the sample. TEM and SEM are high-resolution imaging techniques and can resolve structures with sizes ≤ 1 nm [2]. However, the drawback of these techniques is that the samples need to be analyzed in vacuum. Thus, artifacts due to drying are highly likely to occur [2]. Therefore, in the past, many efforts were done to provide solutions which allow for the characterization of the samples in their original state. One of the first improvements was the use of cryo-TEM, where samples are analyzed via TEM in a frozen state. Newer advances include environmental scanning electron microscopy (ESEM) and the use of WetSEM™ capsules. In ESEM, gun and lenses are operated under vacuum conditions but the sample is kept in an environment at higher pressures of about 0.01–0.06 bar. The WetSEM™ capsules are specimen holders in which liquid or wet samples are filled. Prior to SEM analysis, the capsules are sealed, enabling "in situ imaging of nanoparticles in natural media" [2].

Another microscopic technique for the characterization of nanoparticles is atomic force microscopy (AFM). AFM uses a cantilever equipped with a tip that is used to scan the specimen surface, which can then be used to obtain 3D-images of the topography of the sample. AFM imaging can be performed in liquid media, but however, requires advanced skills if meaningful and precise images are needed. The reasons for this are, that, very often—especially in liquid media—nanoparticles are not bound to a surface and thus tend to stick to the cantilever, which will lead to smearing effects and imaging artifacts [2].

Besides AFM also near-field scanning optical microscopy (NSOM) and confocal laser scanning microscopy (CLSM) can be used for the imaging of nanosized samples. Especially, CLSM is highly suitable, especially, if thick specimens need to be analyzed [4]. New techniques involve fluorescence-lifetime imaging microscopy (FLIM), where an image based on the differences in the exponential decay rate of the fluorescence from a fluorescent sample is created. FLIM and/or the combination of spectroscopic techniques with microscopic techniques, e.g., X-ray spectroscopy, Raman spectroscopy, or Terahertz spectroscopy are novel and emerging techniques that enable detailed insight into the structure of nanoparticles, their interaction with the vehicle, cells, or even skin [5–16]. However, all these emerging techniques require detailed knowledge and expertise. They are not suitable for daily routine and thus are not further discussed in detail in this chapter. Finally, it should be noted that microscopy—independent on which microscopic technique is employed—is typically used to analyze only a few particles. If these data are used for the calculation of other parameters, e.g., size or size distribution, data might not be statistically significant. This can only be overcome by employing automated image analysis, which enables the analysis of several thousand particles and/or by using additional, independent methods to verify the results obtained [2].

10.2.2 Spectroscopy and Spectroscopy-Related Techniques

Various spectroscopic methods are available for the characterization of nanoparticles in dermal formulations. Most frequently, used techniques include mass spectroscopy (MS), UV/Vis spectroscopy, X-ray spectroscopy, small-angle X-ray scattering (SAXS), small angle neutron scattering (SANS), nuclear magnetic resonance spectroscopy (NMR), as well as dynamic and static light scattering techniques (DLS and MALLS). The spectroscopic techniques are used for size characterization (DLS, MALLS, and SAXS) [2, 17, 18], elemental analysis (MS) [2, 19, 20] and for the determination of the crystalline state or structure of the nanoparticles (NMR, SAXS, and SANS) [2, 21–25]. The use of these techniques for the characterization of nanoparticles in dermal formulations is limited, because in most cases, the viscosity of the dermal formulations is too high. This means prior to analysis, nanoparticles must be separated from the vehicle. As mentioned above, any sample preparation must be done with care to avoid changes in the properties of the nanoparticle during sample preparation. Therefore, a careful selection of the separation technique must be included in each analysis protocol.

10.2.3 Separation Techniques

10.2.3.1 Chromatography and Chromatography-Related Techniques

In some cases, it is necessary to separate nanoparticles from their matrix. This can be the case if a precise size characterization of the nanoparticles is required, which will not be possible in the matrix because the matrix disturbs the measurement or if the number (amount) of nanoparticles in the dermal vehicle is below the detection limit of the characterization instrument. Other purposes might be the need for an elemental analysis of the particles, which is important to prove or disprove the presence or absence of specific particles—for example, titanium nanoparticles—in cosmetic formulations. Most relevant separation techniques include size exclusion chromatography (SEC), capillary electrophoresis (CE), hydrodynamic chromatography (HDC) and field-flow-fractionation (FFF) [2]. In recent years, much research has been done to optimize these techniques and to allow for fast and precise separation and characterization of the particles. The latter could be achieved by coupling the separation techniques with appropriate and sensitive analysis tools. Most often employed coupling techniques include UV/Vis or inductively coupled plasma-mass spectrometry (ICP-MS) for elemental analysis. For size analysis, dynamic light scattering or multiple angle light scattering (MALLS) are most often coupled to the separation techniques [2].

10.2.3.2 Filtration and Centrifugation

The spectroscopic techniques are sophisticated. Thus, to save time and costs, in some cases, it might be more appropriate to use more simple techniques. Filtration is a simple approach to separate particles but should be used with caution for the separation of nanoparticles from dermal vehicles. In most cases, the viscosity of the vehicle is too high, thus heating of the formulation is required, which can already lead to changes of the nanoparticles. Additional artifacts can be created due to the formation of filter cakes and the loss of particles that are caught in the filter membrane. They need always to be considered, especially when filtration is used for the quantification of the nanoparticles. Depending on the size of the nanoparticles, different filters and filtration techniques can be employed. Typically, microfiltration with membrane filters with pore sizes > 100 nm can be used for the separation of the nanoparticles. A more exhausting and efficient method is the cross-flow filtration (CFF) method, where samples are recirculated, which reduces the hazard of agglomeration and changes of the particles during the filtration [2].

Another simple technique to separate particles from the vehicle is centrifugation. Depending on the size of the particles, high centrifugal forces are required for the separation of the particles. The required acceleration and the time needed for the separation of the particles depend on the size of the particles, the viscosity of the matrix, and the differences in the densities of the particle and the vehicle and can be calculated by Stoke's law. If small particles need to be separated from high-viscosity matrices, very high acceleration forces and long centrifugation times can be required. In some cases, ultracentrifugation, which can provide acceleration forces to up to 1,000,000 g, might be most appropriate to achieve good separation results [2]. Besides preparative centrifuges, analytical centrifuges (AC) are available. Analytical centrifuges are able to monitor the centrifugation process in real time. This is done either by monitoring the changes in transmission in the cuvette via UV-spectroscopy or via the monitoring of changes in refractive index. In the last years, especially AC has attracted attention in the cosmetic field for daily routine analysis of cosmetic products [26–34].

10.3   Characterization of Nanoparticles in Dermal Formulations in Daily Routine

The use of nanoparticles in cosmetic products is relatively novel and many different nanoparticles can be incorporated in dermal products. For example, carbon black is used in mascara, titanium dioxide and zinc oxide nanoparticles are used in sun-screen products [3] and various nanocarrier systems are used as "taxis" to transport cosmetic actives more efficiently to the site of action. Examples for "nanotaxis" are micelles, liposomes, lipid nanoparticles, polymeric nanoparticles, or nanocrystals (c.f. previous chapters in this book). Each nanoparticle is used for different purposes and in different concentrations. Each carrier system possesses different properties and thus no standard characterization method is available for the characterization of nanoparticles in dermal products. Therefore, it is necessary to establish an individual characterization protocol for each type of product [35].

One type of product is the incorporation of inorganic nanoparticles, e.g., pigments like carbon black or titanium dioxide, into dermal products. The performance of the products will depend on the distribution of the particles within the product, which means that the particles need to be homogeneously distributed within the products. Agglomeration of the particles will lead to a loss of product quality and thus the major aim of the characterization of such products will be to prove homogeneous distribution of the particles within the product.

Another type of product is the use of nanocarriers which are employed to improve the performance of cosmetic actives. Typically, the active is incorporated into the nanocarrier. In this type of products, the quality and performance of the product depend on many parameters. The first prerequisite for good performance is the homogeneous distribution of the nanocarriers within the dermal product, which must be maintained throughout the whole shelf life of the product. The second prerequisite is that the carriers do not undergo significant changes during storage. Possible changes are changes in size or the expulsion of active form the carrier. If changes like this occur, the performance of the product will be impaired. Consequently, to avoid this, the observation of possible changes is highly important for the formulation and production of high-performing "nanoproducts". Reasons for changes in size or drug loading are different for the different carrier systems and thus—as stated above—no standard method is available for monitoring the product quality of these products.

For the establishment of a meaningful characterization protocol, it is, therefore, necessary to consider which parameters will be most affected if changes in the carrier or instability in the product occur. From our experience, we found that in most cases, it will be easier, faster, and more cost-efficient to analyze indirect parameters, i.e., macroscopic changes of the product, instead of changes in the nanocarriers, which typically require sophisticated analytic tools. Parameters that can be easily analyzed are phase separation of the vehicle, changes in color, odor, viscosity, or changes in the product performance, e.g., changes in degradation or dermal penetration of active, etc. Parameters that need relatively simple analytical tools are the microscopic observation of the formulations, observation of chemical stability of the active, or enhanced physical stability testing with centrifuges. These analytical tools are typically available in all formulation units and thus should be used for the characterization of nanocarrier-based formulations, whenever possible.

10.3.1 Light Microscopy

Light microscopy can analyze particles > 500 nm. Hence, the nanoparticles are not visible by this technology [36]. However, possible agglomerates of nanocarriers, large crystals, that might be derived from expulsed active from the nanocarriers that recrystalized in the vehicle due to poor solubility in this environment or interactions of the nanocarriers which lead to changes in the colloidal structure of the vehicle and might cause changes in the appearance of the product (stability, viscosity, etc.), can nicely be observed via light microscopy (Fig. 10.5). In fact, even though nanoparticles as such cannot be seen via light microscopy, instabilities that cause the appearance of larger objects can easily be detected with this technique, thus allowing to discriminate between "good" and "bad" formulations [37]. Light microscopy should, therefore, be the method of choice for the early characterization of nanocosmetics.

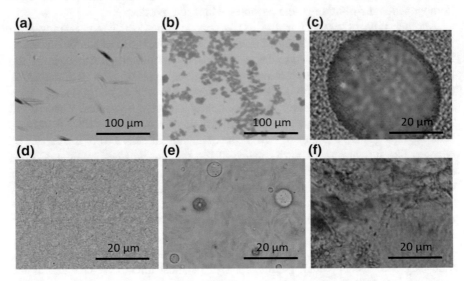

Fig. 10.5 Microscopic images of nanocarriers in dermal formulations: **a** large crystals of recrystalized active, **b** agglomerated nanocrystals in a hydrogel, **c** oily droplet formed upon the addition of perfume oil and adsorbed nanocrystals on the surface of the perfume droplet causing instability of the formulation, **d** homogenously distributed nanocarriers incorporated in a hydrophilic cream, **e** non-homogenously distributed nanocarriers incorporated in a hydrophobic formulations containing no water—nanocarriers (formulated in aqueous suspension) form water droplets in the vehicle, **f** homogenously distributed nanocarriers incorporated in a hydrophobic cream with low water content—nanocarriers are homogeneously distributed in the water phase of the vehicle

10.3.2 Particle Size Analysis

Other methods for characterization—as mentioned above—will strongly depend on the purpose of the analysis. In most cases, size analysis will be the next step in characterization. Size characterization for nanoparticles is most frequently performed by using a combination of microscopic imaging, dynamic light scattering, and laser diffraction. However, a direct analysis of the size of the nanoparticles in the matrix is often not possible. In DLS, where the diffusion constant of the particles is analyzed (c.f. chapter "Characterization of nanoparticles"), the viscosity of the matrix is too high and thus DLS cannot be employed for the characterization of the carriers in their natural environment. Possible methods to measure the nanoparticles in their matrix without prior separation might be the analysis of the particles in the dermal formulation at elevated temperatures. The increase in temperature—due to the melting of the lipidic components of the matrix—decreases the viscosity of the matrix, which might enable the measurement. However, measurements will only be possible if heating will not modify the nanocarriers as such.

Laser diffraction has the advantage of a broad measuring range (typically between 20 nm – 2000 µm), which enables the detection of the small sized nanoparticles besides larger droplets of the vehicle in theory [38–40]. However, LD has the disadvantage, that small particles—especially if the number is small—can be overlooked if larger particles are present within the sample. Thus, LD can be used—however, the feasibility of the method—that means to the possibility to reliably detect the small-sized particles within the matrix and changes over time—need to be evaluated for each sample. Special focus should be put on the sample preparation and the measurement setup, because both, dispersions in aqueous medium and stirring speed, can influence the results obtained [41]. New setups of LD instruments, e.g., small volume dispersion units allow for the measurement of the original samples and might, therefore, be the best method for the characterization of nanoparticles in dermal formulations [42–45]. Also, the analysis method is highly important. Hence, Fraunhofer approximation can only be used for the analysis of microparticles and thus, it is of uppermost importance to use Mie-theory and correct optimal parameters, when nanoparticles are analysed [19, 20]. If complex systems are used which contain nano- and microparticles being composed of different materials, refractive indices of the nanoparticles should be used to obtain more reliable results [46].

10.3.3 Differential Scanning Calorimetry

The detection of nanocarriers in dermal formulations can also be performed by using the differences in the characteristics of the materials. For example, most particles possess different melting peaks than the other excipients that are used in the formulation. Thus, the determination of melting peaks is often used for the detection of nanocarriers. Also, changes over time, which are often correlated to changes in the melting behavior of the particles can be determined by this. The most appropriate technique for this is differential scanning calorimetry (DSC). DSC belongs to the field of thermal analysis methods. It measures emitted or absorbed heat quantities of a sample compared to a reference, giving a curve of heat flux versus temperature or versus time, which allows for the detection of endothermic or exothermic processes, e.g., melting or recrystallization of compounds. DSC is used to observe the crystalline state and the degree of crystallinity of many different nanocarriers, e.g., nanocrystals, liposomes, lipid dispersions, polymers [47–52] in both, original medium and in complex, semi-solid systems, i.e., dermal vehicles [53–60]. Moreover, if lipid-based nanocarriers are used, DSC can also provide information about possible changes in lipid modifications, which are a sensitive marker for the expulsion of actives during the storage and typically cause both—physical instability of the nanocarriers and chemical instability of the incorporated active [61–63]. In fact, DSC—in addition to imaging and size analysis—is a valuable method for the characterization of nanocarrier-based cosmetic products.

10.3.4   X-Ray Diffraction

If more information is needed for the characterization of the formulations, X-ray diffraction should be considered as next possible analysis tool. X-ray is used for structure analysis and the determination of the crystalline state of materials [64–66]. In the development of dermal formulations containing nanoparticles, X-ray is highly attractive, because the analysis can be performed in the original state and without the need for a special sample preparation. Similar to DCS, X-ray analysis— especially wide-angle x-ray scattering (WAXS)—is used to confirm the presence of materials or the maintenance of the crystalline state of the nanocarriers during storage and thus allows for the prediction of the storage stability and for the detection of (unknown) materials, i.e., titanium dioxide in dermal formulations [67–72]. Besides WAXS, also small-angle x-ray scattering (SAXS) can used for the characterization of nanocarrier based formulations [73–76]. The principle of WAXS and SAXS analysis are similar, the only difference between SAXS and WAXS is the distance of the detectors from the sample. In SAXS, the distance is longer and thus the diffraction maxima are observed at smaller angles. In contrast to WAXS, which is mainly used to determine the crystalline state of materials, SAXS is used to determine size, shape or even pore sizes, and structures of the nanoparticles [77–80]. It is a sophisticated method that requires skills and experience but enables to gain various and detailed information about the formulations [72]. Consumer-friendly, combined instruments for SAXS and WAXS measurements are meanwhile available on the market and enable the use of this technique in a broader sense in the future. With an increasing number of cosmetic products containing nanocarriers and with an increasing number of novel and more sophisticated nanocarriers, SAXS/WAXS measurements are believed to become more important and more frequently applied in the next years.

10.3.5   Analytical Centrifugation

Another emerging technique for the characterization of nanocosmetic products is the use of analytical centrifuges, which can be used for various purposes for the characterization of disperse systems [26–33]. For the characterization of nanoparticles, in dermal formulations, it can mainly be used for the determination of the physical stability (homogeneous distribution of the particles within the vehicle without agglomerates) and the determination of size and size distribution of the nanoparticles. However, it can also be used to determine suitable compositions of vehicles (Hansen parameter) which will provide the environment for highly physically stable dispersions, without the formation of agglomerates and finally—if needed—it can also be used for the analysis of physico-chemical properties, e.g., density or shape of the particles [34, 81–83].

10.3.6   Characterization of Nanoparticles in Dermal Formulations for Safety Evaluation

Nanoparticles—due to their small size—possess different properties when compared to larger-sized material. Besides positive properties, possible toxic effects need to be considered. Possible toxic "nano-effects" can be due to improved uptake of the small-sized nanoparticles [84]. This means nanoparticles could reach target sides which are not reached with larger-sized particles. However, nanotoxic effects can only be expected, if the nanoparticles are released from the vehicle and are able to enter the body. If this is not the case, nanotoxic effects are very unlikely to occur [85]. Therefore, to evaluate the nanosafety of nanocosmetics, test methods, that investigate the release of the particles from their vehicles, should be included in such safety investigations. Only if particles enter the body and only if they are released from their vehicles as nanosized particles, they can possess nanotoxic effects.

Today—except from pigments that are used for sunscreen products and decorative cosmetics—only biodegradable nanoacarriers, e.g., liposomes, smartLipids, smartCrystals, or micelles are used for improved delivery of cosmetic actives. If these carriers are released from the vehicle and will be able to enter the body will depend on the vehicle and the type of product (i.e. spray, lipstick or cream). In most cases dermal, semi-solid formulations are applied to skin, from which the nanoparticles are not released. Hence, cosmetics containing innovative nanocarriers for improved delivery of cosmeceutical actives are believed to be safe and to possess no nanotoxicological risk.

10.4   Conclusions

Characterization of nanoparticles in dermal formulations is challenging and should only be done if particles cannot be analyzed in its original—unformulated state. Due to the variety of nanoparticles, vehicles, and types of products, no standard protocol for the characterization of nanoparticles is available and for each formulation—depending on what sort of information is needed—an individual protocol needs to be established. The protocol should contain not only the characterization method and the measurement protocol but should also contain detailed information about the sample preparation. Whenever possible, the sample should be analyzed in its original state, because separation or other preparation methods can lead to dramatic changes and thus to artifacts in the analysis results. Various characterization methods are available for the characterization of nanoparticles and many more are expected to come in the near future. However, none of these methods can be used as a stand-alone method for the characterization of nanoparticles in dermal formulations.

Basic information of the formulation should be obtained with imaging methods, where light microscopy is strongly recommended as a simple and fast screening method for the detection of possible larger particles, i.e., crystals or agglomerates. High-resolution imaging techniques should be employed if detailed information on structure and shape of the nanoparticles and the vehicle are needed. Additional information should be gained by particle size determination. The most appropriate technique will depend on the sample and should be selected individually. If more details are needed, X-ray diffraction and/or differential scanning calorimetry should be employed to gain more insights into the crystalline structure and composition of the nanocarriers. In most cases, these analytical tools are fully sufficient for the development and characterization of nanocarrier-based cosmetic products. In other—more sophisticated—techniques are required, cooperations, e.g., with universities instead of self-acquisition—is strongly recommended. This will save time and costs and will ensure meaningful results at the same time.

Acknowledgements The authors would like to thank Abraham Abraham, Noor Almohsen, Pascal Stahr, and Florian Stumpf for their contribution and help with the preparation of this chapter.

References

1. López-Serrano A, Olivas RM, Landaluze JS, Cámara C. Nanoparticles: a global vision. Characterization, separation, and quantification methods. Potential environmental and health impact. Anal Methods. 2014;6:38–56.
2. Tiede K, Boxall ABA, Tear SP, Lewis J, et al. Detection and characterization of engineered nanoparticles in food and the environment. Food Addit Contam Part A: Chem Anal Control Expo Risk Assess. 2008;25:795–821.
3. Lu P-J, Huang S-C, Chen Y-P, Chiueh L-C, et al. Analysis of titanium dioxide and zinc oxide nanoparticles in cosmetics. J Food Drug Anal. 2015;23:587–94.
4. Möckl L, Lamb DC, Bräuchle C. Super-resolved fluorescence microscopy: nobel prize in chemistry 2014 for Eric Betzig, Stefan Hell, and William E. Moerner. Angew Chem. 2014;53:13972–7.
5. Alexiev U, Volz P, Boreham A, Brodwolf R. Time-resolved fluorescence microscopy (FLIM) as an analytical tool in skin nanomedicine. Eur J Pharm Biopharm. 2017;116:111–24.

6. Alnasif N, Zoschke C, Fleige E, Brodwolf R, et al. Penetration of normal, damaged and diseased skin—an in vitro study on dendritic core-multishell nanotransporters. J Controlled Release. 2014;185:45–50.
7. Balke J, Volz P, Neumann F, Brodwolf R, et al. Visualizing oxidative cellular stress induced by nanoparticles in the subcytotoxic range using fluorescence lifetime imaging. Small. 2018;14:e1800310.
8. Boreham A, Kim T-Y, Spahn V, Stein C, et al. Exploiting fluorescence lifetime plasticity in FLIM: target molecule localization in cells and tissues. ACS Med Chem Lett. 2011;2: 724–8.
9. Boreham A, Brodwolf R, Walker K, Haag R, et al. Time-resolved fluorescence spectroscopy and fluorescence lifetime imaging microscopy for characterization of dendritic polymer nanoparticles and applications in nanomedicine. Molecules. 2016;22.
10. Edlich A, Volz P, Brodwolf R, Unbehauen M, et al. Crosstalk between core-multishell nanocarriers for cutaneous drug delivery and antigen-presenting cells of the skin. Biomaterials. 2018;162:60–70.
11. Ostrowski A, Nordmeyer D, Boreham A, Holzhausen C, et al. Overview about the localization of nanoparticles in tissue and cellular context by different imaging techniques. Beilstein J Nanotechnol. 2015;6:263–80.
12. Radbruch M, Pischon H, Ostrowski A, Volz P, et al. Dendritic core-multishell nanocarriers in murine models of healthy and atopic skin. Nanoscale Res Lett. 2017;12:64.
13. Volz P, Schilrreff P, Brodwolf R, Wolff C, et al. Pitfalls in using fluorescence tagging of nanomaterials: tecto-dendrimers in skin tissue as investigated by Cluster-FLIM. Ann NY Acad Sci. 2017;1405:202–14.
14. Witting M, Boreham A, Brodwolf R, Vávrová K, et al. Interactions of hyaluronic acid with the skin and implications for the dermal delivery of biomacromolecules. Mol Pharm. 2015;12:1391–401.
15. Zhu Y, Choe C-S, Ahlberg S, Meinke MC, et al. Penetration of silver nanoparticles into porcine skin ex vivo using fluorescence lifetime imaging microscopy, Raman microscopy, and surface-enhanced Raman scattering microscopy. J Biomed Optics. 2015;20:51006.
16. Ajito K, Ueno Y, Kim J-Y, Sumikama T. Capturing the freeze-drying dynamics of NaCl nanoparticles using THz spectroscopy. J Am Chem Soc. 2018.
17. Chen ZH, Kim C, Zeng X-B, Hwang SH, et al. Characterizing size and porosity of hollow nanoparticles: SAXS, SANS, TEM, DLS, and adsorption isotherms compared. Langmuir. 2012;28:15350–61.
18. Follens LRA, Aerts A, Haouas M, Caremans TP, et al. Characterization of nanoparticles in diluted clear solutions for Silicalite-1 zeolite synthesis using liquid 29Si NMR, SAXS and DLS. Phys Chem Chem Phys. 2008;10:5574–83.
19. Keck CM. Particle size analysis of nanocrystals: improved analysis method. Int J Pharm. 2010;390:3–12.
20. Keck CM, Müller RH. Size analysis of submicron particles by laser diffractometry–90% of the published measurements are false. Int J Pharm. 2008;355:150–63.
21. Lee Y-T, Li DS, Ilavsky J, Kuzmenko I, et al. Ultrasound-based formation of nano-pickering emulsions investigated via in-situ SAXS. J Colloid Interface Sci. 2019;536:281–90.
22. Sunaina Sethi V, Mehta SK, Ganguli AK, et al. Understanding the role of co-surfactants in microemulsions on the growth of copper oxalate using SAXS. Phys Chem Chem Phys. 2018;21:336–48.
23. Schmitt T, Gupta R, Lange S, Sonnenberger S, et al. Impact of the ceramide subspecies on the nanostructure of stratum corneum lipids using neutron scattering and molecular dynamics simulations. Part I: impact of CERNS. Chem Phys Lipids. 2018;214:58–68.
24. Schroeter A, Stahlberg S, Školová B, Sonnenberger S, et al. Phase separation in ceramide NP containing lipid model membranes: neutron diffraction and solid-state NMR. Soft Matter. 2017;13:2107–19.

25. Sonnenberger S, Eichner A, Schmitt T, Hauß T, et al. Synthesis of specific deuterated derivatives of the long chained stratum corneum lipids EOS and EOP and characterization using neutron scattering. J Labelled Compd Radiopharm. 2017;60:316–30.

26. Caddeo C, Manconi M, Fadda AM, Lai F, et al. Nanocarriers for antioxidant resveratrol: formulation approach, vesicle self-assembly and stability evaluation. Colloids Surf B: Biointerfaces. 2013;111:327–32.

27. Fernandes AR, Ferreira NR, Fangueiro JF, Santos AC, et al. Ibuprofen nanocrystals developed by 22 factorial design experiment: a new approach for poorly water-soluble drugs. Saudi Pharm J. 2017;25:1117–24.

28. Gross-Rother J, Herrmann N, Blech M, Pinnapireddy SR, et al. The application of STEP-technology® for particle and protein dispersion detection studies in biopharmaceutical research. Int J Pharm. 2018;543:257–68.

29. Mäkinen OE, Uniacke-Lowe T, O'Mahony JA, Arendt EK. Physicochemical and acid gelation properties of commercial UHT-treated plant-based milk substitutes and lactose free bovine milk. Food Chem. 2015;168:630–8.

30. Pereira I, Zielińska A, Ferreira NR, Silva AM, et al. Optimization of linalool-loaded solid lipid nanoparticles using experimental factorial design and long-term stability studies with a new centrifugal sedimentation method. Int J Pharm. 2018;549:261–70.

31. Thompson KL, Derry MJ, Hatton FL, Armes SP. Long-term stability of n-alkane-in-water pickering nanoemulsions: effect of aqueous solubility of droplet phase on Ostwald ripening. Langmuir. 2018;34:9289–97.

32. Xu D, Qi Y, Wang X, Li X, et al. The influence of flaxseed gum on the microrheological properties and physicochemical stability of whey protein stabilized β-carotene emulsions. Food Funct. 2017;8:415–23.

33. Zielińska A, Martins-Gomes C, Ferreira NR, Silva AM, et al. Anti-inflammatory and anti-cancer activity of citral: optimization of citral-loaded solid lipid nanoparticles (SLN) using experimental factorial design and LUMiSizer®. Int J Pharm. 2018;553:428–40.

34. Miller R, Lerche D, Schäffler M. (editors). Dispersionseigenschaften, 2D-Rheologie, 3D-Rheologie, Stabilität. Eigenverlag. 2014.

35. Peters R, ten Dam G, Bouwmeester H, Helsper H, et al. Identification and characterization of organic nanoparticles in food. Trends Anal Chem. 2011;30:100–12.

36. Goodhew PJ, Humphreys J, Beanland R. Electron microscopy and analysis. 3rd ed. London/UK: Taylor & Francis; 2001.

37. Aulton ME, Taylor K. Aulton's pharmaceutics: the design and manufacture of medicines. 5th ed. New York, USA: Elsevier; 2018.

38. ISO 13320, Particle size analysis—laser diffraction methods. 2009.

39. Müller RH, Schuhmann R. Teilchengrößenmessung der Laborpraxis. Stuttgart/Germany: Wissenschaftliche Verlagsgesellschaft mbH Stuttgart; 1998.

40. Varenne F, Makky A, Gaucher-Delmas M, Violleau F, et al. Multimodal dispersion of nanoparticles: a comprehensive evaluation of size distribution with 9 size measurement methods. Pharm Res. 2016;33:1220–34.

41. Acar Kübart S, Keck CM. Laser diffractometry of nanoparticles: frequent pitfalls & overlooked opportunities. J Pharm Technol Drug Res. 2013;2:2–17.

42. Malvern Instruments Ltd. Mastersizer 2000—User Manual. Malvern Instruments Ltd., Worcestershire/UK. 2007.

43. Beckman Coulter Inc. LS 13 320 Series—Particle Size Analyzer. Beckman Coulter Inc., Fullerton/USA. 2004.

44. Pathak P, Nagarsenker M. Formulation and evaluation of lidocaine lipid nanosystems for dermal delivery. AAPS PharmSciTech. 2009;10:985–92.

45. Olejnik A, Goscianska J, Zielinska A, Nowak I. Stability determination of the formulations containing hyaluronic acid. Int J Cosmet Sci. 2015;37:401–7.

46. Li X. Nanocrystals for topical delivery: nanocrystals, nanoemulsions & smartLipids. Ph.D. Thesis, Freie Universität Berlin. 2016.

47. Unruh T, Westesen K, Bösecke P, Lindner P, et al. Self-assembly of triglyceride nanocrystals in suspension. Langmuir. 2002;18:1796–800.
48. Souto EB, Müller RH. Investigation of the factors influencing the incorporation of clotrimazole in SLN and NLC prepared by hot high-pressure homogenization. J Microencapsul. 2006;23:377–88.
49. Souto E. SLN and NLC for topical delivery of antifungals. Ph.D-Thesis, Berlin. 2005.
50. Pardeike J. Nanosuspensions of phospholipase A2 inhibitors and coenzyme Q10-loaded nanostructured lipid carriers for dermal application. Ph.D-Thesis, Berlin. 2008.
51. Höhne GWH, Hemminger W, Flammersheim H-J, Theoretical fundamentals of differential scanning calorimeters. In: Höhne GWH, Hemminger W, Flammersheim H-J, editors. Differential scanning calorimetry: an introduction for practitioners. Heidelberg: Springer. p. 21–40.
52. Jenning V, Thünemann AF, Gohla SH. Characterisation of a novel solid lipid nanoparticle carrier system based on binary mixtures of liquid and solid lipids. Int J Pharm. 2000;199: 167–77.
53. Pireddu R, Sinico C, Ennas G, Marongiu F, et al. Novel nanosized formulations of two diclofenac acid polymorphs to improve topical bioavailability. Eur J Pharm Sci. 2015;77: 208–15.
54. Sahoo NG, Kakran M, Shaal LA, Li L, et al. Preparation and characterization of quercetin nanocrystals. J Pharm Sci. 2011;100:2379–90.
55. Lv Q, Yu A, Xi Y, Li H, et al. Development and evaluation of penciclovir-loaded solid lipid nanoparticles for topical delivery. Int J Pharm. 2009;372:191–8.
56. Castelli F, Puglia C, Sarpietro MG, Rizza L, et al. Characterization of indomethacin-loaded lipid nanoparticles by differential scanning calorimetry. Int J Pharm. 2005;304:231–8.
57. Müller RH, Dingler A. The next generation after liposomes: solid lipid nanoparticles (SLN Lipopearls) as dermal carrier in cosmetics. Eurocosmetics. 1998;7(8):19–26.
58. Pardeike J, Schwabe K, Müller RH. Influence of nanostructured lipid carriers (NLC) on the physical properties of the Cutanova Nanorepair Q10 cream and the in vivo skin hydration effect. Int J Pharm. 2010;396:166–73.
59. de Vringer T, de Ronde HAG. Preparation and structure of a water-in-oil cream containing lipid nanoparticles. J Pharm Sci. 1995;84:466–72.
60. Dingler A, Blum RP, Niehus H, Müller RH, et al. Solid lipid nanoparticles (SLN/Lipopearls)-a pharmaceutical and cosmetic carrier for the application of vitamin E in dermal products. J Microencapsul. 1999;16:751–67.
61. Pérez-Monterroza E, Ciro-Velásquez HJ. Study of the crystallization and polymorphic structures formed in oleogels from avocado oil. Revista Facultad Nacional de Agronomía. 2016;69:7945–54.
62. Müller RH, Shegokar R, Keck CM. 20 years of lipid nanoparticles (SLN and NLC): present state of development and industrial applications. Curr Drug Discovery Technol. 2011; 8:207–27.
63. Biltonen RL, Lichtenberg D. The use of differential scanning calorimetry as a tool to characterize liposome preparations. Chem Phys Lipids. 1993;64:129–42.
64. Smyth MS, Martin JHJ. X-ray crystallography. Mol Pathol. 2000;53:8–14.
65. Spieß L, Teichert G, Schwarzer R, Behnken H, Genzel C. Moderne Röntgenbeugung: Röntgendiffraktometrie für Materialwissenschaftler, Physiker und Chemiker. 2nd ed. Wiesbaden/Germany: Vieweg+Teubner Verlag/GWV Fachverlage GmbH Wiesbaden; 2009.
66. Waseda Y, Matsubara E, Shinoda K. X-ray diffraction crystallography. Heidelberg, Germany: Springer; 2011.
67. Wade M, Tucker I, Cunningham P, Skinner R, et al. Investigating the origins of nanostructural variations in differential ethnic hair types using X-ray scattering techniques. Int J Cosmet Sci. 2013;35:430–41.
68. Radtke M. Grundlegende Untersuchungen zur Arzneistoffinkorporation, -freisetzung und Struktur von SLN und NLC. Ph.D. Thesis, Berlin/Germany. 2003.

69. Romero GB, Chen R, Keck CM, Müller RH. Industrial concentrates of dermal hesperidin smartCrystals®—production, characterization & long-term stability. Int J Pharm. 2015;482:54–60.

70. Souto EB, Müller RH. Investigation of the factors influencing the incorporation of clotrimazole in SLN and NLC prepared by hot high-pressure homogenization. J Microencapsul. 2006;23:377–88.

71. Oka T, Miyahara R, Teshigawara T, Watanabe K. Development of novel cosmetic base using sterol surfactant. I. Preparation of novel emulsified particles with sterol surfactant+. J Oleo Sci. 2008;57:567–75.

72. Lu PJ, Fang SW, Cheng WL, Huang SC, et al. Characterization of titanium dioxide and zinc oxide nanoparticles in sunscreen powder by comparing different measurement methods. J Food Drug Anal. 2018;26:1192–200.

73. Marsh JM, Brown MA, Felts TJ, Hutton HD, et al. Gel network shampoo formulation and hair health benefits. Int J Cosmet Sci. 2017;39:543–9.

74. Choudhury P, Kumar S, Singh A, Kumar A, et al. Hydroxyethyl methacrylate grafted carboxy methyl tamarind (CMT-g-HEMA) polysaccharide based matrix as a suitable scaffold for skin tissue engineering. Carbohydr Polym. 2018;189:87–98.

75. Hoppel M, Caneri M, Glatter O, Valenta C. Self-assembled nanostructured aqueous dispersions as dermal delivery systems. Int J Pharm. 2015;495:459–62.

76. Guillot S, Tomsic M, Sagalowicz L, Leser ME, et al. Internally self-assembled particles entrapped in thermoreversible hydrogels. J Colloid Interface Sci. 2009;330:175–9.

77. Barbosa RM, Casadei BR, Duarte EL, Severino P, et al. Electron paramagnetic resonance and small-angle X-ray scattering characterization of solid lipid nanoparticles and nanostructured lipid carriers for dibucaine encapsulation. Langmuir. 2018;34:13296–304.

78. Boge L, Hallstensson K, Ringstad L, Johansson J, et al. Cubosomes for topical delivery of the antimicrobial peptide LL-37. Eur J Pharm Biopharm. 2019;134:60–7.

79. Hunter SJ, Thompson KL, Lovett JR, Hatton FL, et al. Synthesis, characterization, and pickering emulsifier performance of anisotropic cross-linked block copolymer worms: effect of aspect ratio on emulsion stability in the presence of surfactant. Langmuir. 2018.

80. Li T, Senesi AJ, Lee B. Small angle X-ray scattering for nanoparticle research. In Li T, Senesi AJ, Lee B, editors. Chemical. Chemical Reviews 2016. p. 11128–80.

81. Detloff T, Lerche D, Sobisch T. Particle size distribution by space or time dependent extinction profiles obtained by analytical centrifugation. In: Particle & particle systems characterization, p. 184–7.

82. Detloff T, Sobisch T, Lerche D. Particle size distribution by space or time dependent extinction profiles obtained by analytical centrifugation. Part Part Syst Charact. 2006;23:184–7.

83. ISO 13318-2. Determination of particle size distribution by centrifugal liquid sedimentation methods. Part 2: Photocentrifuge method. 2007.

84. European Commission—Scientific Committee on Consumer Products (SCCP), SCCP/1147/07—Opinion on safety of nanomaterials in cosmetic products. 2007.

85. Keck CM, Müller RH. Nanotoxicological classification system (NCS)—a guide for the risk-benefit assessment of nanoparticulate drug delivery systems. Eur J Pharm Biopharm. 2013;84:445–8.

J. Lademann, M. E. Darvin, M. C. Meinke and A. Patzelt

Abstract

The effect of cosmetic formulations containing nanostructures should stabilize the actives, enhance their penetration efficiency or provide a reservoir for the actives. Exceptions are nanostructures in sunscreens and make-ups with glitter effects. They should not penetrate in the viable epidermis. Methods to investigate these promises should provide a depth resolution and, if possible, should be non-destructive. In this chapter, different methods are presented to monitor stratum corneum, hair follicle and transdermal penetration of the actives and/or nanostructures.

Keywords

Sunscreens · Stratum corneum · Hair follicle · Drug delivery · Differential stripping · Microdialysis · Confocal Raman microscopy

J. Lademann (✉) · M. E. Darvin · M. C. Meinke · A. Patzelt
Department of Dermatology, Venerology and Allergology, Center of Experimental and Applied Cutaneous Physiology, Charité-Universitätsmedizin Berlin, Charitéplatz 1, 10117 Berlin, Germany
e-mail: juergen.lademann@charite.de

Freie Universität Berlin, Berlin, Germany

Humboldt-Universität zu Berlin, Berlin, Germany

Berlin Institute of Health, Berlin, Germany

© Springer Nature Switzerland AG 2019
J. Cornier et al. (eds.), *Nanocosmetics*,
https://doi.org/10.1007/978-3-030-16573-4_11

11.1 Introduction

In cosmetic science, products are often developed to improve the skin properties or to protect the skin from environmental pollutants or light. Some of these formulations do contain nanoparticles (NPs) [1, 2] which possess different action mechanisms. In sunscreen products, e.g., NPs such as TiO_2 and ZnO are frequently added as physical filters to increase the sun protection factor (SPF) [3, 4].

Moreover, there is a great interest to develop NPs which are able to increase the drug delivery of active components into and even through the skin barrier [5, 6]. However, NPs are able to overcome the blood-brain barrier and are effectively used for drug delivery in other medical disciplines, e.g. for tumour cell targeting, the application of NP for the transdermal transport still remains a challenge as the NPs are not able to overcome the very potent skin barrier.

Since more than 20 years, there is great scientific effort to develop NP for the application in transdermal drug delivery and despite the huge investments spent for that research, a commercial pharmaceutical product for drug delivery based on NP is still missing.

However, very recent research could demonstrate that there is still a future for NPs for transdermal drug delivery if they are customized especially for the follicular penetration pathway. Current studies impressively demonstrate that NPs can be effectively penetrated into the hair follicles where they are stored over several days. The future prospects for NP in this context are described in this chapter.

11.2 Methods of Analysing the Penetration of Particles into the Skin

Many different types of NPs (1–100 nm) such as polymeric or dendritic NP, solid lipid NP, silver NP, nanocrystals or nanosized (50–5000 nm) components such as nanoemulsions, nanocapsules, nanosomes, niosomes, liposomes are commercially available. Frequently, the NPs are only used as carriers for active substances and may degrade during the drug delivery process. There are different methods available to either track the penetration of the NP or the active. For most methods, a fluorescent labelling of the NP or the active facilitates their detection but there are also methods available, which can determine the penetration of label-free substances.

A further challenge represents the fact that NP preferentially penetrates into the hair follicles meaning that the methods used for the analytic of the penetration need spatial resolution so that a distinction between intercellular and follicular penetration becomes available.

In the following, several methods are presented which are able to track the NP during their penetration process into the skin:

11.2.1   Tape Stripping

Tape stripping is a well-established non-invasive method to investigate the penetration of substances into the stratum corneum and has been described in detail elsewhere [7, 8]. After topical application and penetration of a substance, the stratum corneum is removed together with the applied substance by adhesive tapes layer by layer. Each tape strip can then by analysed with regard to its content of corneocytes and topically applied substance. As analytical methods to determine the amount of corneocytes on each tape strip, spectroscopic determination of the pseudo-absorption [9], infrared spectroscopy [10] and weighing [11] are conceivable. The amount of active agent on each tape strip can either be determined by UV/VIS spectroscopy, HPLC or others depending on the active (Fig. 11.1).

Based on these data, a horny-layer profile can be calculated according to Weigmann et al. [12].

Although the tape stripping procedure does not allow the distinction between the intercellular and the follicular penetration pathways, it can be demonstrated that most of the NPs are located on the skin surface or the upper skin layers at the best,

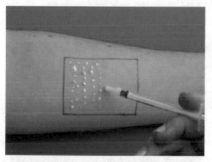
Application of the emulsion

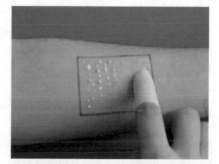
Homogeneous distribution

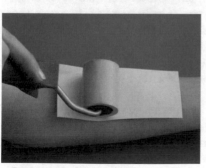
Pressing of the tape by a roller

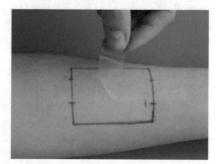
Removing of the adhesive film

Fig. 11.1   Method of tape stripping

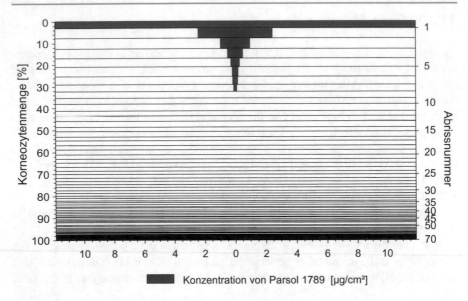

Fig. 11.2 Penetration profile of the UV filter PARSOL

demonstrating that an effective intercellular penetration of NP into the skin is not feasible. A typical penetration profile of the sunscreen's UV filter PARSOL is presented in Fig. 11.2.

11.2.2 Differential Stripping

Due to the need to distinguish between the follicular and intercellular penetration pathway, different methods have been developed to investigate the follicular penetration pathway selectively. Differential stripping allows the determination of the amount of a substance or a NP that has penetrated into the hair follicle, and however is not able to analyse the transfollicular penetration process of actives which are overcoming the follicular barrier and are uptaken by the viable tissue. In principle, differential stripping combines the above-mentioned tape stripping method with a cyanoacrylate skin surface biopsy and has been described in detail elsewhere [13] (Fig. 11.3).

After topical application and penetration of a substance/NP, the stratum corneum is removed completely by tape stripping as described above. Subsequently, a cyanoacrylate glue is applied onto the skin together with an additional tape strip. The cyanoacrylate glue also enters the hair follicles. Once the glue has polymerized, it will be removed together with the tape strip from the skin and contains the content of the hair follicles. By selectively analysing the tape strips and

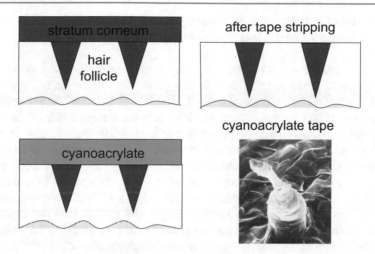

Fig. 11.3 Schematic of the differential stripping method

the cyanoacrylate skin surface biopsies by conventional analytical methods, it can be distinguished which part of substance has penetrated into the stratum corneum and which part into the hair follicles [14].

By applying this technique, it could be demonstrated in previous studies that also the hair follicles represent an important penetration pathway and moreover, a long-term reservoir for topically applied NPs which were partially stored inside the hair follicles for a period of 10 days, whereas the stratum corneum reservoir was already depleted after 24 h.

11.2.3 Microdialysis

To distinguish whether the formulation penetrated into the reservoir of the stratum corneum or the follicular reservoir, which have similar volumes [15], the method of microdialysis was developed and applied [16]. A flexible tube consisting of a porous membrane is led through the stratum corneum into the living tissue. This small tube is floated with a solution, which absorbs the substances that passed the membrane. The concentration of the topically applied substances that were absorbed by the solution is determined again by conventional chemical methods.

Although microdialysis is an invasive procedure, it is the only possibility to efficiently detect penetrated formulations in the living tissue underneath the skin barrier.

11.2.4 Confocal Raman Microscopy and Surface-Enhanced Raman Scattering Microscopy

Much more convenient than tape stripping is the application of a non-invasive in vivo Raman microscopy to analyse the penetration of topically applied substances. A prerequisite for its application is that Raman bands of the particles in the formulation can be detected that are different from the Raman bands of the skin. In case of Raman bands overlapping, non-restricted multiple least square fit and the Gaussian function-based deconvolution methods could be applied for determination of penetration profiles into the stratum corneum [17]. In this case, these bands are detected after application and penetration of the topically applied formulations in different depths of the skin. For this purpose, the focus of the Raman microscope is gradually moved from the skin surface deeper and deeper into the tissue. Figure 11.4a shows the typical Raman spectra of untreated and treated with silver nanoparticles (about 70 nm) coated with poly(vinylpyrrolidone) (PVP) porcine s-tratum corneum (depth 4 μm). Figure 11.4b shows a calculated penetration profile of the PVP-coated silver nanoparticles into the porcine ear skin ex vivo. Analysis were performed using non-restricted multiple least square fit method available in the SkinTools RiverD software. As can be seen from Fig. 11.4b, maximum penetration depth is determined to be approx. 11 μm, i.e. particles do not permeate stratum corneum barrier [18], whose thickness is approx. 19 μm [19].

It is known that silver nanoparticles are able to generate surface-enhanced Raman scattering (SERS) signal, whose intensity is much higher than the Raman signal intensity [20]. Using the SERS microscopy, it was determined by tracking the SERS signal in the skin that silver nanoparticles are able to penetrate until 15,6 μm into the porcine ear skin ex vivo. Thus, even using such a highly sensitive technique, results show that silver nanoparticles do not penetrate through the s-tratum corneum barrier (depth ≈ 19 μm).

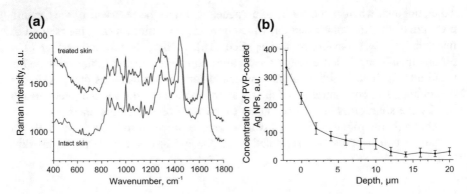

Fig. 11.4 Raman spectra of the untreated and PVP-coated silver nanoparticle-treated porcine ear skin at a depth of 4 μm (**a**) and the penetration profile of PVP-coated silver nanoparticles into the porcine ear skin (**b**) obtained using confocal Raman microscopy. The spectra are taken from [18] with permission

First disadvantage of confocal Raman (SERS) microscopy is that the analysing depths into the skin are limited to approx. 50 μm due to the optical properties of the skin. This is deeper than the skin depth that can be analysed by tape stripping, which is restricted to the thickness of the stratum corneum.

Another disadvantage is that confocal Raman and SERS microscopy do not provide images of the skin but only one depth scan (Z-scan) profile.

The main advantage of confocal Raman (SERS) microscopy is that the kinetics of the penetration of topically applied substances in the stratum corneum can be detected easily at μm resolution. The SERS technique has an additional advantage in possibility to analyse the Raman-inactive metal nanoparticles, such as gold and silver nanoparticles. The non-invasiveness of this method permits the tissue to be analysed at the same position several times.

It was recently shown that confocal Raman microscopy is able to determine in vivo the skin barrier function-related parameters, such as lateral packing order of intercellular lipids [21], hydrogen bounding state of water molecules [22] and keratin folding/unfolding states [23], depth-dependently in the entire stratum corneum at μm resolution. This is promising for future in vivo measurements investigations of the influence of cosmetic formulations (including nanocosmetics) on the stratum corneum-based skin barrier function-related parameters, which is already realized for cosmetic oils [24].

Figure 11.5 shows a confocal Raman microscope applicable for in vivo/ex vivo measurements on the skin (RiverD International B.V., Model 3510, Rotterdam, The Netherlands).

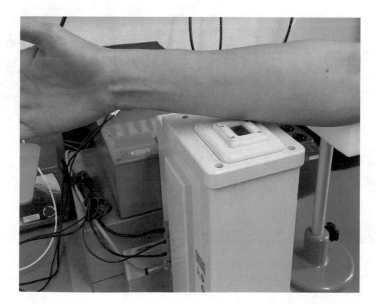

Fig. 11.5 Raman microscope (RiverD International B.V., Model 3510, Rotterdam, The Netherlands) applicable for in vivo/ex vivo measurement of the skin

11.2.5 Two-Photon Tomography with Fluorescence Lifetime Imaging

Another non-invasive technique appropriated for in vivo/ex vivo analysis is two-photon tomography.

It is widely used as an UV filter in sunscreens, ZnO nanoparticles, characterized by the strong second harmonic generation (SHG) signal. Therefore, the two-photon tomography was used for their visualization. This imaging technique is highly sensitive to ZnO nanoparticles (an estimated detection limit is 0.08 fg/m^3), and due to the absence of SHG signal in the stratum corneum, this technique is well suited for determination of the ZnO particles' penetration into the stratum corneum [25]. Typical SHG images of ZnO on the skin surface and in the furrows and wrinkles in vivo are shown in Fig. 11.6. Results show that ZnO nanoparticles do not permeate through the stratum corneum saturating only the outermost layers of stratum corneum, furrows and orifices of the hair follicles.

Additional application of the fluorescence lifetime imaging technique gives an advantage to measure metal nanoparticles, which was shown for measurement of the PVP-coated silver nanoparticle (about 70 nm) penetration into the porcine ear skin ex vivo. The main idea of this method is represented in Fig. 11.7a and lies in obvious differences in mean lifetime distributions for stratum corneum (925–1675 ps) and for silver nanoparticles (50–125 ps), which can be easily detected [18]. Figure 11.7b represents the corresponding penetration profile, which shows the maximum penetration depth into the skin to be approx. 14 μm. These results are

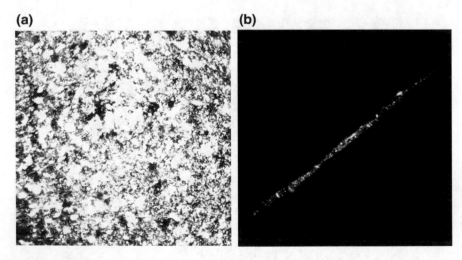

(a)   **(b)**

Fig. 11.6 Distribution of ZnO nanoparticles (around 30 nm) on the skin surface (**a**) and in the furrows and wrinkles (**b**) measured in vivo using two-photon tomography. Image size $130 \times 130 \ \mu m^2$

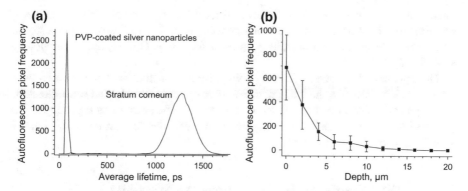

Fig. 11.7 Average fluorescence lifetime distribution of PVP-coated silver nanoparticles (around 70 nm) and the stratum corneum (**a**) and the corresponding penetration profile into the porcine ear skin ex vivo (**b**) measured using two-photon tomography with fluorescence lifetime imaging

in accordance with results obtained using confocal Raman and SERS microscopy, showing absence of permeation of PVP-coated silver nanoparticles through the porcine stratum corneum ex vivo.

It should be noted that due to the higher skin barrier function of human skin in vivo than of porcine skin ex vivo [26], it is expected that penetration of silver nanoparticles into the human stratum corneum in vivo will even lower than shown in ex vivo experiments.

Recent development for the visualization of papillary dermis blood capillaries and extracellular matrix components using two-photon tomography with fluorescence lifetime imaging technique [27], promising future determination of the effect of topical drug delivery in the papillary area.

11.3   Particles in Cosmetic Formulations

11.3.1   Sunscreens

Sunscreens are the only cosmetic products for topical application, which contain particulate ingredients. The standard pigments used for this purpose are TiO_2 and ZnO. The physical UV filter systems exhibit absorption bands and increase the efficacy of sunscreens. These particles are most important because of their reflecting and scattering properties [28]. Like micro-mirrors on the skin, they reflect the incident UV photons of the sun in different directions. Consequently, these pigments have a protective effect as they prevent the photons from penetrating the human skin. After sunscreen application, the particles should therefore remain in the superficial layers of stratum corneum and not penetrate into deeper skin layers. This can be demonstrated easily for TiO_2 particles [29].

For ZnO, the analysis of this penetration process is more complicated as ZnO is not only topically applied via the sunscreen but also a basic element, which is present in the human body in a dissolved form. Although the element is the same, its structure is different for the endogenous and the topically applied form. Therefore, two-photon tomography was used for the ZnO analysis, as previously shown [25].

No ZnO could be detected in the living tissue, so that ZnO particles show a similar distribution like the TiO_2 particles if sunscreens are applied onto the skin.

In general, it can be stated that sunscreens are efficient and safe products. No penetration of organic and filter components through the stratum corneum, i.e. the skin barrier, could be observed.

11.3.2   Disturbance of the Skin Barrier to Stimulate the Penetration of Particulate Formulations

Normally, particles do not pass the intact skin barrier. Thus, in order to stimulate the penetration of particulate formulations, the stratum corneum has to be damaged. This can be done by different methods. One of these methods is the topical application of penetration enhancers [30], another one electrophoresis [31], by means of which the formulations are forced by electrical fields to pass the skin barrier. The most intensive method is the application of needle rollers to prick small holes into the skin [32]. These needles are very short so that they can only damage the stratum corneum, without reaching the nerve cells.

If the skin barrier is damaged by manipulations or injury, it can also be penetrated by particles.

11.3.3   Particles as Carrier Systems for Hair Follicle Targeting

In Sect. 3.2., it was explained that particles do not penetrate through the skin barrier. However, particulate formulations efficiently penetrate into the hair follicles contrary to non-particulate ones. Both the terminal and the vellus hair follicles are provided with a barrier in their upper part, whereas they have tight junctions in their deeper part [33]. This structure seems to be less protective. However, this is not necessary because non-particulate formulations do not penetrate deeply into the hair follicles [34]. On the other hand, particles of approximately 600 nm in diameter are capable of penetrating deeply into the hair follicles until they reach the barrier structure of the tight junctions [35]. This surprising process is due to the structure of the surface of the hairs. The cuticula has a thickness of approximately 600 nm as shown in Fig. 11.8. The moving hair acts as a gearing pump, pushing the parts of the fitting size deeply into the hair follicles [36]. The particle diameter of 600 nm is an optimum value for penetration into the hair follicles. Smaller or larger particles can be applied too, although at lower efficiency.

Figure 11.9 shows the relation between the particle size and the penetration depth in the hair follicle.

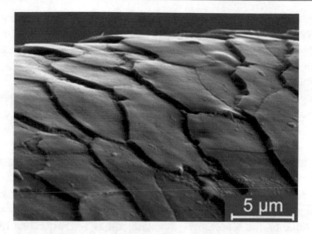

Fig. 11.8 Approximate size of the vellus and terminal hairs' cuticula (600 nm)

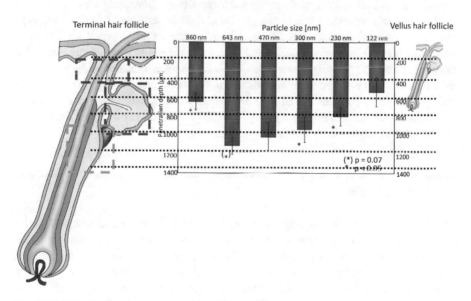

Fig. 11.9 Relation between particle size and penetration depth

Particles of 600 nm are too large to pass the barrier of the tight junctions in the deeper part of the hair follicle. So, in a new drug delivery concept, the particles shall be used as transport systems that will be loaded with the active drug. Once the particles have been penetrated deeply into the hair follicles, the drug has to be released either by diffusion or by a trigger signal. If the drug is not expensive, porous particles can be used, which absorb the formulation during storage. They release the formulation by diffusion after penetration into the hair follicles.

Much more practicable is the stimulated release of the active drug from the particles in the hair follicles. The release can be triggered by an external signal like ultrasound or infrared radiation [37]. More convenient is the use of an internal trigger signal system, e.g. different pH values on the skin surface (pH = 4.5) and deep in the hair follicles (pH = 7.5) (REF).

These drug delivery strategies are still under investigation using, for example, microdialysis.

11.3.4 Nanocrystals for Drug Delivery in Cosmetic Science

Recently, a method was proposed to produce nanocrystals of different size from single substances [38]. Among the substances often used in cosmetic science is caffeine. Stimulating the blood flow, this molecule is an antioxidant [39] important for the prevention of skin ageing [40]. The possibility of enhanced penetration into the stratum corneum was shown for combination of penetration enhancer propylene glycol and caffeine nanocrystals using confocal Raman microscopy [41].

Two types of nanocarriers were produced and incorporated in formulations. While one crystal system was 200 nm in size, the other measured 600 nm in diameter. Two formulations contained the same concentration of caffeine. The crystals were applied onto human skin, and the caffeine concentration was determined in the blood of volunteers at different points in time. Using this new technology, the penetration of caffeine could be increased by one order of magnitude.

These results confirmed the efficacy of the new method for drug delivery based on particles as transport systems.

11.4 Conclusions

In this chapter, different in vivo methods to investigate the penetration of active and/or particles into the skin were presented. If SC penetration is the aim, tape stripping, two-photon tomography with fluorescence lifetime imaging and Raman (SERS) microscopy can be applied. For hair follicle penetration, differential stripping is the method of choice and for transepidermal and transfollicular penetration microdialysis should be applied. With these methods, it could be proven that sunscreens are safe, and particles are efficient systems for carrying active molecules into the hair follicles. It could be shown that particles are excellent structures to target hair follicles. Using these nanosystems, typical cosmetic components like hyaluronic acid, antioxidants and moistening components can efficiently pass the skin barrier and reach their target structures.

Non-invasive methods for measurements of penetration profiles into the skin are continuously developed and optimized making more sensitive detection promising.

References

1. Ahmad U, Ahmad Z, Khan AA, Akhtar J, Singh SP, Ahmad FJ. Strategies in development and delivery of nanotechnology based cosmetic products. Drug Res (Stuttg); 2018.
2. Jose J, Netto G: Role of solid lipid nanoparticles as photoprotective agents in cosmetics: J Cosmet Dermatol. 2018;13, Epub ahead of print] Review. https://doi.org/10.1111/jocd.12504.
3. Martini APM, Maia CP. Influence of visible light on cutaneous hyperchromias: Clinical efficacy of broad-spectrum sunscreens. Photodermatol Photoimmunol Photomed; 2018.
4. Ouyang H, Meyer K, Maitra P, Daly S, Svoboda RM, Farberg AS, Rigel DS. Realistic sunscreen durability: a randomized, double-blinded, controlled clinical study. J Drugs Dermatol. 2018;17:116–7.
5. Tyagi P, Subramony JA. Nanotherapeutics in oral and parenteral drug delivery: key learnings and future outlooks as we think small. J Controlled Release. 2018;272:159–68.
6. Moghadas H, Saidi MS, Kashaninejad N, Nguyen NT. Challenge in particle delivery to cells in a microfluidic device. Drug Deliv Transl Res. 2017.
7. Ilic T, Pantelic I, Lunter D, Dodevic S, Markovic B, Rankovic D, Daniels R, Savic S. Critical quality attributes, in vitro release and correlated in vitro skin permeation-in vivo tape stripping collective data for demonstrating therapeutic (non) equivalence of topical semisolids: A case study of "ready-to-use" vehicles. Int J Pharm. 2017;528:253–67.
8. Wolf M, Halper M, Pribyl R, Baurecht D, Valenta C. Distribution of phospholipid based formulations in the skin investigated by combined atr-ftir and tape stripping experiments. Int J Pharm. 2017;519:198–205.
9. Jacobi U, Weigmann HJ, Ulrich J, Sterry W, Lademann J. Estimation of the relative stratum corneum amount removed by tape stripping. Skin Res Technol. 2005;11:91 6.
10. Voegeli R, Heiland J, Doppler S, Rawlings AV, Schreier T. Efficient and simple quantification of stratum corneum proteins on tape strippings by infrared densitometry. Skin Res Technol. 2007;13:242–51.
11. Weigmann HJ, Lindemann U, Antoniou C, Tsikrikas GN, Stratigos AI, Katsambas A, Sterry W, Lademann J. Uv/vis absorbance allows rapid, accurate, and reproducible mass determination of corneocytes removed by tape stripping. Skin Pharmacol Appl Skin Physiol. 2003;16:217 27.
12. Pelchrzim R, Weigmann HJ, Schaefer H, Hagemeister T, Linscheid M, Shah VP, Sterry W, Lademann J. Determination of the formation of the stratum corneum reservoir for two different corticosteroid formulations using tape stripping combined with uv/vis spectroscopy. J Dtsch Dermatol Ges. 2004;2(11):914–9.
13. Teichmann A, Jacobi U, Ossadnik M, Richter H, Koch S, Sterry W, Lademann J. Differential stripping: Determination of the amount of topically applied substances penetrated into the hair follicles. J Invest Dermatol. 2005;125:264–9.
14. Ossadnik M, Czaika V, Teichmann A, Sterry W, Tietz HJ, Lademann J, Koch S. Differential stripping: introduction of a method to show the penetration of topically applied antifungal substances into the hair follicles. Mycoses. 2007;50:457–62.
15. Otberg N, Richter H, Schaefer H, Blume-Peytavi U, Sterry W, Lademann J. Variations of hair follicle size and distribution in different body sites. J Invest Dermatol. 2004;122:14–9.
16. Zhang HY, Zhang K, Li Z, Zhao JH, Zhang YT, Feng NP. In vivo microdialysis for dynamic monitoring of the effectiveness of nano-liposomes as vehicles for topical psoralen application. Biol Pharm Bull. 2017;40:1996–2000.
17. Yu DG, Ding HF, Mao YQ, Liu M, Yu B, Zhao X, Wang XQ, Li Y, Liu GW, Nie SB, Liu S, Zhu ZA. Strontium ranelate reduces cartilage degeneration and subchondral bone remodeling in rat osteoarthritis model. Acta Pharmacol Sin. 2013;34:393–402.
18. Zhu YJ, Choe CS, Ahlberg S, Meinke MC, Alexiev U, Lademann J, Darvin ME. Penetration of silver nanoparticles into porcine skin ex vivo using fluorescence lifetime imaging microscopy, raman microscopy, and surface-enhanced raman scattering microscopy. J Biomed Opt. 2015;20.

19. Choe C, Lademann J, Darvin ME. Confocal raman microscopy for investigating the penetration of various oils into the human skin in vivo. J Dermatol Sci. 2015;79:176–8.
20. Huang H, Shi H, Feng S, Lin J, Chen W, Huang Z, Li Y, Yu Y, Lin D, Xu Q, Chen R. Silver nanoparticle based surface enhanced raman scattering spectroscopy of diabetic and normal rat pancreatic tissue under near-infrared laser excitation. Laser Phys Lett 2013;10.
21. Choe C, Lademann J, Darvin ME. A depth-dependent profile of the lipid conformation and lateral packing order of the stratum corneum in vivo measured using raman microscopy. Analyst. 2016;141:1981–7.
22. Choe C, Lademann J, Darvin ME. Depth profiles of hydrogen bound water molecule types and their relation to lipid and protein interaction in the human stratum corneum in vivo. Analyst. 2016;141:6329–37.
23. Choe C, Schleusener J, Lademann J, Darvin ME. Keratin-water-nmf interaction as a three layer model in the human stratum corneum using in vivo confocal raman microscopy. Scientific Reports 2017:7.
24. Choe C, Schleusener J, Lademann J, Darvin ME. In vivo confocal raman microscopic determination of depth profiles of the stratum corneum lipid organization influenced by application of various oils. J Dermatol Sci. 2017;87:183–91.
25. Darvin ME, Konig K, Kellner-Hoefer M, Breunig HG, Werncke W, Meinke MC, Patzelt A, Sterry W, Lademann J. Safety assessment by multiphoton fluorescence/second harmonic generation/hyper-rayleigh scattering tomography of zno nanoparticles used in cosmetic products. Skin Pharmacology and Physiology. 2012;25:219–26.
26. Choe CS, Schleusener J, Lademann J, Darvin ME. Human skin in vivo has a higher skin barrier function than porcine skin ex vivo—comprehensive raman microscopic study of the stratum corneum. J Biophotonics. 2018; Accepted 2018; https://doi.org/10.1002/jbio.201700355.
27. Shirshin EA, Gurfinkel YI, Priezzhev AV, Fadeev VV, Lademann J, Darvin ME. Two-photon autofluorescence lifetime imaging of human skin papillary dermis in vivo: Assessment of blood capillaries and structural proteins localization. Scientific Reports 2017;7.
28. Zastrow L, Meinke MC, Albrecht S, Patzelt A, Lademann J. From uv protection to protection in the whole spectral range of the solar radiation: new aspects of sunscreen development. Adv Exp Med Biol. 2017;996:311–8.
29. Lademann J, Weigmann HJ, Rickmeyer C, Barthelmes H, Schaefer H, Mueller G, Sterry W. Penetration of titanium dioxide microparticles in a sunscreen formulation into the horny layer and the follicular orifice. Skin Pharm Appl Skin Phys. 1999;12:247–56.
30. Nokhodchi A, Shokri J, Dashbolaghi A, Hassan-Zadeh D, Ghafourian T, Barzegar-Jalali M. The enhancement effect of surfactants on the penetration of lorazepam through rat skin. Int J Pharm. 2003;250:359–69.
31. Dai Y, Zhang Q, Jiang Y, Yin L, Zhang X, Chen Y, Cai X. Screening of differentially expressed proteins in psoriasis vulgaris by two-dimensional gel electrophoresis and mass spectrometry. Exp Ther Med. 2017;14:3369–74.
32. Stahl J, Wohlert M, Kietzmann M. Microneedle pretreatment enhances the percutaneous permeation of hydrophilic compounds with high melting points. Bmc Pharm Toxicol. 2012;13.
33. Mathes C, Brandner JM, Laue M, Raesch SS, Hansen S, Failla AV, Vidal S, Moll I, Schaefer UF, Lehr CM. Tight junctions form a barrier in porcine hair follicles. Eur J Cell Biol. 2016;95:89–99.
34. Tachaprutinun A, Meinke MC, Richter H, Pan-In P, Wanichwecharungruang S, Knorr F, Lademann J, Patzelt A. Comparison of the skin penetration of garcinia mangostana extract in particulate and non-particulate form. Eur J Pharm Biopharm. 2014;86:307–13.
35. Patzelt A, Richter H, Knorr F, Schafer U, Lehr CM, Dahne L, Sterry W, Lademann J. Selective follicular targeting by modification of the particle sizes. J Control Release. 2011;150:45–8.

36. Patzelt A, Richter H, Dähne L, Walden P, Wiesmüller KH, Wank U, Sterry W, Lademann J. Influence of the vehicle on the penetration of particles into the hair follicles. Int J Pharm. 2011;3:307–14.
37. Mak WC, Richter H, Knorr F, Patzelt A, Darvin ME, Rühl E, Renneberg R, Lademann J. Triggered drug release from nanoparticles for follicular targeting using ira irradiation. Acta Biomater. 2016;30:388–96.
38. Pyo SM, Hespeler D, Keck CM, Muller RH. Dermal miconazole nitrate nanocrystals - formulation development, increased antifungal efficacy & skin penetration. Int J Pharm. 2017;531:350–9.
39. Bors L, Bajza A, Kocsis D, Erdo F. Caffeine: traditional and new therapeutic indications and use as a dermatological model drug. Orv Hetil. 2018;159:384–90.
40. Türkoğlu M, Uğurlu T, Gedik G, Yılmaz AM, Süha YA. In vivo evaluation of black and green tea dermal products against uv radiation. Drug Discov Ther. 2010;4:362–7.
41. Ascencio SM, Choe C, Meinke MC, Muller RH, Maksimov GV, Wigger-Alberti W, Lademann J, Darvin ME. Confocal raman microscopy and multivariate statistical analysis for determination of different penetration abilities of caffeine and propylene glycol applied simultaneously in a mixture on porcine skin ex vivo. Eur J Pharm Biopharm. 2016;104:51–8.

Characterisation of Nanomaterials with Focus on Metrology, Nanoreference Materials and Standardisation

12

Kirsten Rasmussen, Agnieszka Mech and Hubert Rauscher

Abstract

Nanomaterials (NMs) may be seen as a particular group of chemicals that are defined by their size. This chapter provides an overview of NMs and requirements under the Cosmetic Products Regulation (CPR), which was the first piece of EU legislation that explicitly defined NMs. A publicly available "catalogue of nanomaterials used in cosmetic products placed on the EU market" by the European Commission provides an overview of NMs used in cosmetic products. The CPR introduced legal requirements for NMs to ensure their safety, including data for physicochemical characterisation, and labelling of cosmetic products for their content of NMs. The Scientific Committee on Consumer Safety (SCCS) evaluates the safety of NMs in cosmetic products and their ingredients, and the SCCS has issued two guidance documents supporting the assessment of NMs. Those documents include listings of physicochemical properties proposed by the SCCS for identification and characterisation of NMs. In general, physicochemical characterisation of a NM is essential to precisely identify and characterise that material. Furthermore, this information is needed prior to further testing for assessing potential toxicological effects to ensure that

Disclaimer: The views expressed in this review are that of the authors and do not represent the views and/or policies the European Commission. The authors are not responsible for any use which might be made of this text.

K. Rasmussen (✉) · A. Mech · H. Rauscher
European Commission, Joint Research Centre, Directorate F, Health, Consumers and Reference Materials, Via E. Fermi 2749, 21027 Ispra, VA, Italy
e-mail: kirsten.rasmussen@ec.europa.eu
URL: https://ec.europa.eu/jrc/en

A. Mech
e-mail: agnieszka.mech@ec.europa.eu

H. Rauscher
e-mail: hubert.rauscher@ec.europa.eu

© Springer Nature Switzerland AG 2019
J. Cornier et al. (eds.), *Nanocosmetics*,
https://doi.org/10.1007/978-3-030-16573-4_12

conclusions from (eco)toxicological tests will be relevant for other experimental contexts and possibly allow linking the NM's physicochemical properties with identified adverse effects. Characterisation along the life cycle is desirable as some physicochemical properties depend on the immediate surroundings. Different datasets for physicochemical characterisation of NMs have been suggested both in a regulatory and research context, and an overview of datasets is presented here. They all include properties that are beyond the standard dataset for chemicals and are thought to be of particular relevance for NMs. To achieve a comprehensive characterisation and to compensate for the weaknesses of individual methods, the measurement of a property should, where possible, be made using several methods. The chapter highlights the need to develop and agree on methods for physicochemical characterisation; regulatory methods are developed by the OECD test guidelines programme and ISO complements such methods by providing standardised measurement techniques.

Keywords
Nanomaterial · Cosmetic products · Physicochemical properties

12.1   Introduction

The use of nanomaterials (NMs) is expected to increase as illustrated by an estimate of the value of the global NM market, which was at $14,741 million in 2015 and is expected to arrive at $55,016 million by 2022 [50]. As the use of NMs has increased, so have the concerns over possible adverse effects and the adequacy of current legislation to address NMs. Hence, over the past 20 years, much research into NM safety and methodologies for their safety assessment, including data requirements, has been performed. Furthermore, some legislation, including the EU Cosmetic Products Regulation [41, 42], has been amended to specifically address NMs, whereas the necessity for such an amendment is still being evaluated for other legislation.

NMs may be viewed as a particular group of chemicals that are defined by their size. Lövestam et al. [24] highlight that the only common feature among all NMs is the nanoscale. Their small size may lead to special properties not present in a corresponding macro-sized material, and therefore, NMs are much investigated as well as used in innovative products. Furthermore, even if a material does not exhibit truly "new" features at the nanoscale, it can have properties that are clearly different from those of the macroscale material just because of its reduced size. Besides being of scientific interest, this calls for a dedicated risk assessment in a regulatory context.

The international communities of legislators and scientists have, via work in the Organisation for Economic Co-operation and Development's (OECD) Working Party on Manufactured Nanomaterials (WPMN), developed their visions for characterisation data relevant for safety assessment [30–32] and have proposed a first set of methods for characterising NMs [38].

12.2   What Are Nanomaterials? Definition of the Term "Nanomaterial" for Cosmetic Products and Other Legislation

The Cosmetic Products Regulation (CPR) [41, 42] is the first piece of EU legislation that explicitly addressed NMs. The CPR introduced a definition of the term "nanomaterial" and outlined specific requirements for NMs in terms of data requirements, content labelling and evaluation of cosmetic products that contain NMs to ensure that NMs are specifically addressed when evaluating the safety of cosmetic products.

The NM definition in the CPR states (Article (2) Definitions, paragraph 1 (k)): "'*nanomaterial*' *means an insoluble or biopersistant and intentionally manufactured material with one or more external dimensions, or an internal structure, on the scale from 1 to 100 nm;*". Paragraph 3 furthermore states: "*In view of the various definitions of nanomaterials published by different bodies and the constant technical and scientific developments in the field of nanotechnologies, the Commission shall adjust and adapt point (k) of paragraph 1 to technical and scientific progress and to definitions subsequently agreed at international level. That measure, designed to amend non-essential elements of this Regulation, shall be adopted in accordance with the regulatory procedure with scrutiny referred to in Article 32 (3)*".

In addition to the NM definition given in CPR, other regulatory definitions of "nanomaterial" exist. Rauscher et al. [39] give an overview of these. The European Commission's Recommendation on the Definition of Nanomaterial [6], in the following called the "EC Recommendation", is especially relevant here as it was proposed in order to have a broadly applicable definition of "nanomaterial" and later to harmonise that term across EU legislation. The EC Recommendation is currently being reviewed (http://ec.europa.eu/info/law/better-regulation/initiatives/ares-2017-4513169) and will possibly be revised. Thereafter, it is expected that EU legislation, in which a definition of the term "nanomaterial" is relevant, will be reviewed and aligned with the future EC Recommendation. The EC Recommendation differs from the CPR definition in several important points: it addresses any NM, including incidental and naturally occurring NMs, whereas the CPR definition addresses specifically "intentionally manufactured materials". The CPR includes a criterion that the NM is "insoluble" or "biopersistent", which is not required in the EC Recommendation. The EC Recommendation also specifies that "... *50% or more of the particles in the number size distribution, one or more external dimensions is in the size range 1 nm-100 nm. ...*" and states that the threshold of "*.... 50% may be replaced by a threshold between 1 and 50% ...*", under specific conditions. These two provisions are not included in the CPR.

The main piece of legislation addressing almost all chemicals placed on the market, the Regulation concerning the Registration, Evaluation, Authorisation and Restriction of Chemicals (REACH) [40], defines nanoforms on the basis of the EC Recommendation [6] in a recent amendment to the REACH annexes [7]. The European Chemicals Agency (ECHA) is responsible for the implementation of REACH and has developed guidance as well as a "best practices" document for the registration of NMs [12], more guidedance for nanoforms is under development. The regulation on classification, labelling and packaging (CLP) [4], which applies the same definition of the term "substance" as REACH, also applies to all chemicals, and it does not mention NMs specifically.

Other EU legislations that include definitions of NM are the Biocidal Products Regulation (BPR) [44] and the food legislation. The former uses a definition which is based on the EC Recommendation with sector-specific modifications. For food legislation, the term "nanomaterial" as defined in the Novel Foods Regulation [45] is applicable also in the Regulation on the Provision of Food Information to Consumers [43]; an overview of how NMs are defined and addressed in legislation for foodstuff is found in Rasmussen et al. [36]. Neither the food legislation nor BPR will be further discussed here.

Table 12.1 gives an overview of EU legislation on chemicals, which is relevant in the context of NMs, including the wording defining "nanomaterial" where relevant.

As this chapter concerns cosmetic products, the CPR definition of NM is the one used; Article 16 of the CPR concerns NMs and specific data requirements and is cited in full in Box 1.

Table 12.1 Selected EU legislation relevant for nanomaterials

Legislation	Regulatory definition of nanomaterial
Regulation (EC) No 1907/2006 concerning the Registration, Evaluation, Authorisation and restriction of Chemicals (REACH), establishing a European Chemicals Agency [40], and Commission Regulation (EU) 2018/1881 [7]	On the basis of the Commission Recommendation of 18 October 2011 on the definition of nanomaterial, a nanoform is a form of a natural or manufactured substance containing particles, in an unbound state or as an aggregate or as an agglomerate and where, for 50% or more of the particles in the number size distribution, one or more external dimensions is in the size range 1 nm–100 nm, including also by derogation fullerenes, graphene flakes and single wall carbon nanotubes with one or more external dimensions below 1 nm. For this purpose, "particle" means a minute piece of matter with defined physical boundaries; "agglomerate" means a collection of weakly bound particles or aggregates where the resulting external surface area is similar to the sum of the surface areas of the individual components and "aggregate" means a particle comprising of strongly bound or fused particles.
Regulation (EC) No 1272/2008 on Classification, Labelling and Packaging (CLP) [4]	No definition Same substance definition as REACH; nanomaterials are implicitly covered
Regulation (EC) No 1223/2009 on Cosmetic Products [41]	*Article 2* **Definitions** (k) an insoluble or biopersistent and intentionally manufactured material with one or more external dimensions, or an internal structure, on the scale from 1 to 100 nm
Regulation (EU) No 528/2012 on Biocidal Products [44]	*Article 3* **Definitions** (z) "nanomaterial" means a natural or manufactured active substance or non-active substance containing particles, in an unbound state or as an aggregate or as an agglomerate and where, for 50% or more of the particles in the number size distribution, one or more external dimensions is in the size range 1–100 nm Fullerenes, graphene flakes and single-wall carbon nanotubes with one or more external dimensions below 1 nm shall be considered as nanomaterials For the purposes of the definition of nanomaterial, "particle", "agglomerate" and "aggregate" are defined as follows: – "Particle" means a minute piece of matter with defined al boundaries – "Agglomerate" means a collection of weakly bound particles or aggregates where the resulting external surface area is similar to the sum of the surface areas of the individual components – "Aggregate" means a particle comprising strongly bound or fused particles

(continued)

Table 12.1 (continued)

Legislation	Regulatory definition of nanomaterial
Regulation (EC) No 1107/2009 on the placing of plant protection products on the market [42]	No definition
Regulation (EU) 2015/2283 on Novel Foods [45]	*Article 3* **Definitions** (f) "Engineered nanomaterial" means any intentionally produced material that has one or more dimensions of the order of 100 nm or less or that is composed of discrete functional parts, either internally or at the surface, many of which have one or more dimensions of the order of 100 nm or less, including structures, agglomerates or aggregates, which may have a size above the order of 100 nm but retain properties that are characteristic of the nanoscale. Properties that are characteristic of the nanoscale include: (i) Those related to the large specific surface area of the materials considered; and/or (ii) Specific physicochemical properties that are different from those of the non-nanoform of the same material
Regulation (EU) No 1169/2011 on the Provision of Food Information to Consumers [43]	Yes. Refers to the NM definition in the Novel Foods Regulation
Regulation (EU) No 10/2011) on plastic materials and articles intended to come into contact with food [6]	No definition Regulation (EU) No 10/2011 provides some specifications for engineered NMs
Regulation (EU) 2017/745 on Medical Devices [46]	Yes, same as the EC Recommendation

EC Recommendation [5]

2. "Nanomaterial" means a natural, incidental or manufactured material containing particles, in an unbound state or as an aggregate or as an agglomerate and where, for 50% or more of the particles in the number size distribution, one or more external dimensions is in the size range 1 nm–100 nm. In specific cases and where warranted by concerns for the environment, health, safety or competitiveness, the number size distribution threshold of 50% may be replaced by a threshold between 1 and 50%

3. By derogation from point 2, fullerenes, graphene flakes and single-wall carbon nanotubes with one or more external dimensions below 1 nm should be considered as nanomaterials

4. For the purposes of point 2, "particle", "agglomerate" and "aggregate" are defined as follows:

(a) "Particle" means a minute piece of matter with defined physical boundaries

(b) "Agglomerate" means a collection of weakly bound particles or aggregates where the resulting external surface area is similar to the sum of the surface areas of the individual components

(c) "Aggregate" means a particle comprising of strongly bound or fused particles

Box 1. Article 16 "Nanomaterials" of the Cosmetics Regulation
Article 16
  Nanomaterials

1. For every cosmetic product that contains nanomaterials, a high level of protection of human health shall be ensured.
2. The provisions of this article do not apply to nanomaterials used as colourants, UV filters or preservatives regulated under Article 14, unless expressly specified.
3. In addition to the notification under Article 13, cosmetic products containing nanomaterials shall be notified to the Commission by the responsible person by electronic means six months prior to being placed on the market, except where they have already been placed on the market by the same responsible person before 11 January 2013.

In the latter case, cosmetic products containing nanomaterials placed on the market shall be notified to the Commission by the responsible person between 11 January 2013 and 11 July 2013 by electronic means, in addition to the notification in Article 13. The first and the second subparagraphs shall not apply to cosmetic products containing nanomaterials that are in conformity with the requirements set out in Annex III.

The information notified to the Commission shall contain at least the following:

(a) the identification of the nanomaterial including its chemical name (IUPAC) and other descriptors as specified in point 2 of the Preamble to Annexes II to VI;
(b) the specification of the nanomaterial including size of particles, physical and chemical properties;
(c) an estimate of the quantity of nanomaterial contained in cosmetic products intended to be placed on the market per year;
(d) the toxicological profile of the nanomaterial;
(e) the safety data of the nanomaterial relating to the category of cosmetic product, as used in such products;
(f) the reasonably foreseeable exposure conditions.

The responsible person may designate another legal or natural person by written mandate for the notification of nanomaterials and shall inform the Commission thereof.

The Commission shall provide a reference number for the submission of the toxicological profile, which may substitute the information to be notified under point (d).

4. In the event that the Commission has concerns regarding the safety of a nanomaterial, the Commission shall, without delay, request the SCCS to

give its opinion on the safety of such nanomaterial for use in the relevant categories of cosmetic products and on the reasonably foreseeable exposure conditions. The Commission shall make this information public. The SCCS shall deliver its opinion within six months of the Commission's request. Where the SCCS finds that any necessary data is lacking, the Commission shall request the responsible person to provide such data within an explicitly stated reasonable time, which shall not be extended. The SCCS shall deliver its final opinion within six months of submission of additional data. The opinion of the SCCS shall be made publicly available.

5. The Commission may, at any time, invoke the procedure in paragraph 4 where it has any safety concerns, for example, due to new information supplied by a third party.

6. Taking into account the opinion of the SCCS, and where there is a potential risk to human health, including when there is insufficient data, the Commission may amend Annexes II and III.

7. The Commission may, taking into account technical and scientific progress, amend paragraph 3 by adding requirements.

8. The measures, referred to in paragraphs 6 and 7, designed to amend non-essential elements of this Regulation, shall be adopted in accordance with the regulatory procedure with scrutiny referred to in Article 32(3).

9. On imperative grounds of urgency, the Commission may use the procedure referred to in Article 32(4).

10. The following information shall be made available by the Commission:

(a) By 11 January 2014, the Commission shall make available a catalogue of all nanomaterials used in cosmetic products placed on the market, including those used as colourants, UV filters and preservatives in a separate section, indicating the categories of cosmetic products and the reasonably foreseeable exposure conditions. This catalogue shall be regularly updated thereafter and be made publicly available.

(b) The Commission shall submit to the European Parliament and the Council an annual status report, which will give information on developments in the use of nanomaterials in cosmetic products within the Community, including those used as colourants, UV filters and preservatives in a separate section. The first report shall be presented by 11 July 2014. The report update shall summarise, in particular, the new nanomaterials in new categories of cosmetic products, the number of notifications, the progress made in developing nanospecific assessment methods and safety assessment guides, and information on international cooperation programmes.

> 11. The Commission shall regularly review the provisions of this Regulation concerning nanomaterials in the light of scientific progress and shall, where necessary, propose suitable amendments to those provisions.
>
> The first review shall be undertaken by 11 July 2018.

According to Article 16, the cosmetic products containing nanomaterials (other than colourants, preservatives and UV filters and not otherwise restricted by Regulation (EC) No 1223/2009) are subject to an additional procedure and are not a part of the standard notification (Article 13). They require a specific notification to the Cosmetic Products Notification Portal (CPNP) 6 months before being placed on the market (Art. 16 (3)). If the EC has concerns about the safety of a NM to be used as ingredient in cosmetic products, the Scientific Committee on Consumer Safety (SCCS) is requested to give its opinion on the safety of the NM (e.g. [54]). The SCCS has issued, among others, two guidance documents aimed at the assessment of NMs: Guidance on Risk Assessment of Nanomaterials (SCCS/1484/12) [52] and Memorandum "Relevance, Adequacy and Quality of Data in Safety Dossiers on Nanomaterials" (SCCS/1524/13) [53].

Certain groups of substances, i.e. colourants, preservatives and UV filters, including those that are nanomaterials, must be authorised by the EC priori to their use in cosmetic products and the allowed substances and nanomaterials are listed in the dedicated Annexes of CPR.

Currently, EC has authorised three UV filters to be used in cosmetic products in form of a nanomaterial: titanium dioxide, zinc oxide and tris-biphenyl triazine, and one colourant: "carbon black (nano)" for use in cosmetic products.

Furthermore, Article 19 of the CPR concerns product labelling and states "*All ingredients present in the form of nanomaterials shall be clearly indicated in the list of ingredients. The names of such ingredients shall be followed by the word 'nano' in brackets*". Also, Annex I on Cosmetic Product Safety Report states "*… Particular consideration shall be given to any possible impacts on the toxicological profile due to - particle sizes, including nanomaterials …*". The Preamble to Annexes II to VI states: "*(3) Substances listed in Annexes III to VI do not cover nanomaterials, except where specifically mentioned*".

The first version of the living and informative "Catalogue of nanomaterials used in cosmetic products placed on the EU market" required according to Article 16(10) was published mid-2017 by the Commission at http://ec.europa.eu/docsroom/documents/24521. The information in this catalogue is based on information submitted to the Cosmetic Products Notification Portal (CPNP). Of the NMs in the CPNP, 12 NMs are stated to be colourants, 6 NMs are stated to be UV filters, and 25 NMs are stated to have other functions. CPR explicitly requires authorisation of colourants and UV filters.

12.3   Relevant Properties and Parameters

Physicochemical properties of NMs and of materials with the same composition, but with larger dimensions, can differ for two main reasons: (1) NMs have a relatively larger specific surface area (SSA) which makes them chemically more reactive (per unit mass or amount). Moreover, the atomic surface structure can be different from a macroscale material, with potentially more reactive sites, kinks and edges, and such differences can lead to changes in the surface reaction kinetics. In some cases, such differences even make materials that are inert in their larger form reactive when produced in their nanoscale form. (2) With decreasing particle size, effects due to electron confinement, including quantum effects, can begin to dominate the behaviour of matter at the nanoscale, and such effects become more pronounced as particle size decreases. Electron confinement effects may concern, for example, the optical, electrical and magnetic behaviour of materials.

In a regulatory context, the defining physicochemical property of any NM is "particle size", often supplemented by "particle size distribution". The CPR and certain other regulations do not explicitly specify a particle size distribution or a cut-off value for the relative amount of particles that are in the size range between 1 and 100 nm. Nevertheless, for the practical identification of NMs and when the applicable NM definition does not have a quantitative criterion for the particle size distribution, there must anyhow be an agreement on some specification of the particle size distribution, e.g. based on guidance, best practices or taken from the EC Recommendation. Under this assumption, in the European Union knowledge of the values of these two parameters is necessary for a given material to decide whether it is regarded as a NM in a regulatory context, or not (e.g. [5, 41]). Hence, particle size and particle size distribution would be the first physicochemical properties to measure, in order to decide whether the material is a NM or not. Materials can be nanoscale in one dimension (e.g. very thin platelets), in two dimensions (e.g. nanofibres) or in all three dimensions (e.g. nanoparticles). Depending on the applicable definition of NMs, additional defining criteria may be relevant, e.g. "insoluble" and "intentionally manufactured".

Also, non-regulatory definitions of NMs exist, and for example, ISO has introduced the collective term "nano-object" to describe materials that are nanoscale in one, two or three dimensions [16].

Depending on the exact information needs, additional physicochemical properties may be relevant for materials identified as nanomaterials according to a regulatory definition of NM (see Table 12.1 for examples of such definitions). Sets of physicochemical properties thought to be relevant for the safety assessment of a NM in a regulatory context have been proposed by the OECD [28, 38] and numerous other sources. Stefaniak et al. [55] analysed the datasets proposed by 28 different sources, such as committees [e.g. the Scientific Committee for Emerging and Newly Identified Health Risks (SCENIHR)], organisations (e.g. OECD) and researchers (e.g. [21]). They provide an overview of all physicochemical parameters proposed for NMs considered relevant for different purposes, including safety assessment. Only one common parameter, "surface area (specific)", was found among all the sources, while several parameters (defect density, hardness, magnetic properties and optical properties) were listed in only one or few sources. The consolidated list of parameters from Stefaniak et al. contains some additions compared to the WPMN parameters, e.g. solubility (biological), stability, defect density, hardness, magnetic and optical properties. The list of relevant parameters to be characterised links to the purpose of characterisation, e.g. research or addressing regulatory requirements. Table 12.1 gives an example of this. It lists the physico-chemical properties suggested by the SCCS [52, 53, 58] for NMs when used as cosmetic ingredients (columns 1 and 2), by the OECD WPMN (column 3) and by Stefaniak et al. (column 4). As seen from Table 12.1, there is significant overlap of parameters between the sources.

It is important to perform the physicochemical characterisation on a representative sample of the NM. If possible, different stages of the life cycle of the material should also be characterised. For characterising and testing NMs in complex media, a number of requirements, challenges and solutions need to be considered.

The measurement of each property listed in Table 12.2 can be performed by choosing wisely among available methods. Rasmussen et al. [38] give an overview, see Table 12.3, of which methods are available for characterising NMs for the WPMN endpoints, and as seen, a variety of methods are available for measuring the endpoints for pristine materials. These methods may be applicable only to certain materials, and they may have various degrees of strengths and limitations. It should be noted that the field of NM characterisation is progressing fast and additional methods or combinations of methods are under development or already used for specific purposes.

Under CPR the information requirements include "the specification of the nanomaterial including size of particles, physical and chemical properties" (Box 1). The guidance [58] states as one requirement "Relevant physical and chemical specifications" and [53] states "It is important that a safety dossier on nanomaterial (s) contains sufficient data and supporting information to enable adequate risk assessment. The dataset should be complete in relation to physicochemical properties, exposure, toxicological effects, and safety evaluation, as indicated in the SCCS Nano-Guidance (SCCS/1484/12)". CPR thus states the need to provide characterisation data, however, without listing the exact requirements. Thus, the all information possibilities listed in Table 12.2 should be considered when characterising NMs.

In this context, it should be noted that the sample preparation and dispersion media are important parameters as they can affect the outcomes of testing NMs that are insoluble or sparingly soluble in water and other media, which are used in (eco)toxicological tests. Test item preparation and dispersion (including stability) in appropriate media are thus a critical issue. Thus, sample characterisation should be performed for a number of the different stages of testing NMs. The OECD prepared a Guidance Document on Sample Preparation and Dosimetry [29], which refers and applies to water-insoluble manufactured NMs, as it was considered unlikely that soluble NMs would need a sample preparation different from what is applied to

Table 12.2 Suggested physicochemical endpoints for nanomaterials

SCCS [52, 53, 58]	SCCS explanation/details (2012)	WPMN endpoints	Stefaniak et al.
Chemical Identity	Information on structural formula(e)/molecular structure(s) of the constituents of nanomaterial must be provided, along with chemical and common names, and CAS and EINECS numbers (where available)	*Nanomaterial information/identification* (not listed separately) – Nanomaterial name (from list) – CAS number – Structural formula/molecular structure – Major commercial uses – Method of production (e.g. precipitation, gas phase)	
Chemical composition	Information on full chemical composition of the nanomaterial must be provided. This should include purity, nature of impurities, coatings or surface moieties, doping material, encapsulating materials, processing chemicals, dispersing agents, and other additives or formulants, e.g. stabilisers	Composition of NM being tested (incl. degree of purity, known impurities or additives)	Elemental/molecular composition (bulk)
Catalytic activity	Information on the chemical reactivity of the nanomaterial core material or surface coating must be provided. Information on photocatalytic activity and radical formation potential of relevant materials must also be provided	Known catalytic activity Radical formation potential Photocatalytic activity	
Surface characteristics	Detailed information on nanomaterial surface must be provided. This should include information on surface charge (zeta potential), morphology/topography, interfacial tension, reactive sites, as well as any chemical/biochemical modifications or coatings that could change the surface reactivity or add a new functionality	Description of surface chemistry (e.g. coating or modification) Zeta potential (surface charge)	Surface chemistry Surface morphology/structure Surface charge
Morphology	Information on the physical form and crystalline phase/shape must be provided. The information should indicate whether the nanomaterial is present in a particle-, tube-, rod-shape, crystal or	Basic morphology Crystallite size and crystalline phase Shape	Morphology/shape/form Crystal structure

(continued)

Table 12.2 (continued)

SCCS [52, 53, 58]	SCCS explanation/details (2012)	WPMN endpoints	Stefaniak et al.
	amorphous form. Also, whether the nanomaterial is in the form of primary particulates or agglomerates/aggregates. Information should also indicate whether the nanomaterial preparation is in the form of a powder, solution, suspension or dispersion. Aspect ratio of nanomaterial should be calculated and provided (for fibre/tube like material)		
Size	Information on primary and secondary particle size, particle number size distribution and particle mass size distribution must be provided. Product specification and any batch-to-batch variation during manufacturing must be indicated. The use of more than one method (one being electron microscopy based imaging) for determination of size parameters has been recommended by EFSA (2011) and OECD (2010). This must also be a prerequisite for nanoscale cosmetic ingredients	Particle size and particle size distribution	Particle size Particle size distribution
Surface area	Information on BET specific surface area of the nanomaterial, and volume specific surface area (VSSA) must be provided. At the moment, the VSSA is only applicable to nanomaterials in powder formulation.	Specific surface area	Surface area (specific)
Solubility	Information on solubility of the nanomaterial in relevant solvents and partitioning between aqueous and organic phase (e.g. log K_{ow} for organic nanomaterials, and surface modified inorganic nanomaterials) must be provided. Dissolution rates in relevant solvent for soluble and partially soluble nanomaterials should also be provided. Information on hygroscopicity of powders should also be provided	Water solubility/dispersibility Octanol–water partition coefficient, where relevant	Solubility (water) Solubility (biological)
Physical aggregation state		Agglomeration/aggregation	Agglomeration/aggregation state

(continued)

Table 12.2 (continued)

SCCS [52, 53, 58]	SCCS explanation/details (2012)	WPMN endpoints	Stefaniak et al.
Dustiness	Information on dustiness of dry powder products must be provided	Dustiness	
Redox potential	Information on oxidation state and redox potential (for inorganic materials) must be provided, including the conditions under which redox potential was measured should be documented	Redox potential	Surface reactivity
Concentration	Information on concentration in terms of particle mass and particle number per volume must be provided for dispersions and per mass for dry powders		Particle concentration
Stability	Data on stability/dissociation constant of the nanomaterial in relevant formulation/media must be provided	Dispersion stability in water	Stability
Density and pour density	Information on density/porosity of granular materials and pour density must be provided	Pour density Porosity	Density Porosity (specific)
pH	pH of aqueous suspension must be provided		
Viscosity	Information on viscosity of liquid dispersions must be provided		
Other aspects	UV absorption (extinction coefficient), light reflection	Other relevant information (where available)	Dispersibility (dry/wet)
		Representative TEM picture(s)	Defect density
			Hardness
			Magnetic properties
			Optical properties (refractive index)

Characterisation at several stages

Cosmetic ingredients
The characterisation of the nanomaterial needs to be carried out at the raw material stage, in the cosmetic formulation, and during exposure for toxicological evaluations [58]

OECD WPMN
Depending on the material this could be relevant ([28, page 39], Additional Material Characterisation Considerations)

Table 12.3 Overview of commonly used methods for characterisation of physicochemical properties of nanomaterials

Method	Physicochemical characteristics analysed/parameters	Type of material	Strengths	Limitations
SEM (scanning electron microscopy)	Size and size distribution, Shape, Aggregation, Dispersion (cryo-SEM), Crystal structure (lattice system, electron backscattering detection SEM: crystallinity)	Inorganic, organic, carbon-based, biological, complex materials (coated, core–shell), Spherical, equiaxial particles, tubes, flakes, rods and fibres	High-resolution, down to sub-nanometre	Conductive materials (sample or coating) required for high resolution, Dry samples required, Sample analysis in non-physiological conditions, Expensive equipment, Cryogenic method required for most NP-biomaterials
TEM (transmission electron microscopy)	Size and size distribution, Shape, Aggregation, Dispersion (environmental TEM), Crystal structure	Inorganic, organic, carbon-based, biological, complex materials (coated, core–shell), Spherical, equiaxial particles, tubes, flakes, rods and fibres	Direct measurement of the size/size distribution and shape of nanomaterials (higher resolution than SEM, down to sub-nanometre), Several analytical methods can be coupled with TEM for analysis of chemical composition and electronic structure	Very thin samples are required, To achieve statistic reliability several scans of a one sample which means time-consuming and expensive measurements, Samples measured in non-physiological conditions, Damage/change of sample, Expensive equipment
STM (scanning tunnelling microscopy)	Size and size distribution, Shape, Aggregation, Structure, Surface charge density (mapping)	Inorganic, organic, carbon-based, biological, complex materials (coated, core–shell), Spherical, equiaxial particles, tubes, flakes, rods and fibres	Direct measurement, High spatial resolution down to atomic scale	Conductive samples only
AFM (atomic force microscopy)	Size and size distribution, Shape, Structure, Surface properties (modified AFM), Surface charge density (mapping)	Inorganic, organic, carbon-based, biological, complex materials (coated, core–shell), Spherical, equiaxial particles, tubes, flakes, rods and fibres	3D sample surface topography, Sub-nanoscale topographic Resolution, Direct measurement of samples in dry, aqueous or ambient environment	Difficult and time-consuming sample preparation, Tip effects often disturb measurements, Analysis in general limited to the external of nanomaterials
DLS (dynamic light scattering) in combination with electrophoretic light scattering	Hydrodynamic size, Distribution/particle size distribution, Zeta potential	Inorganic, organic, carbon-based, biological, complex materials (coated, core–shell), Spherical and equiaxial particles	Non-destructive/invasive method, Measures in many liquid media (solvent of interest), Hydrodynamic sizes accurately determined for monodisperse samples, Modest cost of apparatus. Rapid technique	A small number of large particles may shield small particles, Not suitable for polydisperse sample measurements, Limited size resolution, Model for data evaluation assumes spherical particles

(continued)

Table 12.3 (continued)

Method	Physicochemical characteristics analysed/parameters	Type of material	Strengths	Limitations
MALS (multi-angle light scattering)	Particle size distribution Mass distribution	Inorganic, organic, carbon-based, biological, complex materials (coated, core–shell) Spherical, equiaxial particles, tubes, flakes, rods and fibres	Good resolution Can correctly characterise non-spherical particles Can be coupled with separating methods, e.g. FFF.	Sample dispersion should be transparent and do not absorb the light wavelength Typically large samples volume—1 ml
SAXS (small-angle X-ray scattering)	Size/size distribution SSA Shape Crystal structure	Inorganic, organic, carbon-based, biological, complex materials (coated, core–shell) Spherical, equiaxial particles to a certain extent also: tubes, flakes, rods and fibres	Non-destructive method Simple sample preparation Possible for amorphous materials and sample in solution.	Low resolution Not suitable for polydisperse samples
TOF-MS (time-of-flight mass spectrometry)	Particle size distribution	Inorganic, organic, carbon-based, complex materials (coated, core–shell) Spherical, equiaxial particles. Non-spherical particles.	Fast analysis Broad total range of measurements (0.1–7 µm)	The size of non-spherical particles is underestimated Suspension of particles may be difficult or impossible to measure. High-resolution analysis is limited
SIMS (secondary ion mass spectrometry)	Composition Surface chemistry	Inorganic, organic, carbon-based, biological, complex materials (coated, core–shell) Spherical, equiaxial particles to a certain extend also: tubes, flakes, rods and fibres	Despite the fact that it has 200-nm spatial resolution, it can be used for nanomaterials chemical identification Depth resolution up to 1 nm for inorganic and up to 10 nm organic materials	Samples analysed are in vacuum Quantification may be challenging
XRD (X-ray diffraction)	Size, shape, and crystal structure for crystalline materials	Inorganic, carbon-based, complex materials (coated, core–shell not organic)	Well-established technique High spatial resolution at atomic scale	Limited to crystalline materials Low intensity compared to electron diffraction
CHDF (capillary hydrodynamic fractionation)	Particle size distribution	Inorganic, organic (particulate and non-particulate), carbon-based, complex materials (coated, core–shell) Spherical, equiaxial particles	Relatively fast analysis time (7–10 min)	Aqueous emulsifier medium only Poor resolution (40% difference between sizes) Capillary plugging Non-spherical particles may not be correctly measured

(continued)

Table 12.3 (continued)

Method	Physicochemical characteristics analysed/parameters	Type of material	Strengths	Limitations
FFF (field flow fractionation)	Particle size distribution (in combination with a detection method)	Inorganic, organic, carbon-based, complex materials (coated, core–shell); Spherical, equiaxial particles	Very good resolution with small particles, narrow peaks as little as 5–6% different in size; Certified reference materials are available	Relatively complex algorithm for size separation. Relatively long run times (>1 h); Assumption of spherical particles; It does not determine particles in aggregates and agglomerates
SMPS (scanning mobility particle spectrometry)	Particle size; Particle size distribution	Inorganic, organic, carbon-based, complex materials (coated, core–shell); Spherical, equiaxial particles	Most widely used technique for aerosols; Can also be used on site	Only aerosol samples in size range 2.5–1000 nm with resolution 2 nm; Particles need to be charged
Electrophoretic mobility	Zeta potential	Inorganic, organic, carbon-based, biological, complex materials (coated, core–shell); Spherical, equiaxial particles, tubes, flakes, rods and fibres	Simultaneous measurement of many particles (using ELS)	Not very precise method; Very diluted sample required
BET (Brunauer–Emmett–Teller method)	SSA; Porosity (if extended pressure range and modified data analysis are applied)	Inorganic, carbon-based, complex materials (coated, core–shell); Spherical, equiaxial particles	Certified reference materials are available; Fast and relatively cheap method	Particles and non-particulate porous materials cannot be distinguished; Materials must be free of any volatile compounds; Long measurement time
sp-ICP-MS (single particle-inductively coupled plasma-mass spectrometry)	Particle size distribution; Surface chemistry	Inorganic, complex materials (coated, core–shell if contains detectable element); Spherical, equiaxial particles	Measures individual particles; Method is chemically specific, rapid, cost-efficient	Can detect only one element at a time; Accurate size determination limited to spherical particles only; The method does not resolve particles in aggregates and agglomerates
IR/FTIR (infrared/fourier-transform infrared spectroscopy)	Structure; Molecular composition; Surface properties (FTIR)	Inorganic, organic, carbon-based, biological, complex materials (coated, core–shell); Spherical, equiaxial particles, tubes, flakes, rods and fibres	Fast and inexpensive measurement; Minimal or no sample preparation requirement	Interference and strong absorbance of H_2O (IR); Relatively low sensitivity in nanoscale analysis; In complex samples, analysis may be very challenging; FTIR mostly for dry powder samples. Molecules must be active in IR region. Detection limits routine: 2%, special techniques: 0.01%

(continued)

Table 12.3 (continued)

Method	Physicochemical characteristics analysed/parameters	Type of material	Strengths	Limitations
Raman spectroscopy	Composition Surface chemistry Particle size (for certain materials)	Inorganic, organic, carbon-based, biological, complex materials (coated, core–shell) Spherical, equiaxial particles, tubes, flakes, rods and fibres	Commonly available, easy and cheap technique	It can determine size for certain materials only, e.g. CNT, thickness of graphene flakes It cannot be used for metals and alloys Very sensitive to impurities. Samples heating through intense laser radiation can lead to sample decomposition
UV-Vis (ultraviolet–visible)	Composition Surface chemistry Optical properties Particle size (for certain materials)	Inorganic, organic, carbon-based, biological, complex materials (coated, core–shell). Only certain elements	Easy and cheap technique Both solid and liquid samples can be analysed	Very material dependent. Size determination possible only for plasmonic samples. Only for diluted samples (not in powders), e.g. thin films and dispersions Samples in powders require reflectance accessory
NMR (nuclear magnetic resonance)	Size (indirect analysis) Structure Composition Purity Conformational change	Inorganic, organic, carbon-based, biological, complex materials (coated, core–shell). Only for certain elements	Non-destructive/non-invasive method Sample preparation easy	Low sensitivity Time-consuming Relatively large amount of sample required Only for certain elements
EELS (electron energy-loss spectroscopy)	Composition Surface chemistry Electronic properties	Inorganic, organic, carbon-based, complex materials (coated, core–shell) Spherical, equiaxial particles, tubes, flakes, rods and fibres	Depth resolution up to 10 nm Can be used in combination with TEM	Tends to work best at relatively low atomic number Need very thin specimens <30 nm. Intensity week for energy losses >300 eV
XRF (X-ray fluorescence)	Composition	Inorganic, organic, carbon-based, complex materials (coated, core–shell) Spherical, equiaxial particles, tubes, flakes, rods and fibres	Despite the fact that it has 100 nm spatial resolution, it can be used for nanomaterials chemical identification	Fairly high limits of detection Possibility of matrix effects
XPS (X-ray photoelectron spectroscopy)	Surface chemistry composition	Inorganic, organic, carbon-based, complex materials (coated, core–shell) Spherical, equiaxial particles, tubes, flakes, rods and fibres	Very sensitive for surface chemistry	Only detects the elements in a very thin layer close to surface (about 10 nm) Very sensitive, requires careful sample preparation Ultrahigh Vacuum

(continued)

Table 12.3 (continued)

Method	Physicochemical characteristics analysed/parameters	Type of material	Strengths	Limitations
EDX ((EDS)-energy dispersive X-ray spectrometry)	Composition Surface chemistry	Inorganic, organic, carbon-based, complex materials (coated, core–shell) Spherical, equiaxial particles, tubes, flakes, rods and fibres	Gives information on elements from beryllium to einsteinium Detection limit 0.1 atomic %	Used in conjunction with TEM or SEM Not real surface science technique Not suitable for non-conductive samples Elements of low atomic number can be difficult to detect (overlap with X-ray peaks)
TGA/DSC (thermogravimetric analysis/differential scanning calorimetry)	Composition Thermal properties Crystalline phase	Inorganic, organic, carbon-based, complex materials (coated, core–shell) Spherical, equiaxial particles, tubes, flakes, rods and fibres	DSC can operate from—180 °C to 1750 °C	Large samples required: 1–50 mg Destructive method
ICP-MS (inductively coupled plasma–mass spectrometry)	Composition Surface chemistry	Inorganic, organic, carbon-based, complex materials (coated, core–shell) Spherical, equiaxial particles, tubes, flakes, rods and fibres	Very sensitive: limits of detection 1 ng/l	Lower sensitivity towards some of the lighter elements, such as H, I, O, N, C, S
ICP-OES (inductively coupled plasma–optical emission spectrophotometry)	Composition Surface chemistry	Inorganic, organic, carbon-based, complex materials (coated, core–shell) Spherical, equiaxial particles, tubes, flakes, rods and fibres	Can basically characterise all types of material	2–3 orders of magnitude lower sensitivity than ICP-MS

Reproduced from Rasmussen et al. [37], with permission. © Elsevier chemicals in general. In addition to the Guidance Document on Sample Preparation and Dosimetry [29], a number of specific protocols have emerged (NIST, link http://www.nist.gov/mml/np-measurement-protocols.cfm, PROSPEcT [35], Nanogenotox [25], and ENPRA (Jacobsen et al. 2010) and NanoDefine [57]) and recommendations [56]. These may become references for future studies. Hartmann et al. [14] reviewed several dispersion protocols for dispersing NMs in aqueous media, analysing which aspects of dispersion procedures could be harmonised. Currently, a detailed generic dispersion protocol for all types of NMs is not available. Taking into consideration the specificities of NMs and test methods requirements, it is unlikely that such a protocol will be developed in the near future.

12.4 Metrology, Harmonisation and Standardisation Needs and Initiatives for Nanomaterials

Through applying metrology to nanoscience and nanotechnology, the methods and techniques for characterising NMs can be improved and developed. Metrology has been defined as "2.2 (2.2) metrology science of measurement and its application. NOTE Metrology includes all theoretical and practical aspects of measurement, whatever the measurement uncertainty and field of application" [3]. In their brief description of metrology as the science of measurement, Howarth and Redgrave [15] outline three types of activities: "(1) the definition of internationally accepted units of measurement (e.g. the metre); (2) the realisation of units of measurement by scientific methods (e.g. the realisation of a metre through the use of lasers); and (3) the establishment of traceability chains by determining and documenting the value and accuracy of a measurement and disseminating that knowledge (e.g. the documented relationship between the micrometre screw in a precision engineering workshop and a primary laboratory for optical length metrology)". They furthermore note that metrology can be divided into three relevant categories: (1) scientific, (2) industrial and (3) legal metrology. To ensure the quality of a wide range of activities and processes, inputs such as metrological activities, calibration, testing and measurements are needed. Additionally, traceability is gaining importance on a par with the measurement itself, and at each level of the traceability chain, recognition of metrological competence can be established through mutual recognition agreements or arrangements, and accreditation and peer review.

As a consequence of their size, NMs pose unique measurement challenges for characterisation (see, e.g., [22, 47]), as also reflected in the lack of harmonised, standardised and (regulatory) recognised methods for NMs, including methods for physicochemical properties, to obtain quality data. For cosmetic ingredients, SCCS also noted [58] these issues, e.g. that toxicological hazards of chemical substances are measured and conventionally expressed in weight or volume units (such as mg/kg, or mg/l), which may not be appropriate for NMs [51]. Appropriate dose metrics for NMs are currently being discussed. Until there is agreement on suitable dose metrics, it is important that tests on NMs are evaluated using more than one dose metrics, e.g. weight/volume concentration, particle number concentration and surface area. Furthermore, SCCS noted [58] that although most analytical methods used routinely for chemical substances have not yet been validated for NMs, a careful choice from these method(s) should provide sufficient means to gather adequate characterisation data. An additional issue is that many of the physico-chemical properties relevant for NMs are method-defined; i.e., the value of the property measured is determined by how (by which method) it is measured [48]. Thus, the use of more than one method generally adds more confidence to the measured values, and the SCCS recommended (SCCS/1484/12) applying more than one method for the measurement of particle size distribution and particle imaging.

Two types of harmonised and standardised methods are applied for testing and characterising chemicals, including NMs: the regulatory test guidelines (harmonised methods) that are developed in the OECD Test Guidelines Programme (TGP), and the methods that are developed by standardisation bodies, notably the International Organization for Standardization (ISO) via its Technical Committee (TC) 229 "Nanotechnologies"; in Europe the European Committee for Standardization (CEN) is developing European standards. These organisations have published several documents describing methods for characterising NMs.

OECD Test Guidelines

Methods recognised worldwide for the regulatory testing of chemicals are developed and agreed in the OECD TGP. When generating data by applying the OECD test guidelines (and good laboratory practice), the data fall under a legally binding instrument, the Mutual Acceptance of Data (MAD) agreement [8] that facilitates the international acceptance of information for the regulatory safety assessment of chemicals. With regard to nanomaterials, an analysis of the OECD test guidelines [27] concluded that the physicochemical characterisation methods described in the OECD test guidelines for chemicals are largely not, in their current form, directly applicable to NMs. Furthermore, the WPMN suggests a number of additional physicochemical properties for the characterisation of NMs compared to that for chemicals (see Table 12.2), as necessary information for the toxicological assessment of NMs. In order to assure that the OECD test guidelines are relevant for and applicable to NMs, the OECD is currently reviewing and adapting existing test guidelines, as necessary, as well as developing new ones to ensure that all relevant endpoints are covered. Table 12.4 gives an overview of the currently ongoing or planned work on OECD test guidelines.

ISO Documents

ISO/TC 229 works on standardising methods for characterisation of NMs which is largely done in ISO/TC 229 "Nanotechnologies", though also other TCs under ISO work on relevant documents. Table 12.5 gives an overview of methods developed, or in development, for characterisation of NMs, as well as documents for

Table 12.4 Overview of OECD Test Guidelines (TGs) or Guidance Documents (GDs) for physicochemical methods proposed by the OECD WPMN to be developed to ensure the availability of regulatory test methods also for nanomaterials (February 2018)

Title of OECD test guideline/guidance document to be revised or developed
New TG on dissolution rate of nanomaterials in aquatic environment
Test Guideline or Guidance Document on Particle size and size distribution
New GD (Decision-Tree) on the agglomeration and dissolution behaviour of nanomaterials in aquatic media
New Test Guideline on Determination of the Specific Surface Area of Manufactured Nanomaterials
Identification and quantification of the surface chemistry and coatings on nano- and microscale materials
Determination of solubility and dissolution rate of nanomaterials in water and relevant synthetic biologically mediums, Denmark
New TG on Determination of the Dustiness of Manufactured Nanomaterials
GD on Aquatic (Environmental) Transformation of Nanomaterials
GD on Biopersistent/Biodurable manufactured nanomaterials

Table 12.5 ISO/TC 229. Characterisation methods published or under development (January 2018)

ISO/TC 229 published methods for characterisation of NMs
ISO 9276-6:2008 Representation of results of particle size analysis—Part 6: Descriptive and quantitative representation of particle shape and morphology
ISO/TS 10798 Characterization of single wall carbon nanotubes using scanning electron microscopy and energy dispersive X-ray spectrometry analysis
ISO/TS 11308 Characterization of single wall carbon nanotubes using thermogravimetric analysis
ISO/TS 12025 Quantification of nano-object release from powders by generation of aerosols
ISO/TS 13278 Determination of elemental impurities in samples of carbon nanotubes using inductively coupled plasma mass spectrometry
ISO/TS 16195 Nanotechnologies—Guidance for developing representative test materials consisting of nano-objects in dry powder form
ISO/TS 10797 Nanotechnologies—Characterization of single-wall carbon nanotubes using transmission electron microscopy
ISO/TS 10867:2010 Nanotechnologies—Characterization of single-wall carbon nanotubes using near infrared photoluminescence spectroscopy
ISO/TS 10868:2017 Nanotechnologies—Characterization of single-wall carbon nanotubes using ultraviolet-visible-near infrared (UV-Vis-NIR) absorption spectroscopy
ISO/TR 10929:2012 Nanotechnologies—Characterization of multiwall carbon nanotube (MWCNT) samples
ISO/TS 11251:2010 Nanotechnologies—Characterization of volatile components in single-wall carbon nanotube samples using evolved gas analysis/gas chromatograph-mass spectrometry
ISO/TR 11811:2012 Nanotechnologies—Guidance on methods for nano- and microtribology measurements

(continued)

ISO/TS 11888:2017 Nanotechnologies—Characterization of multiwall carbon nanotubes—Mesoscopic shape factors

ISO/TS 11931:2012 Nanotechnologies—Nanoscale calcium carbonate in powder form—Characteristics and measurement

ISO/TS 11937:2012 Nanotechnologies—Nanoscale titanium dioxide in powder form—Characteristics and measurement

ISO/TS 13278:2017 Nanotechnologies—Determination of elemental impurities in samples of carbon nanotubes using inductively coupled plasma mass spectrometry

ISO/TR 18196:2016 Nanotechnologies—Measurement technique matrix for the characterization of nano-objects

ISO/TS 19590:2017 Nanotechnologies—Size distribution and concentration of inorganic nanoparticles in aqueous media via single particle inductively coupled plasma mass spectrometry

ISO/TR 19716:2016 Nanotechnologies—Characterization of cellulose nanocrystals

ISO/TS 16195:2013 Nanotechnologies—Guidance for developing representative test materials consisting of nano-objects in dry powder form

ISO/TS 17200:2013 Nanotechnology—Nanoparticles in powder form—Characteristics and measurements

ISO/TS 17466:2015 Use of UV-Vis absorption spectroscopy in the characterization of cadmium chalcogenide colloidal quantum dots

ISO/TC 229 published methods for characterisation of test suspensions containing NMs

ISO 10808:2010 Nanotechnologies—Characterization of nanoparticles in inhalation exposure chambers for inhalation toxicity testing

ISO/TR 13014:2012 (and ISO/TR 13014:2012/Cor 1:2012) Nanotechnologies—Guidance on physico-chemical characterization of engineered nanoscale materials for toxicologic assessment

ISO/TR 16196:2016 Nanotechnologies—Compilation and description of sample preparation and dosing methods for engineered and manufactured nanomaterials

ISO/TR 16197:2014 Nanotechnologies—Compilation and description of toxicological screening methods for manufactured nanomaterials

ISO/TS 19337:2016 Nanotechnologies—Characteristics of working suspensions of nano-objects for in vitro assays to evaluate inherent nano-object toxicity

ISO/TC 229 work in progress, or proposed, on methods for characterisation of NMs

\ISO/TR 19733 Matrix of characterization and measurement methods for graphene

ISO/TR 20489 Separation and size fractionation for the characterisation of metal based nanoparticles in water samples

ISO/TS 19805 On-line/off-line techniques for characterizing size distribution of airborne nanoparticle populations

ISO/19749 Determination of size and size distribution of nano-objects by scanning electron microscopy

ISO/21363 Nanotechnologies—Protocol for particle size distribution by transmission electron microscopy

ISO/TS 21362 Nanotechnologies—Application of Field Flow Fractionation to the Characterization of Nano-Objects

21356 Nanotechnologies—Structural characterization of graphene

21346 Nanotechnologies—Characterization of cellulose elementary fibril samples

21357 Nanotechnologies—Measurement of average nanoparticle size and assessment of agglomeration state by static multiple light scattering (SMLS) in concentrated media

(continued)

Table 12.5 (continued)

ISO/TC 229 work in progress, or proposed, on methods for characterisation of NMs
Nanotechnologies—Identification and quantification of airborne nano-objects in a mixed dust industrial environment
Nanotechnologies—3-D Tomography in a transmission electron microscope (TEM)
23151 Nanotechnologies—Particle size distribution for cellulose nanocrystals
IEC 62565-3-1 Nanomanufacturing—Material specifications—Part 3-1: Graphene—Blank detail specification
IEC/NP 62607-6-3 Nanomanufacturing—Key control characteristics—Part 6-3: Graphene—Characterization of graphene domains and defects

characterisation of suspensions tested. As seen from this list, methods may be very specific and address one technique applied to one type of material.

12.5 Use of Reference Materials

To support metrology for NMs, reference NMs are being developed for use in method validation and quality control, so that the results obtained by testing different sub-samples, at different places and times and with different methods, can be meaningfully compared. For specific purposes, these materials also need to have reliable, assigned property value(s). A dedicated ISO Committee deals with reference materials (RM) (ISO/REMCO) and addresses all generic measurement and testing needs across the main scientific disciplines physics, chemistry and biology, and it liaises with several international organisations (e.g. the World Health Organization, and the International Union of Pure and Applied Chemistry). REMCO introduced the following generic reference material definition (2015): a reference material is a "material, sufficiently homogeneous and stable with respect to one or more specified properties, which has been established to be fit for its intended use in a measurement process". For each of the measurands, corresponding with the properties of interest, homogeneity and stability must be determined, and a RM is a RM only for these specified properties. ISO 17034 [18] describes the conditions and terms for producing and using RMs that meet these requirements. "Certified reference material" (CRM) are RMs with known property value(s), and the term is proposed by ISO/REMCO with the following definition [17]: "reference material (RM) characterised by a metrologically valid procedure for one or more specified properties, accompanied by an RM certificate that provides the value of the specified property, its associated uncertainty and a statement of metrological traceability". CRMs are required for many applications within calibration, validation and quality assurance, and they are accompanied by a certificate containing all the relevant information. ISO 17034 and Guide 35 [18, 19], for example, provide metrologically valid procedures for the production and certification of reference materials. As CRMs allow the calibration of instruments and quality control of methods and laboratories based on metrological traceability [3], and thus ensure comparability of test results [11], they are fundamental to quality assurance.

Reference materials are also used to assure metrological traceability when testing NMs. Certain physicochemical properties relevant for NMs are method-defined, meaning that the value of the property measured depends on the method by which it is measured [48], and hence also, the reference property is linked to the method applied. This can be illustrated by the size of a particle which, for particles with a regular shape, can be defined by one value (e.g. the diameter of a sphere) or a small number of values (e.g. the length, width and height of a cuboid). For particles with irregular shapes, however, so-called apparent size values are obtained, which depend both on how the particle size is measured and on the evaluation of the measurement. Furthermore, in practice, size values are often derived from other measurands (e.g. light scattering intensity) or from other quantities (e.g. of a sedimentation or diffusion rate), and thus, the result of the measurement depends on the applied method, even for near-spherical particles. When the certified property values are method-defined, this has important implications for the use of test results, as in principle results cannot be converted from one method to another, e.g. size as measured by DLS to size as measured by SAXS. To partially overcome this, the JRC has developed the certified reference materials ERM-FD100 [2] and ERM-FD304 [13]. ERM-FD100 is a CRM of silica nanoparticles suspended in aqueous solution, with four different certified particle size values. Each of these values corresponds to a different measurement technique: electron microscopy (EM), centrifugal liquid sedimentation (CLS), small-angle X-ray scattering (SAXS) and dynamic light scattering (DLS). In addition to its certified property values, ERM-FD100 provides a tool enabling laboratories to ensure that the results of their DLS, CLS, EM and SAXS measurements of (near-)spherical nanoparticles are traceable to the SI unit metre and thereby comparable with other results obtained with the same methods. ERM-FD304 is also a colloidal silica with monomodal size distribution, but its polydispersity is higher than the polydispersity of ERM-FD100, and thus, the differences between the certified values for different method-defined particle size methods are more pronounced for ERM-FD304 than for ERM-FD100.

The OECD TGs also use the terms "reference substance" and "reference chemical", but these terms have a different meaning than those used by ISO. In the OECD TGs, the terms refer to materials required in the process of hazard identification, especially when evaluating responses in test systems [34] to compare the observed effects of an unknown substance to the known effects of a known substance in biological test systems. A wider knowledge than the elemental or molecular composition may be relevant for reference substances to be used for testing NMs for safety assessment, as NMs may have (eco)toxicological and other properties that depend also on material-specific parameters such as size, shape and surface area. The variations in these parameters could have a larger influence on the test results than variations in chemical composition within or between batches or between producers of nominally the same material. According to OECD MAD [8], new tests should be performed applying GLP ([10, 9, 26, 33]) and the principles of OECD Guidance Document 34 "on the validation and international acceptance of new or updated test methods for hazard assessment" [34].

Validation and, for OECD test guidelines, regulatory acceptance are important aspects when developing test methods. Validation is supported by (C)RMs, which are commonly used and required by international standards to ensure the comparability of measurement data obtained in each step of the method development and implementation cycle. Currently, a limited range of (C)RMs for NMs is available.

As a first step towards developing nano-(C)RMs, the European Commission's Joint Research Centre (JRC) established the JRC Repository for Representative Nanomaterials [60]. The repository contains industrially sourced representative test materials, each from a single batch. Roebben et al. [49] define representative test materials (RTMs) as follows: "a representative test material is a material from a single batch, which is sufficiently homogeneous and stable with respect to one or more specified properties and which implicitly is assumed to be fit for their intended use in the development of test methods which target properties other than the properties for which homogeneity and stability have been demonstrated." Each material from this repository originates from the same lot and batch of the respective industrial product, and thus when used for testing, one source of uncertainty is eliminated when comparing data obtained, e.g. from different laboratories.

12.6 Summary of Open Issues and Needs in This Field

In the EU, there are currently several regulatory definitions of the term "nanomaterial" in different legislative areas. The Cosmetic Products Regulation specifically defines "nanomaterial" differently as the later EC Recommendation, which is intended to be broadly applicable across different sectors. The EC Recommendation is currently being reviewed, which may lead to its revision. The intention behind having a broadly applicable definition is harmonisation of the regulatory use of the term "nanomaterial" across sectors (with the possibility of sector-specific provisions). This is to minimise legal uncertainty that would otherwise be caused if materials may be considered NMs under one piece of legislation, but not under another.

The only common property of all NMs is "size"/"size distribution" at the nanoscale, and agreed methods are needed to reliably measure "size"/"size distribution". Currently, the full characterisation needs for NMs are being discussed, and the set of physicochemical parameters may depend on the purpose of the characterisation, e.g. research, a specific toxicity test or regulatory requirements. Thus, also the characterisation needs in a regulatory context are still to be clarified.

Generally, methods for characterising NMs are becoming available, but there is a further need to standardise these methods to ensure their technical and regulatory acceptance, repeatability and validation and to agree on limits of application for each method. Regulatory need, usefulness and limitations of the test method are aspects of its relevance. The latter can be defined as a description of a relationship of the test to the effect (or property) of interest and whether it is meaningful and useful for a particular purpose. Relevance includes the extent to which the test correctly measures or predicts the (biological) effect or property of interest; i.e., it also incorporates consideration of the accuracy of a test method. Reliability is a measure of the extent that a test method can be performed reproducibly within and between laboratories over time, when performed using the same protocol. It is assessed by calculating intra- and inter-laboratory reproducibility and intra-laboratory repeatability. Validation is the process by which the reliability and relevance of a particular approach, method, process or assessment are established for a defined purpose [34]. Based on these definitions, most of the methods currently available are relevant and reliable, but not validated.

Many of the different regulatory areas have guidance on information needs which also address characterisation requirements for regulatory purposes, because physicochemical characterisation of NMs is a prerequisite for assessing their (eco)toxicological effects. Furthermore, safety evaluation and research on the safety of NMs are meaningful only if the NMs to be studied are well characterised, and the measured properties can be linked unequivocally with a specific NM. Several sets of physicochemical properties, which agree to some extent, have been suggested in various contexts as relevant for the (eco)toxicological assessment of NMs and may be used as a basis for deriving a future definitive list of relevant (regulatory) physicochemical properties. For CPR, the Scientific Committee on Consumer Safety has issued two guidance documents that also further detail the characterisation requirements for NMs used in cosmetic products. The testing chain needs to be fully transparent, traceable, including exact identification and characterisation of the tested material, over actually used SOPs to evaluation of the data. Methods are available, but given the complexity of NMs, there is no single characterisation technique that can be used to characterise one property for all NMs. Additionally, different methods may give different results for identical NM samples, and hence, where possible, more than one method should be applied for measuring a specific property to obtain a comprehensive characterisation and to compensate for the weaknesses of individual methods.

Nanoparticles may consist of several layers, and the outer may be functionalised, inducing different properties to the surface compared to the core of the particle, thus influencing the behaviour of the particle; it is an important descriptor for NMs (e.g. [12]). When performing physicochemical characterisation of NMs, a layered particle design needs to be considered. It is not always possible to directly measure the property of interest, e.g. size or surface charge, and often, the instrument's measurand needs to be converted to obtain information on that property, thus introducing an additional source of uncertainty. An example is DLS for which the intensity of scattered light is measured, which is then converted to the desired particle size information. In the conversion step, it is assumed that the measured particles are spherical, and when the real particles have other shapes, it results in errors in the determined particle sizes. Therefore, a careful interpretation of the converted results is needed, taking into account uncertainties, limits and bias originating both from the indirect measurement and the conversion [1].

Whereas methods for characterisation of pristine NMs are being developed and in certain cases become more and more mature, methods that can characterise NMs at points of interaction with organisms are still in their infancy and more research and development is needed. One important issue is the formation of a corona of material, e.g. proteins, from the dispersion or biological medium surrounding the NMs [23], and further investigation is needed on the degree to which this needs to be characterised.

Sample preparation and dispersion in test media are a critical issue both for the reliable physicochemical characterisation of NMs and for interpreting results of toxicity tests. At this time, no universal dispersion protocol exists that would be applicable to all types of NMs in all types of media.

To ensure correct data interpretation and repeatability of test results, appropriately collected data from sample preparation to measurement should be documented in as much detail as possible. Several data formats targeting the testing of NMs have been developed, and the use of widely accepted formats to report data is advantageous for the comparability and sharing of the results. Such formats include ISA-TAB-Nano, a specification for sharing NM research data in spreadsheet-based format [59], and the NANoREG templates for data reporting [61]. The OECD has suggested harmonised templates for some physicochemical endpoints relevant for NMs (OECD website). Furthermore, data management and analysis infrastructures have been developed building on such standardised formats (e.g. [20]). Additionally, a data completeness check is highly recommended and should be regarded as an extension of NM characterisation.

For regulatory testing, validated OECD test guidelines targeting the special requirements for physicochemical characterisation of NMs are still lacking, and thus, each NM is currently tested on a case-by-case basis.

The traditional measurand for the dose in (eco)toxicological testing and for safety assessment is milligram of chemical per kilogram of test organism. More than a decade ago, SCENIHR noted that for nanoparticles, "for the determination of dose–response relationships, special attention should be given to the expression of the metrics of the nanoparticle dose since mass concentration is not necessarily the best description of dose for these materials and number concentration and surface area are likely to be more appropriate" [51], an observation that is still valid.

The following main challenges have been identified in relation to assessing the safety of NMs: understanding which physicochemical properties most significantly influence biological effects, thus being most relevant for the safety of NMs, is still incomplete. Many physicochemical properties, e.g. the zeta potential and the state of agglomeration, depend on the experimental conditions. Potentially also time-dependent, these properties may change with contact time with the media and with the ageing of the NM, and moreover, there is often batch-to-batch variability of NMs. As highlighted by OECD and other organisations, there is a lack of guidelines specifically suited for physicochemical characterisation of NMs.

As far as possible, data generation, either by direct measurement or by derivation of a quantity from measured data, should be performed according to agreed standards (e.g. by ISO) or test guidelines (e.g. by OECD). However, the majority of these test guidelines and standards for determining certain physicochemical properties were developed for chemicals, and they may not be applicable to NMs. For addressing regulatory information requirements, the OECD has started to adapt existing OECD TGs to NMs and to develop new TGs, including TGs for physicochemical endpoints. ISO is also very active in standardising techniques for measuring the physicochemical properties of (certain groups of) NMs. In addition to standardised and agreed procedures, reference materials and certified reference materials are needed to ensure the traceability and reliability of measurements and the quality of the data, as well as to improve the comparability of data generated by different laboratories. It is essential that the chain of data collection from sample preparation to measurement is documented in as much detail as possible, regardless of whether international standards, test guidelines or even case-specific protocols are followed.

High quality of the data is important, and to understand its quality, it is useful to assess the measured data and whether the experimental protocols were followed appropriately. Data quality is linked to compliance with both GLP (OECD website) and with harmonised and standardised test guidelines and protocols. The application of the agreement on Mutual Acceptance of Data (MAD) requires data quality. Thus, sufficient metadata should be provided for appropriate completeness and quality of any measured data. NM identification and characterisation based on physicochemical measurements are only meaningful if the corresponding experimental protocols are adequately described, as many of the physicochemical properties of NMs are context-dependent. (Future) availability of standardised methods, procedures and guidance will support the application of MAD and improve capacity to meaningfully assess the safety of NMs.

Thus, many of the issues for physicochemical characterisation of NMs are not specific to cosmetic products, but are general issues independent from the use of the NM in question.

References

1. Babick F, Mielke J, Wohlleben W, Weigel S, Hodoroaba VD. How reliably can a material be classified as a nanomaterial? Available particle-sizing techniques at work. J Nanoparticle Res. 2016;18:158. https://doi.org/10.1007/s11051-016-3461-7.
2. Braun A, Franks K, Kestens V, Roebben G, Lamberty MA, Linsinger T. Certification of equivalent spherical diameters of silica nanoparticles in water—Certified reference material ERM®-FD100. EUR 24620 EN; 2011. https://doi.org/10.2787/33725.
3. Bureau International des Poids et Mesures (BIPM). International vocabulary of metrology—basic and general concepts and associated terms (VIM). 3rd ed. 2008 version with minor corrections; 2008. Available at https://www.bipm.org/en/publications/guides/.

4. Classification, Labelling and Packaging (CLP) Regulation. Regulation (EC) No 1272/2008, OJ L 353, 31 Dec 2008.
5. Commission Recommendation of 18 October 2011 on the definition of nanomaterial. Official Journal of the European Union. 2011/696/EU. p. 38–40.
6. Commission Regulation (EU) No 10/2011 of 14 January 2011 on plastic materials and articles intended to come into contact with food. OJ L328; 2011. p. 20–29.
7. Commission Regulation (EU) 2018/1881 of 3 December 2018 amending Regulation (EC) No 1907/2006 of the European Parliament and of the Council on the Registration, Evaluation, Authorisation and Restriction of Chemicals (REACH) as regards Annexes I, III,VI, VII, VIII, IX, X, XI, and XII to address nanoforms of substances OJ L 308, 4.12.2018, p. 1–20.
8. Decision of the Council concerning the Mutual Acceptance of Data in the Assessment of Chemicals. 12 May 1981—C(81)30/FINAL; 1981. Available at http://www.oecd.org/env/ehs/mutualacceptanceofdatamad.htm.
9. Directive 2004/9/EC on the inspection and verification of good laboratory practice (GLP). OJ L 50; 2004. p. 28–43.
10. Directive 2004/10/EC on the harmonisation of laws, regulations and administrative provisions relating to the application of the principles of good laboratory practice and the verification of their applications for tests on chemical substances, OJ L 50; 2004. p. 44–59.
11. Emons H, Linsinger TPJ, Gawlik BM. Reference materials: terminology and use. Can't one see the forest for the trees? Trends Anal Chem. 2004;23:442–9.
12. European Chemicals Agency. How to prepare registration dossiers that cover nanoforms: best practices; 2017. Available at https://echa.europa.eu/documents/10162/13655/how_to_register_nano_en.pdf/f8c046ec-f60b-4349-492b-e915fd9e3ca0.
13. Franks K, Braun A, Charoud-Got J, Couteau O, Kestens V, Lamberty A, Linsinger TPJ, Roebben G. Certification report. Certification of the equivalent spherical diameters of silica nanoparticles in aqueous solution certified reference material ERM®-FD304. Available at http://publications.jrc.ec.europa.eu/repository/bitstream/JRC67374/reqno_jrc67374_ermfd304_final%20report%20pdf%20version.pdf.
14. Hartmann NB, Jensen KA, Baun A, Rasmussen K, Rauscher H, Tantra R, Cupi D, Gilliland D, Pianella F, Riego Sintes JM. Techniques and protocols for dispersing nanoparticle powders in aqueous media—is there a Rationale for Harmonization? J Tox Environ Health, Part B. 2015;00:1–28.
15. Howarth P, Redgrave F. Metrology in short; 2008. Available at http://resource.npl.co.uk/international_office/metrologyinshort.pdf.
16. ISO/TS 80004-2:2015(en) Nanotechnologies—Vocabulary—Part 2: Nano-objects. Geneva: ISO; 2015.
17. ISO Guide 30:2015. Reference materials—selected terms and definitions. Geneva: ISO.
18. ISO 17034:2016 General requirements for the competence of reference material producers. Geneva: ISO; 2016.
19. ISO Guide 35:2017. Reference materials—guidance for characterization and assessment of homogeneity and stability. Geneva: ISO.
20. Jeliazkova N, Chomenidis C, p Doganis P, Fadeel B, Grafström R, Hardy B, Hastings J, Hegi M, Jeliazkov V, Kochev N, Kohonen P, Munteanu CR, Sarimveis H, Smeets B, Sopasakis P, Tsiliki G, Vorgrimmler D and Willighagen E. The eNanoMapper database for nanomaterial safety information. Beilstein J Nanotechnol. 2015;6:1609–34.
21. Johnston H, Brown D, Kermanizadeh A, Gubbins E, Stone V. Investigating the relationship between nanomaterial hazard and physicochemical properties: Informing the exploitation of nanomaterials within therapeutic and diagnostic applications. J Control Release. 2012;164:307–13.
22. Linsinger TPJ, Roebben G, Solans C, Ramsch R. Reference materials for measuring the size of nanoparticles. Trends Analyt Chem. 2011;30:1827.

23. Lundqvist M, Stigler J, Elia G, Lynch I, Cedervall T, Dawson KA. Nanoparticle size and surface properties determine the protein corona with possible implications for biological impacts. PNAS. 2008;105(38):14265–70. https://doi.org/10.1073/pnas.0805135105).

24. Lövestam G, Rauscher H, Roebben G, Sokull-Klüttgen B, Gibson N, Putaud J-P, Stamm H. Considerations on a definition of nanomaterial for regulatory purposes. European Commission Joint Research Centre. Luxembourg: Publications Office of the European Union; 2010. ISBN 978-92-79-16014-1. EUR 24403 EN.

25. Nanogenotox dispersion protocol. Available at http://www.nanogenotox.eu/files/PDF/Deliverables/nanogenotox%20deliverable%203_wp4_%20dispersion%20protocol.pdf.

26. OECD. Harmonised Templates for reporting chemical test summaries: OHT 101-113: Additional physicochemical properties of nanomaterials, http://www.oecd.org/ehs/templates/templates.htm.

27. OECD. Preliminary Review of OECD Test Guidelines for their Applicability to Manufactured Nanomaterials (2009). ENV/JM/MONO(2009)21. Available at URL: http://www.oecd.org/officialdocuments/publicdisplaydocumentpdf/?doclanguage=en&cote=env/jm/mono(2009)21.

28. OECD. Guidance Manual for the Testing of Manufactured Nanomaterials: OECD's Sponsorship programme; first revision. ENV/JM/MONO(2009)20/REV Available at URL: http://www.oecd.org/officialdocuments/publicdisplaydocumentpdf/?cote=env/jm/mono%282009%2920/rev&doclanguage=en.

29. OECD. Guidance on Sample Preparation and Dosimetry for the Safety testing of Manufactured Nanomaterials. ENV/JM/MONO(2012)40; 2012.

30. OECD. Report of the OECD Expert Meeting on the Physical Chemical Properties of Manufactured Nanomaterials and Test Guidelines. ENV/JM/MONO(2014)15; 2014. Available at URL: http://www.oecd.org/officialdocuments/publicdisplaydocumentpdf/?cote=ENV/JM/MONO(2014)15&docLanguage=En.

31. OECD. Physical-Chemical Properties of Nanomaterials: Evaluation of Methods Applied in the OECD-WPMN Testing Programme. ENV/JM/MONO(2016)7; 2016.

32. OECD. Physical-Chemical Parameters: Measurements and Methods Relevant for the Regulation of Nanomaterials. OECD Workshop Report. ENV/JM/MONO(2016)2.

33. OECD. Good Laboratory Practice. IEC/NP 62607-6-3 Nanomanufacturing—Key control characteristics—Part 6-3: Graphene—Characterization of graphene domains and defects. http://www.oecd.org/chemicalsafety/testing/goodlaboratorypracticeglp.htm.

34. OECD Series on Testing and Assessment Number 34 Guidance document on the validation and international acceptance of new or updated test methods for hazard assessment. ENV/JM/MONO(2005)14; 2005.

35. PROSPEcT Protocol for Nanoparticle Dispersion. 2010. Available at http://www.nanotechia.org/sites/default/files/files/PROSPECT_Dispersion_Protocol.pdf,.

36. Rasmussen K, Rauscher H, Gottardo S, Hoekstra E, Schoonjans R, Peters R, Aschberger K. Regulatory status of nanotechnologies in food in the EU. In: Fadeel B, Pietroiusti A, Shvedova AA. Nanomaterials for food applications. Elsevier; 2019.

37. Rasmussen K, Rauscher H, Mech A. Physicochemical characterisation of nanomaterials. In: Fadeel B, Pietroiusti A, Shvedova AA. Adverse effects of engineered nanomaterials. 2nd ed. Published by Academic Press of Elsevier; 2017. ISBN: 978-0-12-809199-9.

38. Rasmussen K, Rauscher H, Mech A, Riego Sintes J, Gilliland D, González M, Kearns P, Moss K, Visser M, Groenewold M, Bleeker EAJ. Physico-chemical properties of manufactured nanomaterials—characterisation and relevant methods. An outlook based on the OECD Testing Programme. Regul Toxicol and Pharmacol. 2018;92:8–28.

39. Rauscher H, Roebben G, Amenta V, Boix Sanfeliu A, Calzolai L, Emons H, Gaillard C, Gibson N, Linsinger T, Mech A, Quiros Pesudo L, Rasmussen K, Riego Sintes J, Sokull-Klüttgen B, Stamm H, Towards a review of the EC recommendation for a definition of the term "nanomaterial"; Part 1: Compilation of Information Concerning the Experience with the Definition; 2014. European Commission, Joint Research Centre from http://publications.jrc.ec.europa.eu/repository/bitstream/JRC89369/lbna26567enn.pdf.

40. Regulation (EC) No 1907/2006 of the European Parliament and of the Council of 18 December 2006 concerning the Registration, Evaluation, Authorisation and restriction of Chemicals (REACH), establishing a European Chemicals Agency. OJ L396 (1); 2006. 1–849.
41. Regulation (EC) No 1223/2009 of the European Parliament and of the Council of 30 November 2009 on Cosmetic Products, O. J., L342, 59; 2009.
42. Regulation (EC) No 1107/2009 concerning the placing of plant protection products on the market and repealing Council Directives 79/117/EEC and 91/414/EEC. OJ L 309; 2009. p. 1–50.
43. Regulation (EU) No 1169/2011 on the provision of food information to consumers. OJ L 304/18; 2011.
44. Regulation (EU) 528/2012 concerning the making available on the market and use of biocidal products, OJ L 167/1; 2012.
45. Regulation (EU) 2015/2283 on novel foods, OJ L 327; 2015.
46. Regulation (EU) 2017/745 on medical devices, amending Directive 2001/83/EC, Regulation (EC) No 178/2002 and Regulation (EC) No 1223/2009 and repealing Council Directives 90/385/EEC and 93/42/EEC. OJ L 117/1; 2017.
47. Roebben G, Emons H, Reiners G. Nanoscale reference materials. In: Murashov V, Howard J, editors. Nanostructure science and technology. New York: Springer; 2011. p. 53–75.
48. Roebben G, Linsinger T, Lamberty A, Emons H. Metrological traceability of the measured values of properties of engineering materials. Metrologia. 2010;47:S23–31.
49. Roebben G, Rasmussen K, Kestens V, Linsinger TPJ, Rauscher H, Emons H, H Stamm. Reference materials and representative test materials: the nanotechnology case. J Nanopart Res. 2013;15:1455.
50. Sahu YS. Nanomaterials market by type (carbon nanotubes, fullerenes, graphene, nano titanium dioxide, nano zinc oxide, nano silicon dioxide, nano copper oxide, nano cobalt oxide, nano iron oxide, nano manganese oxide, nano zirconium oxide, nano silver, nano gold, nano nickel, quantum dots, dendrimers, nanoclay and nanocellulose) and end-user (paint & coatings, adhesives & sealants, healthcare & life science, energy, electronics & consumer goods, personal care, and others)—global opportunity analysis and industry forecast; 2016. p. 2014–2022. https://www.alliedmarketresearch.com/nano-materials-market.
51. Scientific Committee on Emerging and Newly Identified Health Risks (SCENIHR), Opinion on the appropriateness of the Risk Assessment methodology in accordance with the technical guidance documents for new and existing substances for assessing the risks of nanomaterials; 2007. Available at URL: http://ec.europa.eu/health/ph_risk/committees/04_scenihr/docs/scenihr_o_010.pdf.
52. Scientific Committee on Consumer Safety. Guidance on the Safety Assessment of Nanomaterials in Cosmetics. SCCS/1484/12. 2012. Available at http://ec.europa.eu/health/scientific_committees/consumer_safety/docs/sccs_s_005.pdf.
53. Scientific Committee on Consumer Safety. Memorandum on "Relevance, Adequacy and Quality of Data in Safety Dossiers on Nanomaterials". SCCS/1524/13, revision of 27 March 2014. URL: https://ec.europa.eu/health/scientific_committees/consumer_safety/docs/sccs_o_142.pdf.
54. Scientific Committee on Consumer Safety. Opinion on Titanium Dioxide (nano form) as UV-Filter in sprays. Opinion adopted by written procedure on 19 January 2018. SCCS/1583/17; 2017. https://ec.europa.eu/health/sites/health/files/scientific_committees/consumer_safety/docs/sccs_o_206.pdf.
55. Stefaniak AB, Hackley VA, Roebben G, Ehara K, Hankin S, Postek MT, Lynch I, Fu W-E, Linsinger TPJ, Thünemann AF. Nanoscale reference materials for environmental, health and safety measurements: needs, gaps and opportunities. Nanotoxicology. 2013;7(8):1325–37.
56. Tantra R, Sikora A, Hartmann N, Riego Sintes J. Comparison of the effects of different protocols on the particle size distribution of TiO_2 dispersions. Particuology. 2014;19:35–44.

57. Technical Report D2.3: Standardised dispersion protocols for high priority materials groups; 2016. Available at http://www.nanodefine.eu/publications/reports/NanoDefine_TechnicalReport_D2.3. pdf.
58. The SCCS Notes of Guidance for the Testing of Cosmetic Ingredients and their Safety Evaluation. 9th revision. SCCS/1564/15, revised version of 25 April 2016. Available at http:// ec.europa.eu/health/scientific_committees/consumer_safety/docs/sccs_o_190.pdf.
59. Thomas DG, Gaheen S, Harper SL, Fritts M, Klaessig F, Hahn-Dantona E, Paik D, Pan S, Stafford GA, Freund ET, Klemm JD, Baker NA. ISA-TAB-Nano: a specification for sharing nanomaterial research data in spreadsheet-based format. BMC Biotechnol. 2013;13:2.
60. Totaro S, Cotogno G, Rasmussen K, Pianella F, Roncaglia M, Olsson H, Riego Sintes J, Crutzen H. The JRC nanomaterials repository: a unique facility providing representative test materials for nanoEHS research. Regulatory Toxicol Pharmacol. 2016;81:334–40.
61. Totaro S, Crutzen H, Riego Sintes J. Data logging templates for the environmental, health and safety assessment of nanomaterials; EUR 28137 EN; 2017. https://doi.org/10.2787/505397.

Production of Nanocosmetics

13

Carsten Schilde, Jan Henrik Finke, Sandra Breitung-Faes, Frederik Flach and Arno Kwade

Abstract

For successful product development and production of nanocosmetics, the interplay between material properties, formulation, process equipment, and process parameters must carefully be understood. Agglomeration or aggregation state of the starting material and its breakage behavior in combination with the desired product fineness define the necessary process equipment and its applied process parameters. Formulation with additives is crucial for the process performance as well as for product stability. Equipment should be operated in optimal parameter settings to combine the advantages of low energy consumption, high productivity, high product quality, and low product contamination. To realize complex nanocosmetic products, such as nanoparticle-loaded nanoemulsion droplets, process integration, and application of microsystem technology can assist.

Keywords

Wet dispersion · Wet milling · Emulsification · Process intensification · Process models

C. Schilde · J. H. Finke · S. Breitung-Faes · F. Flach · A. Kwade (✉)
Institute for Particle Technology and Center of Pharmaceutical Engineering PVZ, Technische Universität Braunschweig, Volkmaroder Straße 5, 38104 Brunswick, Germany
e-mail: a.kwade@tu-braunschweig.de

© Springer Nature Switzerland AG 2019
J. Cornier et al. (eds.), *Nanocosmetics*,
https://doi.org/10.1007/978-3-030-16573-4_13

13.1 Introduction

Nanosized particles and structures out of them get more and more important for cosmetic products. At first, the special optical behavior of nanosized zinc oxide and titanium oxide particles was used as sun creams. Moreover, submicron and nano-sized color pigments are applied in several cosmetic products. In addition to solid, suspended nanoparticles, emulsion droplets, and liposomes in the nanometer range are frequently applied in cosmetics. Especially, the combination of dispersing and emulsifying processes is of interest for complex, innovative products.

As feed material for the dispersing and milling processes either clusters of nanoparticles which are synthesized by precipitation, pyrolysis, and other means or coarse primary particles can be used. We distinguish between primary particles and secondary particles, which can be aggregates, agglomerates, and flocculates of primary particles. In aggregates, the already nanosized primary particles are connected by solid bonds, especially also at flat surfaces, in agglomerates by weak pointwise solid and especially physical bonds (e.g., strong van der Waals forces), and in flocculates by weak particle–particle interactions. Consequently, the strength of the particles and structures regarding fracture decreases from primary particles over aggregates and agglomerates to flocculates. For the fracture of primary particles and very strong aggregates with large-area solid bonds, real grinding or milling processes are required. For this purpose, today, mainly stirred media mills are applied. However, for strong aggregates of organic crystals, also high-pressure homogenizers might work. Against that, for the dispersing of normal aggregates, agglomerates, and flocculates, various dispersing machines are sufficient. The main criterion for the usability of a certain machine is the stress intensity which can be achieved to stress and, if the stress intensity is high enough, to break the primary and secondary particles. However, in all cases special care must be taken to stabilize the nanosized particles and structures after their fracture to avoid re-agglomeration, which is especially true for emulsions.

Accordingly, the following sub-chapters distinguish between dispersing of aggregates/agglomerates, fragmentation of droplets, and real breakage of primary particles and very strong aggregates with large-area solid bonds. The different dispersing/emulsification/milling machines and the stresses and their intensities acting in these machines are explained before examples of the application of these machines are shown. Moreover, a special case of high-pressure microsystems with different process layouts for dispersion and emulsification and their combination in a continuous process chain is presented. In the overall context, the importance of stabilization is explained, too. Based on this general description in an extra sub–chapter, the real grinding of nanoparticles with stirred media mills is discussed based on zinc oxide as exemplary material. An overview on common nanoparticles and their application in different cosmetic products is given in the following table (review articles: [1, 2]).

Inorganic nanoparticles	Application	Product features	Particle size
Titanium dioxide; zinc oxide	Sunscreen, day cream	Protection against UV radiation, transparent (smaller 50 nm), white pigment (larger 100 nm), antifungal, antibacterial, anti-inflammatory	15–400 nm
Carbon black	Mascara, eyeliner, kajal, eye shadow, eyebrow pencils, nail polishes	Black color pigment	15–500 nm primary particles
Silica; alumina	Toothpastes, decorative cosmetics, make-up, powders, hair colorations, hair-styling products, moisturizers	Abrasive/friction agent, filler, flow additive, drying agent, thickener, anti-aging, carrier system for active ingredients	5–50 nm primary particles
Gold nanoparticles	Cream, lotion, face pack, deodorant, antiaging creams	Antifungal, antibacterial, acceleration of blood circulation, anti-inflammatory, antiseptic, elasticity of skin, delaying aging process, vitalizing skin metabolism	5–400 nm
Silver nanoparticles	Deodorant	Antifungal, antibacterial, anti-inflammatory	5–200 nm
Hydroxyapatite nanoparticles	Toothpastes	Remineralization	Smaller 50 nm
Organic nanoparticles	Application	Product features	Particle size
Nanospheres	Antiwrinkle creams, moisturizing creams, antiacne creams	Deliver of active ingredients into deep skin layers	10–200 nm
Dendrimers	Hair care, skin care, nail care, shampoos, sunscreen, hair-styling gels, antiacne products	Controlled release of active ingredients from the inner core and the surface	2–20 nm
Nanoemulsion	Deodorants, sunscreens, shampoos, lotions, nail enamels, conditioners, hair serums	Transparent or translucent, low viscosity, high kinetic stability, high interfacial area, high solubilization capacity	50–200 nm
Liposomes	Anti-aging creams, moisturizing creams, sunscreen, beauty creams, shampoo, conditioner, antiperspirants, body sprays, deodorants, lipsticks	Delivery of fragrances, botanicals, and vitamins; softening and conditioning; treatment of hair loss	20 nm to several micrometers
Niosomes	Antiwrinkle creams, skin whitening, moisturizing cream, hair repairing shampoo, conditioner	Delivery of hydrophobic and hydrophilic compounds; enhanced skin penetration of ingredients	100 nm to 2 μm

13.2   Dispersing Nanoparticles for Nanocosmetics

In contrast to the synthesis of highly defined nanoparticles under laboratory conditions, as a rule in industrial scale nanoparticles are produced not as single primary
particles but as agglomerates or even strongly bonded aggregate structures due to
economic reasons (especially to ensure high production capacities and high solid
concentrations). Inorganic nanoparticles being applied in cosmetics, e.g., titanium
oxide, silica, alumina, zinc oxide, or even carbon black nanoparticles, are typically
produced in large scale via precipitation processes followed by drying or via
pyrolysis. As a result from pyrolysis, fractal aggregates with fused primary particles
[3] or from other synthesis processes with subsequent drying steps, spherical
aggregates with primary particles bonded via strong solid bonds at a high coordination number are produced [4]. Hence, the strength of these aggregates is far above
the strength of agglomerates and flocculates which are held together by attractive
neutral or polar particle–particle interaction forces [5]. This leads to a high resistance against fragmentation of the entire particulate structure and requires an intense
dispersion to achieve the beneficial size-related application properties in nanocosmetics. The dispersion process of aggregates is characterized via the three partial
steps: 1. wetting of the nanoparticulate structure, 2. breakage of the solid bonds to
obtain separated primary particles, and 3. steric or electrostatic stabilization to avoid
re-agglomeration [6, 7] (Fig. 13.1).

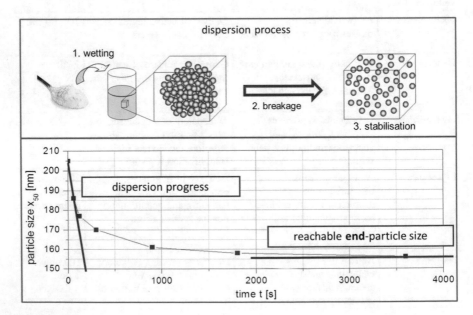

Fig. 13.1 Classification of the dispersing process steps and exemplary progress in median
particle size by dispersing nanoparticulate alumina (original graphic by Schilde [8, 9])

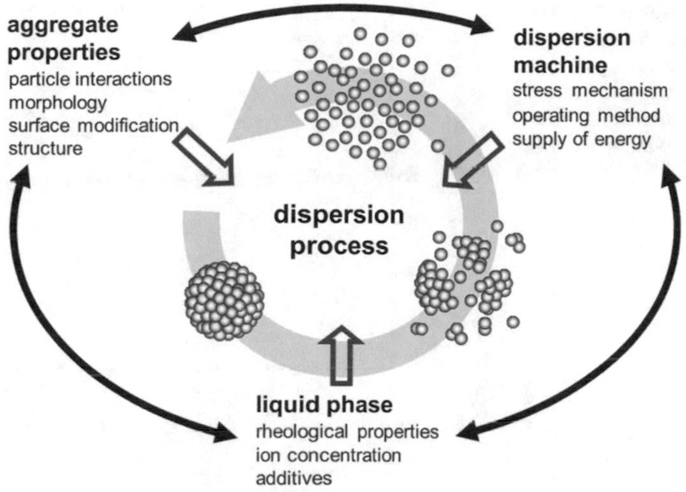

Fig. 13.2 Parameters affecting the dispersion process (original graphic by Schilde [8])

These three steps depend on the properties of the nanoparticulate aggregates, the fluid phase as well as on the characteristics and operation of the dispersing machine (see Fig. 13.2) [8, 10]. The strength of the nanoparticulate aggregates depends on the material and the aggregate structure, e.g., the size distribution, morphology of the primary and secondary particles, the average coordination number of primary particles, the fractal dimension, the surface modification and strength of solid bonds. The fluid phase influences the wetting and stabilization behavior as well as the stress intensity transferred by a dispersing machine. In the context of the fluid phase, the rheological properties, the ion concentration, and the additives are of special importance for wetting, stabilization, and stressing nanoparticulate aggregates. Finally, the dispersion machine determines the stress mechanism, the stress intensity, and frequency acting on the particulate structure. Besides the dispersion machine itself, operation and process parameters substantially determine the stress frequency and intensity.

Wetting of nanoparticulate powders
First step for a sufficient dispersing of nanoparticulate structures is the wetting of the large and often chemically modified particle surface and, thus, the displacement of the gaseous phase. This requires a good wettability of the nanoparticle surface with the fluid phase which can be characterized by a small contact angle of the solid–liquid interface [11]. The contact angle depends on the hydrophilic–hydrophobic behavior of both, the fluid and solid phase, as well as on other properties, e.g., type of additives and the system temperature. Besides the wetting itself, the suspension stability is significantly affected by the contact angle. The characterization of the wetting behavior for particulate materials, especially nanoparticles, is a major challenge, due to powder handling issues and the large specific surface area.

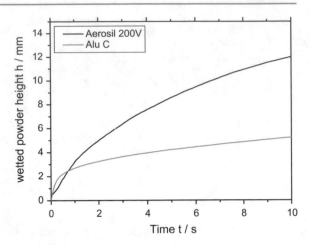

Fig. 13.3 Wetting kinetics as determined by the immersion test of pyrogenic nanoparticulate silica and alumina aggregates in water (original graphic by Schilde et al. [3])

Hence, various methods to characterize the wetting behavior particulate and nanoparticulate materials have been established, e.g., a modified Wilhelmi plate method or modified Washbourne method, or dispersion tests of nanoparticulate powders [10, 12–14]. As an example, Fig. 13.3 exemplarily shows the wetting kinetics of pyrogenic produced silica and alumina nanoparticles in water as fluid phase. The results of the immersion test show a slower wetting of the pyrogenic silica due to a more porous collective structure of the silica [3].

Stress mechanism, frequency, and intensity

Commonly, different devices, e.g., dissolvers, high-pressure and ultrasonic homogenizers, kneaders, three roller mills as well as stirred media mills, are used to disperse nanoparticulate aggregates. The dispersing devices differ in their stress mechanism, stress frequency, and stress intensity. Although often different stress mechanisms are acting on the aggregates during the dispersing process, usually one mechanism dominates the dispersion progress. According to Rumpf, apart from chemical and thermal stressing of particles, different mechanical stress mechanism can be differentiated in three basic mechanisms (see Fig. 13.4) [5, 6, 15]:

1. Shear or compression of the nanoparticulate aggregates between two surfaces with a stress intensity which is inversely proportional to the third power of the particle diameter, x, and proportional to the energy charged by the two surfaces [5, 16–18]:

$$SI \propto \frac{1}{V_{agg}} \propto \frac{1}{x^3}$$

2. Shear or compression stress of nanoparticulate aggregates on one surface via impact with a stress intensity which is proportional to the ratio of the kinetic energy of the particle and its mass. Hence, the stress intensity depends on the square of the relative velocity, v_{rel}, between the particle and surface:

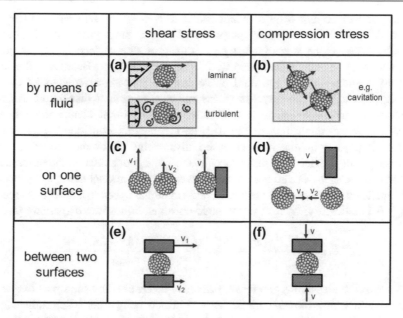

Fig. 13.4 Different stress mechanisms in dispersion processes (original graphic by Schilde [6])

$$SI \propto v_{rel}^2$$

3. Laminar or turbulent shear stress, τ, by the fluid phase with a stress intensity which depends on the particle size and flow regime. For dispersing at particle Reynold's numbers far below 1 and, thus, far away from the transition section for turbulent fluid flow, the stress intensity can be described by the model of Rumpf for stressing particles in a laminar (viscous) shear flow [9, 19] as function of the dynamic viscosity η and the shear rate γ:

$$SI \propto \tau = 2,5 \cdot \gamma \cdot \eta \quad \text{(sphere)}$$

In case of high particle Reynolds numbers, and thus, dispersing in turbulent fluid flow, the stress intensity acting on the particles can be characterized via the Kolmogorov theory of turbulence [20]. For dispersing nanoparticles, the model of the lower dissipation area (ratio of the particle size and the Kolmogorov dissipation length scale, λ, below a value of 3) is applicable. The stress intensity is a function of the fluid density, ρ_L, and the product of the specific power input ε and the kinematic viscosity v_L of the fluid to the power of 0.5 [9].

$$SI \propto \tau = \text{const} \cdot \rho_L \cdot (\varepsilon \cdot v_L)^{1/2} \quad \text{for } \frac{x}{\lambda} < 3$$

For shear or compression of nanoparticulate aggregates between two surfaces, the stress intensity is inversely proportional to the third power of the particle diameter, whereas the stress intensity via impact on one surface or via laminar or turbulent shear stress is independent of the particle size. However, the stress intensity is influenced by the fluid phase and the process parameters of the dispersing process. Additionally, the stress intensity within a dispersion device is subjected to a machine-dependent stress intensity distribution. Hence, the frequency of various stress intensities in the dispersion machine considerably affects the progress in product quality with increasing dispersing time or specific energy input of the dispersing device. The mass-specific energy, E_m, depends on the difference of the total power input, P, and the no-load power and the product mass, m_{solid} (as well as, if applicable, torque on the rotor, M, and rotational speed, n). The mass-specific energy is practically used for the energetic consideration of the dispersing process:

$$E_m = \int \frac{(P - P_0) \cdot dt}{m_{solid}} = \int \frac{2 \cdot \pi \cdot n \cdot (M - M_0) \cdot dt}{m_{solid}}$$

Whereas the dispersing progress is primarily determined by the stress frequency, the minimum achievable end particle size at long dispersing times is determined by the aggregate strength itself and the maximum stress intensity transferred via the dispersing machine. Typically, the aggregate strength increases with decreasing aggregate size [21, 22], mainly because aggregates break off first at the weak bondings. For this reason, the maximum stress intensity which can be transferred by the dispersion machine is very important as far as the product fineness is concerned. Figure 13.5 schematically shows the maximum stress intensity transferred by the dispersing machine and the corresponding aggregate strength as function of the aggregate size. For stress intensities higher than the aggregate strength, a successful reduction of the nanoparticulate aggregate size takes place. Since the stress intensity between two surfaces, e.g., between grinding media within a stirred media mill, increases with decreasing aggregate size, high stress intensities and high product finenesses can be

Fig. 13.5 Comparison of the aggregate strength with the stress intensities transmitted by the fluid phase and grinding media (original graphic by Schilde [6])

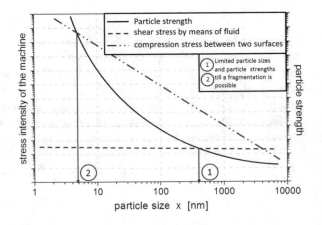

realized. In contrast, stressing nanoparticles by laminar or turbulent shear flow in a fluid, the stress intensity is independent of particle size and therefore limited. Very fine particles can only be dispersed by extremely high shear stresses.

Devices for dispersing nanoparticulate aggregates

The product fineness obtained during dispersing nanoparticulate aggregates is determined by the strength of the aggregates, by the properties of the fluid phase (e.g., its viscosity) as well as by the stress mechanism, the stress intensity, and the stress frequency. The latter three are determined by the dispersing machine and operating parameters, whereas the strength of the aggregates depends on its elementary structure, the properties of the solid bonds and particle interaction forces. Stress mechanisms and intensities for dispersing machines which were typically used in industry to produce nanoparticulate suspensions based on pyrogenic silica, titanium oxide, alumina, or carbon black are briefly summarized in the following [3] (see Fig. 13.6):

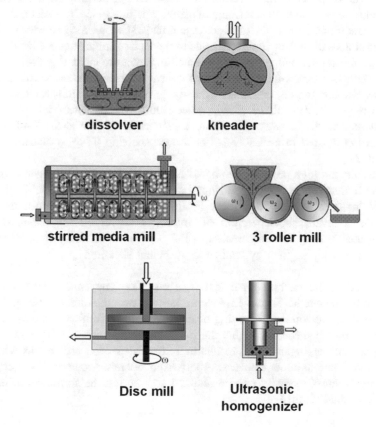

Fig. 13.6 Schematic illustration of different dispersing machines (original graphic by Schilde et al. [3])

- Dissolver: In a dissolver, particles are mainly stressed by fluid shear flow. Partially, a few larger aggregates in the micrometer size range are stressed via shear and impact on a surface, i.e., the dissolver disk. The stress intensity acting on nanoparticles is independent of particle size for laminar and turbulent fluid flow. Since the strength of nanoparticle aggregates increases with decreasing aggregate size, a dispersing limit arises for long dispersing times.
- Three-roller mill and kneader: In a three-roller mill and in a kneader, nanoparticles are stressed by high shear gradients in small dispersing gaps at high fluid viscosities due to high solid concentrations. The particle Reynolds numbers are far below 1, and thus, the stress intensity can be described by the model of Rumpf for stressing particles in a laminar high viscous shear flow [9, 19]. For long dispersing times, a dispersing limit arises.
- Stirred media mill: In a stirred media mill, the product particles are stressed in between grinding beads. The stress intensity is proportional to the kinetic energy of the grinding media and inversely proportional to the third power of the particle diameter [5, 6]. This drastic increase in stress intensity with decreasing particle size leads to the dispersion and even comminution of nanoparticles down to their primary particle size or even to the breakage of primary particles (see later section) in optimum conditions down to the true breakage limit [23]. However, in a wide variety of applications a dispersing limit arises due to the lack of sufficient stabilization when the specific surface and the particle interaction forces are increasing [24].
- Disk mill: In a disk mill, the stress mechanism is based on shear stress in a fluid phase. For this reason, the stress intensity for dispersing nanoparticulate aggregates is independent of the particle size. Due to a high rotational speed and a small gap width between the disks, significantly higher shear gradients are realized compared to a dissolver. For long dispersing times, a dispersing limit arises [3].
- Ultrasonic and high-pressure homogenizer: In high-pressure homogenizers, high shear gradients are realized due to small dispersing gaps and high fluid velocities. Additionally, cavitation takes place in ultrasonic and high-pressure homogenizers. As a result, shock waves and microjets acting on the nanoparticulate aggregates leading to their break-up [25]. The stress intensity by cavitation is considerably higher than by laminar or turbulent shearing.

By varying the operating parameters, stress frequency and intensity can be considerably increased. Figure 13.7 exemplarily shows the effect of tip speed and, thus, the shear rate on the dispersing of nanoparticulate alumina in a dissolver [5]. The higher the tip speed, the higher the stress frequency and intensity acting on the nanoparticulate aggregates. This results in a faster dispersion progress as well as in higher maximum product fineness. Apart from operation parameters, very high increases in stress intensity can be realized by varying the formulation, e.g., an increase in fluid viscosity.

Fig. 13.7 Effect of the tip speed on the product fineness for dispering nanoparticulate alumina in the dissolver (original graphic by Schilde et al. [5])

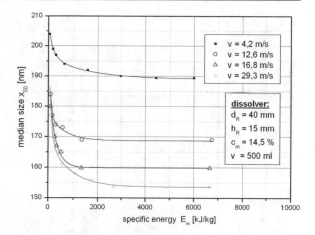

The kinetics of the median aggregate size as function of the dispersion time or specific energy input, E_m, are commonly described via different grinding and dispersing models according to Walker [26], Biedermann and Henzler [27], Winkler [28, 29], and Schilde et al. [9]. The volume related specific energy, E_V, is used more rarely [3].

$$\int_0^{E_{m,1}} dE_m = -C \cdot \int_{x_0}^{x_1} \frac{dx}{x^n} \Rightarrow x_1(E_m) = \left(\frac{E_{m,1} \cdot (n-1)}{C} + \frac{1}{x_0^{n-1}} \right)^{-\frac{1}{n-1}}$$ where : $n < 1$	Walker "general grinding law" [9, 26]
$$x = C \cdot \left(1 + A \cdot a \cdot \left(\frac{P}{V_C} \right)^b \cdot t \right)^{-\frac{1}{a}}$$	Biedermann and Henzler empiric power law [9, 27]
$$x = x_0 - x_0 \cdot p_{\text{dispersion}} = x_0 \cdot \left(1 - \left(1 - e^{-k \cdot \frac{V_{eff}}{V} \cdot t} \right) \cdot \left(1 - e^{-a \cdot \frac{P}{\rho \cdot V}} \right) \right)$$	Winkler "dispersing model" [9, 28, 29]
$$x(E_m) = x_0 + (x_{\text{end}} - x_0) \cdot \frac{E_m}{E_m + K_E}$$ where $x(K_E) = \frac{x_0 - x_{\text{end}}}{2}$	Schilde et al.'s stress frequency- and intensity-related model [9]

In Fig. 13.8, the different dispersing kinetics are exemplarily fitted to the dispersion of pyrogenic nanoparticles in different dispersing machines, i.e., kneader, three roller mill, dissolver, stirred media mill. Hereby, the models of Walker and of Biedermann and Henzler are not able to describe the attainment of a plateau due to reaching the respective dispersing limit. Thus, these models are more appropriate for well-stabilized grinding processes, where the "real grinding limited" [21] is not reached until long grinding times or high specific energies. Thus, both models fit well to the dispersion progress in a stirred media mill. In contrast, the models of Winkler and of Schilde et al. are able to fit dispersing processes, where a dispersing

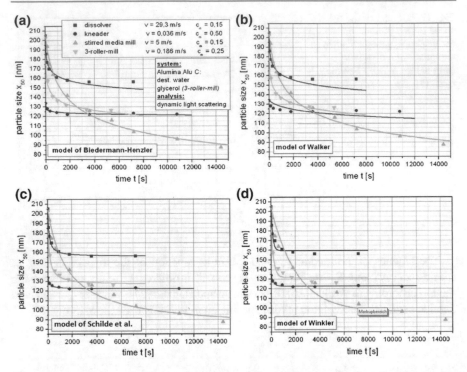

Fig. 13.8 Application of different dispersing models to dispersing kinetics of nanoparticulate alumina in different devices (original graphic by Schilde et al. [9])

limit is reached, even after short dispersing times. The kinetic of Schilde et al. is described by a hyperbolism function, which can theoretically be completely described with two data points. Thus, a dispersing process can accurately be described with only a few data points [24]. In this kinetic, x is the particle size as function of the dispersing time or specific energy input and x_{start} is the aggregate size of the feed particles at the beginning of the dispersing. The parameter x_{end} is the maximum reachable product fineness and is a function of the stress intensity acting on the aggregates. The parameter K_E is a function of the stress frequency in the dispersion process and has a significant influence for short dispersing times/specific energy inputs [9].

In order to compare the dispersing efficiency of different devices based on the achieved product fineness, Fig. 13.9 exemplarily shows the dispersion of pyrogenic alumina and silica nanoparticles in water as a function of the applied specific energy [3]. Within a dissolver and a disk mill, the nanoparticulate aggregates are stressed by high shear rates leading to a considerable energy dissipation into heat due to friction inside the fluid. The realized stress intensities are comparatively low resulting in high median aggregate sizes. Since a dissolver shows a broad distribution in stress intensities, the energetic efficiency of the dispersing process is poor compared with the disk mill. Additionally, higher shear rates are applied in the disk

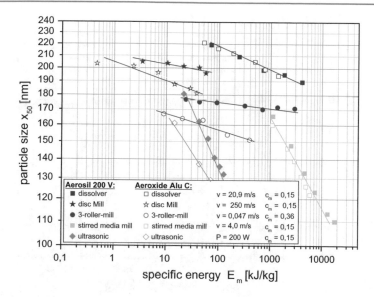

Fig. 13.9 Median particle size during dispersing of nanoparticulate alumina and silica in different dispersing machines (original graphic by Schilde et al. [3])

mill leading to slightly higher product finenesses. Typically, higher stress intensities can be applied in three roller mills and kneaders, since processing at much higher viscosities in laminar shear flow is possible. Moreover, especially in three roller mills high shear rates can be achieved by small roller distances. For this reason, high solid concentrations can be realized which enhance the dispersion efficiency [3, 9, 10] at high suspension viscosities or particle concentrations. In case of dispersing highly fractal silica nanoparticles with very small primary particle sizes, this effect is less pronounced. Fine nanoparticulate aggregates can be produced via ultrasonic homogenizers due to the high stress intensities induced via cavitation. However, this production method is restricted to laboratory and small technical scale. Typically, the increase in aggregate strength with decreasing size [6] restricts the maximum product fineness. In stirred media mills, the stress intensity strongly rises with decreasing aggregate size. Hence, the lowest particle sizes can be achieved by dispersing in stirred media mills.

The polydispersity of the produced particle size distributions in the machines differs substantially (see Fig. 13.10). Since the application properties are often related to the aggregate size, e.g., for good transparency of zinc oxide- or titanium oxide-based suntan lotions, the absence of coarse aggregates in the suspension can be of major importance. Figure 13.10 shows the value of $(x_{90} - x_{10})/x_{50}$ as characteristic parameter for the width of the particle size distribution. Reason for different polydispersity of the product suspensions is that the machines offer different ratios of the active dispersion volume to the total suspension volume or the machines are operated in different modes (circuit or passage operation).

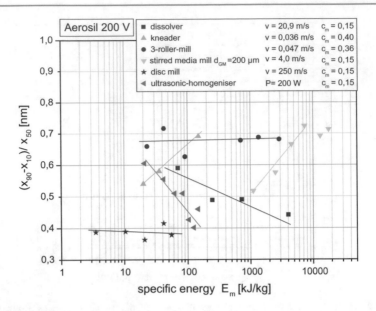

Fig. 13.10 Polydispersity during dispersing of nanoparticulate silica in different dispersing machines (original graphic by Schilde et al. [3])

Stabilization of nanoparticulate aggregates

Stabilization is the last step in the dispersion process to avoid re-agglomeration of nanoparticles and nanoparticulate aggregates. The re-agglomeration is induced by Brownian motion or convective fluid flow, the large specific surface area as well as an increased ratio of attractive particle interactions to weight forces. In principle, the stabilization mechanisms can be classified in electrostatic stabilization, steric stabilization, and electro-steric stabilization (see Fig. 13.11) [30, 31]. For the electrostatic stabilization, the attractive van der Waals and repulsive electrostatic particle–particle interactions were calculated based on the considerations of the DLVO theory [32–35]. If the total interaction potential is positive, an increased charge density on the particle surface leads to the compensation with counterions and the formation of a double layer. These double layers between nanoparticulate aggregates repulse each other and can be influenced by adjusting the ion concentration, particle surface modification, and pH value. As a measurement value for the electrostatic stabilization, the zeta potential is measured. Depending on the characterization method, a zeta potential above an absolute value of 30–50 can empirically be associated with a sufficient electrostatic stabilization [36–38]. Figure 13.12 exemplarily shows the interaction potential and median aggregate size of dispersed pyrogenic titanium oxide at different pH* values. A decrease in the interaction potential leads to an increase in median particle size caused by re-agglomeration.

In a significant number of processes, the electrostatic stabilization mechanism is not applicable, e.g., at high ion concentrations. Thus, organic polymer additives were physi- or chemisorbed at the particle surface. In contrast to electrostatic

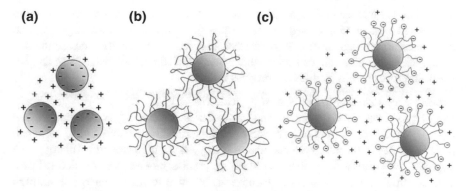

Fig. 13.11 Stabilization mechanisms: **a** electrostatic stabilization; **b** steric stabilization; **c** electro-steric stabilization (original graphic by Schilde [8])

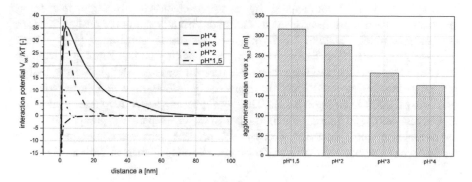

Fig. 13.12 Interaction potential and median aggregate size of dispersed pyrogenic titanium oxide at different pH* values (original graphic by Barth et al. [39])

stabilization mechanisms, the steric stabilization is typically investigated empirically. The organic polymer additives are acting as spacers between the particles which avoid re-agglomeration due to a reduced attractive van der Waals force at a certain distance from the particle surface. Several entropic solvency theories exist for the quantitative prediction of the steric stabilization [40, 41].

13.3 Combination of Dispersing and Emulsifying Processes

For the manufacture of complex cosmetic products, often the combination of dispersion and emulsification processes is necessary. Information on sole emulsification processes can be found elsewhere (Chap. 3.2, [42–44]). Especially for the production of nanocosmetics, this dispersing and emulsifying process chain can open new possibilities such as in case of sun cream production. Here,

Villalobos-Hernandez et al. applied a process chain of titanium dioxide dispersion in a mixed wax matrix, subsequent pre-emulsification into an aqueous phase, and high-pressure homogenization to submicron droplets to drastically increase the sun protection factor [45, 46]. To achieve this synergistic effect, they batch-wise applied separate process steps in different process equipment.

Modular, continuous microsystem approach
In order to reduce loss on batch handling, process time, energy consumption, and overall costs, a continuous process chain for the manufacture of such particle-loaded emulsion systems was developed. A high-pressure microsystem approach was chosen to enable low hold-up, precise process control, defined process conditions, and low educt consumption, but also to realize energy input high enough to yield submicron product sizes in only one run through the system. Additionally, its modularity assures flexibility toward different starting material and desired product properties as well as enhanced cleanability of the system. The microsystems were manufactured by a micro-electrical discharge machining approach that was custom-adapted to facilitate virtually complete design freedom for microchannels manufactured in stainless steel substrates [47]. These microsystems were realized by simply stacking and clamping separate micro-modules, which facilitate a single defined process step, i.e., dispersion, pre-emulsification, or emulsification, respectively, providing a leak-proof connection up to 2000 bar (Fig. 13.13). The two phases, aqueous phase containing emulsifiers and lipid phase containing particles, are separately pumped into the overall microsystem, processed through each process steps and continuously leave the setup as nanoemulsion of lipid droplets containing dispersed particles [48].

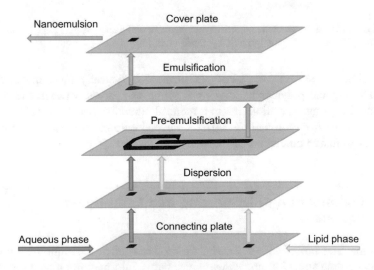

Fig. 13.13 Integrated overall microsystem for continuous production of nanoemulsion droplets with internally dispersed solid particles; assembled by clamping, pressure resistance of 2000 bar, respective micromodules possess measures of 10 × 30 mm, typical channel widths: 1000 μm, typical orifice widths: 50–120 μm, typical channel depths: 30–50 μm [48]

This modular setup additionally provides the advantage that each process step can be characterized individually to identify crucial design rules and improve microchannel performance. In our case, the dispersion and emulsification micromodules were extensively investigated to gain in-depth process knowledge and trace back the influence of geometric parameters on fluid dynamics and their effect on particle deagglomeration and emulsification via high-pressure homogenization in microchannels [49, 50]. Process understanding was drastically supported by computed fluid dynamics (CFD) simulations of the flow in microchannel geometries as well as the coupling of these to aggregate simulations. The stress intensities acting on particles along their streamlines through microsystems were analyzed in CFD and differentiated regarding tensile, shear, turbulent, and compressive stresses. In addition, the strength of the aggregate particles was simulated to capture effects of stress intensities and stress duration by the surrounding flow fields on aggregates and their breakage probability [51, 52]. Additionally, these simulation data were complemented and validated by micro-particle image velocity (μPIV) measurements, displaying the actual flow pattern in microsystems. CFD as well as μPIV were in good accordance with one another. They also showed flow instabilities at the exit of orifice structures that could be traced back to cavitation phenomena which were also visualized by a new method applying a μPIV setup (Fig. 13.14, left) [52, 53]. Experiments showed that cavitation did not contribute to either the dispersion or the emulsification. More precisely, it perturbs the respective size reduction processes, yielding higher sizes with high cavitation under the same energy input (Fig. 13.15) [54]. This clearly proved that certain back pressures lead to process optima regarding product particle size [54, 55].

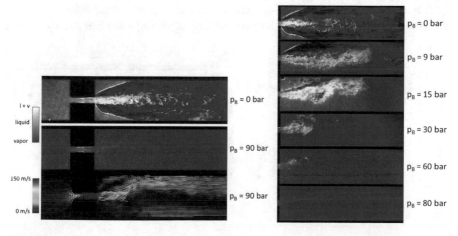

Fig. 13.14 Left: Microscopic images of fluid flow (water stained with Rhodamin B) in an orifice (width 80 μm, depth 50 μm) microsystem at a pressure drop of 200 bar over the orifice, without counter-pressure (top) and with counter-pressure (Th = 0.3; middle); μPIV flow vectors (bottom); right: effect of rising counter-pressure on cavitation at constant pressure drop over orifice of 200 bar [54]

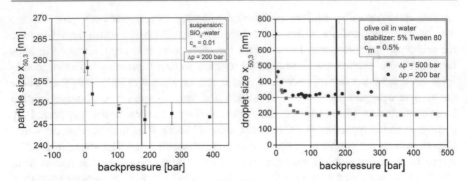

Fig. 13.15 Effect of back pressure on particle sizes of silica dispersions (left) and olive oil emulsions (right) [54]

Nanoemulsification and dispersing performance

Especially for the emulsification step via high-pressure homogenization, a certain backpressure needs to be applied to avoid a free liquid jet behind the homogenizing structure. In order to achieve this aim, multiple orifice micromodules were studied (Fig. 13.16; multiple orifices are in a consecutive arrangement). Single orifice micromodules showed the highest droplet sizes. This inferior emulsification performance can be traced back to the dominance of coalescence as deducible from curve progression. Double orifices which automatically provide a backpressure for the first orifice performed best while triple orifices again displayed higher droplet sizes, nonetheless smaller than those of single orifices. This order of results can be explained by the suppression of coalescence in multiple orifice systems and the counteracting distribution of energy input over the number of orifices [49, 55].

A closer investigation of double orifices further fostered these initial findings, showing that double orifices can be improved by applying a smaller followed by a

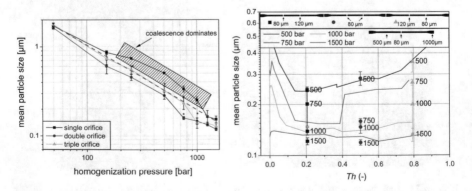

Fig. 13.16 Left: Effect of number of consecutive orifices (width 80 μm, depth 50 μm) on nanemulsification performance; right: correlation of particle sizes achieved with internally applied (by double orifice combination) and externally applied (by a consecutive valve) back pressures, expressed as Thoma numbers (Th, counter-pressure/total pressure loss). Modified after [55]

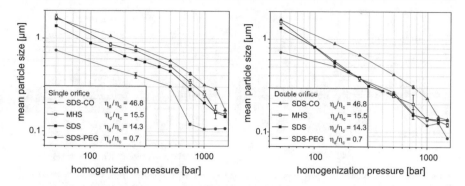

Fig. 13.17 Effect of number of orifices (left: single orifice; right: double orifice) and formulation parameters on nanoemulsion particle size (SDS—sodium dodecyl sulfate, CO—castor oil, MHS—macrogol-15-hydroxystearate, PEG—polyethylene glycol) [55]

larger orifice. In this case, the first orifice yields a high pressure loss and, by that, high energy input effective for droplet stressing and disruption. A lower pressure loss results from the latter orifice due its larger width, yielding a preferred back pressure in the range of Thoma numbers (counter-pressure/total pressure drop) of 0.1–0.4 [55]. Between these orifices, a turbulent mixing zone without cavitation is achieved that keeps droplet contacts short and, by that, reduced coalescence probability. Further, a period of time is provided to facilitate emulsifier transport to newly created interfaces by diffusion and enhanced convective mixing to achieve stabilization of emulsion droplets against coalescence. By this, multiple orifice emulsification microsystems can enable the production of smaller droplet sizes with the same energy input and make the resulting droplet size less dependent on the diffusion and adsorption kinetics of the emulsifier and the viscosity ratios of the phases, as compared with single orifice systems (Fig. 13.17) [55].

Otherwise for dispersing processes, multiple orifices are less effective for particle size reduction with a set overall pressure loss and energy input. This originates from a difference in dominance of dispersing and re-agglomeration processes as compared with emulsifying processes: in dispersing, the aggregate strength and, hence, the pinpointed energy dissipation dominates over the re-agglomeration, while for emulsification, re-agglomeration (coalescence) crucially dominates product qualities [54]. Accordingly, single orifices are to be applied for dispersing while double orifices with a bespoke relation of orifice widths should be applied for emulsifying process steps.

Application of the whole process chain
The full assembly of the overall microsystem for continuous production of complex, colloidal carrier systems (compare Fig. 13.13) was studied with nanoemulsions for performance and design (combination) characterization and with reference formulations from literature for solid lipid and wax nanoparticle production [45, 46, 56]. In this assembly, the composition of the final product as well as the product particle size is determined by the combination of various geometries for the

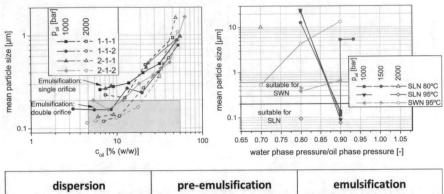

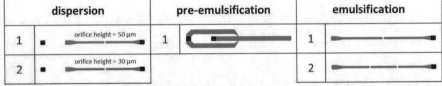

Fig. 13.18 Top left: Effect of pressure drop and microsystem assembly on crucial product parameters (disperse phase content and mean particle size) for nanoemulsions; top right: Effect of the ratio of water phase to oil phase pressure on solid lipid and solid wax nanoparticles (SLN and SWN, respectively) mean particle sizes; bottom: microsystems used for studies of overall microsystems [48]

respective process steps and the insertion pressures of both phases. As an example, by elevating the lipid phase pressure, the content of lipid in the product will rise and the size of the nanoparticle aggregates will decrease, but the lipid product particle size will most likely rise as well as the energy input (by the insertion of the continuous phase) per unit volume of the disperse phase is lower. Studies of various pressure and geometry combinations showed that the desirable ranges of particle size and lipid phase concentration can be achieved for nanoemulsions (Fig. 13.18, left). For that, it is necessary to apply an efficient double orifice microsystem (2) in the emulsification process step. The slight variation of the dispersion microsystem did not measurably influence the overall performance.

For the production of solid lipid or wax nanoparticles, it was necessary to elevate the working temperature of all process steps including feed reservoirs, facilitating the processing as emulsions with a subsequent solidification of the disperse phase by free cooling to ambient temperature. Results show that the product particle sizes more crucially depend on the ratio of pressures between the aqueous and the lipid phase (Fig. 13.18, right). It is necessary to apply a sufficiently high pressure of the aqueous phase to facilitate the disruption of the continuous lipid flow into discrete droplets in the pre-emulsification micromodule (Fig. 13.13, middle). Additionally, smaller particle sizes can be achieved by applying a higher process temperature [48].

On the whole, an overall, modular microsystem was developed, facilitating the continuous production of nanoemulsions that contain nanoparticles dispersed in their internal phase. Such systems can improve the production processes of nanocosmetics by boosting product quality and performance while keeping production costs low, as compared with conventional batch processes.

13.4 Production of Nano-suspensions via Wet Grinding of Primary Particles and Strong Aggregates

The production of nano- and submicron particles by particle breakage is a common way to obtain products with certain properties, like special color effects or UV-protection. Therefore, the coarse particles are dispersed in a fluid phase and ground especially by media milling, where the colliding media supply mechanical stress to fracture particles. For real grinding in production scale, usually stirred media mills are used. They are available in size ranges from grinding chamber volumes of some milliliters up to several cubic meters. They offer the possibility to work in the continuous operation mode, whereas for nanomilling processes they are usually operated in a circuit with a stirred vessel due to the relatively long milling times. Typical other media milling types are planetary ball mills or vibratory mills, with closed grinding chambers for a batch-wise production. They are usually used in laboratory scale to determine formulation parameters. Construction schemes of the mentioned mills can be found elsewhere [57]. For the so-called real grinding, i.e., the breakage of primary particles and dense aggregates with very strong solid bridges stressing by a surrounding fluid as it is realized in most of the dispersing machines does not provide the required stress intensity, i.e., energy related to stress particle mass or compression and shear stress [3, 6, 8].

The focus of this sub-chapter lies on the optimization possibilities to design a good nanomilling process using a stirred media mill. For this purpose, the nanomilling of zinc oxide is considered. Zinc oxide plays a major role in the chemical, cosmetic, and pharmaceutical industry, because it can be used as UV-absorber in sunscreens, as catalyst or as pharmaceutical ingredient for wound healing creams, patches, or bandages. It is a white crystalline powder with a density of 5.61 g/cm^3. The unground material of the study described in the following had a median particle size of around 800 nm (purchased from Sigma Aldrich). It was suspended in water and stabilized with TODS (trioxydecanoic acid) for preventing agglomeration of particles. The suspension was ground in a laboratory stirred media mill, whereas the process parameters stirrer tip speed as well as grinding media size and type were varied. Besides the median particle size, UV/VIS spectroscopy measurements were carried out to determine the UV-absorption of the material. The absorption spectra can be directly linked to the particle size. Figure 13.19 shows the dependency of the suspensions extinction in relation to the process time. It gets obvious that with increasing grinding time the transparency in the visible light range increases and the absorption of the UV-light remains high.

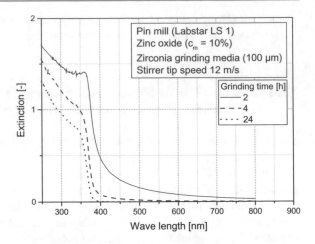

Fig. 13.19 UV/VIS measurement of zinc oxide suspension after different grinding times

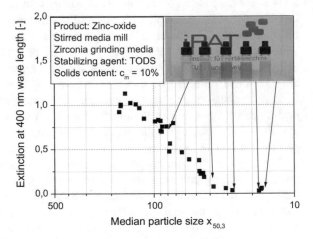

Fig. 13.20 Correlation of the extinction at a wavelength of 400 nm to the median particle size

The boundary between transparency and absorption is located in the wavelength range of 350–400 nm. The extinction at a wavelength of 400 nm was connected to the measured particle, and Fig. 13.20 presents the dependency between absorption and particle size.

With decreasing particle size, the extinction at a wavelength of 400 nm decreases. The samples with particle sizes below 20 nm showed good transparency, although they have a light yellow touch. If the grinding time is increased further, the particle size stays more or less constant, due to reaching the grinding limit of zinc oxide [58]. The transparency decreases at this stage, because the concentration of wear particles increases with increasing energy input.

In order to obtain nanoparticles by real grinding, very long grinding times and therewith high specific energy inputs are necessary. The choice of process parameters gets important at that stage. Figure 13.21 is obtained by plotting the

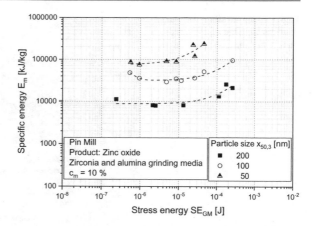

Fig. 13.21 Basic stress model applied on the zinc oxide grinding

specific energy as a function of the stress energy for certain particle sizes. The stress energy represents more or less the energy which is provided by the grinding media at each collision and is a function of the grinding media density ρ_{GM}, size d_{GM} and the stirrer tip speed v_t.

$$SE_{GM} = d_{GM}^3 \cdot v_t^2 \cdot \rho_{GM}$$

The background and definitions to the model of stress energy can be found elsewhere [59–63]. From Fig. 13.21, it gets obvious, that the same particles' size, e.g., 100 nm, can be obtained with different stress energies, but the specific energy to achieve the particle size achieves its minimum value of around 30.000 kJ/kg for the given mill at the optimum stress energy and can rise up to more than the threefold value if too high or low stress energies are set. A minimum specific energy ensures a minimum of product contamination and also the possibility to achieve a maximum in production capacity.

In analogy to the specific energy, the grinding media wear is an important value because on the one hand worn grinding media have to be replaced by new ones and on the other hand the wear particles are part of the product suspension after the grinding process and may have impact on product quality attributes like optical or other properties. The increase in grinding media wear during the grinding process is proportional to the energy introduced into the grinding chamber. Thus, a similar plot results if the product contamination by grinding media wear is plotted as function of the stress energy for constant particle size. This is shown in Fig. 13.22.

It can be concluded that the optimization of stress energy unifies two advantages: The specific energy and the grinding media wear reach a minimum value. In order to reach grinding times as short as possible, the power draw of the mill should be increased to the maximum by increasing the stirrer tip speed. At the same time, the specific energy required for grinding should be kept at the minimum value by adjusting the stress energy close to the optimum, for which usually small grinding

Fig. 13.22 Grinding media wear as function of the stress energy for different particle sizes

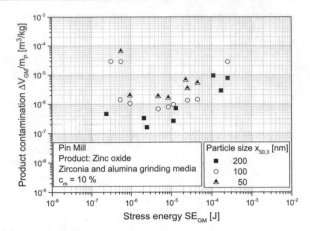

media sizes are more favorable. Overall, one should be aware that high power consumptions are linked to an increase in temperature, which might have an influence on the product quality.

Other examples for the application of stirred media milling can be found in color cosmetics, usually inorganic or organic color pigments are ground to enhance functional and skin feeling properties of cosmetic formulations, especially for decorative applications. Stirred media milling is used for tailoring particle size distributions of organic pigments, and raw materials are often received as micron-sized fractions of aggregated pigments. The choice of milling parameters and their optimization follows the same principles like presented for zinc oxide as representative of inorganic particles. Figure 13.23 presents the correlation between stress energy and required specific energy input to obtain a median product particle size of 250 nm of an organic pigment by stirred media milling. It can be seen that the required specific energy decreases with decreasing stress energies, and the optimum is assumed to be located at even smaller, not investigated stress energies. Furthermore, the level of product contamination from grinding media abrasion is proportional to the specific energy input. Thus, similar dependencies were proven to be valid for nanomilling of organic particles [64].

Several studies have shown that the grinding process of organic particles can be overlaid by different mechanisms, such as agglomeration, crystallization, or degradation [65–67], as it was especially found for pharmaceutical active ingredients. It was also proven that grinding media wear can induce agglomeration of fine ground particles [68]. In general, the aspect of colloidal stability is of high importance, an adequate product formulation is required to prevent the product particles from agglomeration. Apart from worse application properties, agglomerated particles would have a negative impact on the milling efficiency. Elevated suspension viscosities due to agglomeration lead to an increase in power draw and lower effective stress energies [69], thus grinding at high viscosities is less efficient.

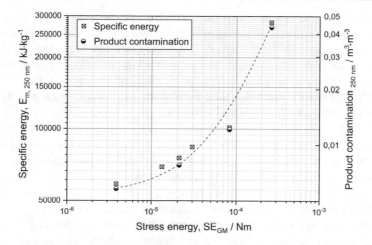

Fig. 13.23 Specific energy input to obtain a product particle size of 250 nm and resulting product contamination as a function of stress energy

13.5   Conclusions

In principle, several different possibilities exist to achieve nanosized pigments for cosmetic products. The right choice of the dispersing or milling method, respectively, mainly depends on the raw material properties and the demanded final product quality. Moreover, also economic criteria affect the final decision of a proper dispersing/milling method. First of all, the method must be able to fragment the primary particle (in case of real breakage), the aggregate, or the agglomerate. Depending on the strength of the raw material particles, only certain dispersing/milling methods are appropriate. It can be stated that generally the higher the strength of the raw material particles, the less processes are able to fragment these particles. Primary particles without integrated fracture planes (like adhered particles in an aggregate) can only be processed down to the nanometer range in mills with moving grinding media, especially in stirred media mills. The more fragile the aggregates and agglomerates are, the more dispersing processes are able to disperse these secondary particles. However, the higher the demanded product fineness is, the higher the acting stress intensity (in most dispersing devices the acting shear stresses) of the dispersing process must be. Beside stirred media mills, in which the stress intensity automatically increases with decreasing particle size, high-pressure homogenizers, three roller mills and kneaders, latter two operated at very high viscosities, are able to produce agglomerate and aggregate size in the range of 100 nm.

Besides choosing an appropriate dispersing machine, stress intensity can also be adjusted by the operating parameters of the machine. For example, if a dissolver, a rotor–stator–dispersing machine, a three roller mill or a kneader is employed, the stirrer speed and/or the viscosity can be increased. In case of a high-pressure dispersing system, especially the operating pressure and with that the fluid velocity inside the system can be enhanced. Depending on the dispersing device and operating parameters used, a different steepness of the particle size distribution can result at the same median particle size. Especially, high-pressure systems and three roll mills with narrow residence time distributions can produce steep particle size distributions with relatively defined maximum particle sizes [1, 6, 7].

For most cosmetic product that contain lipid ingredients besides inorganic materials, it may be advantageous to combine dispersion and emulsification processes to save time and lower costs, but also to improve product performance by creating new submicron structures. In the case of multiple process steps, continuous process chains are of special interest, such as a continuous micro-process chain for the production of such complex systems, as described above.

Besides the achievement of the demanded product quality, other criteria can play an important role in choosing a certain milling or dispersing device. One important issue may be a minimum of product contamination. Another criterion can be a fast and easy transfer from the laboratory to the production scale. For this purpose especially, high-pressure micro-systems can also be favorable. Last but not least, the overall production capacity can lead to a certain milling or dispersing machine. For example, stirred media mills are available in very different scales up to tons per hour, whereas for example ultrasonic homogenizers are only available in relatively small scale and consequently less used in production.

Literature

1. Raj S, et al. Nanotechnology in cosmetics: opportunities and challenges. J Pharm Bioallied Sci. 2012;4(3):186.
2. Kaul S, et al. Role of nanotechnology in cosmeceuticals: a review of recent advances. J Pharm 2018;2018.
3. Schilde C, et al. Efficiency of different dispersing devices for dispersing nanosized silica and alumina. Powder Technol. 2011;207:353–61.
4. Schilde C, Kwade A. Measurement of the micromechanical properties of nanostructured aggregates via nanoindentation. J Mater Res. 2012;27(4):672–84.
5. Schilde C, Breitung-Faes S, Kwade A. Dispersing and grinding of alumina nano particles by different stress mechanisms. Ceram Forum Int. 2007;84(13):12–17.
6. Möller A. Modelling the deagglomeration behavior of nanocrystalline Al_2O_3- and ZrO_2-powders. Darmstadt: Technische Universität Darmstadt, Technische Universität Darmstadt; 2000.
7. Parfitt GD. Dispersion of powders in liquids. Elsevier Publishing; 1969.
8. Schilde C. Structure, mechanics and fracture of nanoparticulate aggregates. Braunschweig: Institute of Particle Technology, TU Braunschweig; 2012.
9. Schilde C, Kampen I, Kwade A. Dispersion kinetics of nano-sized particles for different dispersing machines. Chem Eng Sci. 2010;65(11):3518–27.
10. Schilde C, et al. Effect of fluid-particle-interactions on dispersing nano-particles in epoxy resins using stirred-media-mills and three-roll-mills. Compos Sci Technol. 2010;70(4):657–63.
11. Young T. An essay on the cohesion of fluids. Philos Trans R Soc Lond. 1805;95:65–87.

12. Buckton G, Newton JM. Assessment of the wettability and surface energy of pharmaceutical powder. Int J Pharm. 1986;47:121–8.
13. Stiller S. Pickering-Emulsionen auf Basis anorganischer UV-filter. Braunschweig: Institut für Pharmazeutische Technologie, TU Braunschweig; 2003.
14. Fuji M, et al. Effect of wettability on adhesion forces between silica particles evaluated by atomic force microscopy measurement as a function of relative humidity. Langmuir. 1999;15:4584–9.
15. Rumpf H. Die Einzelkornbeanspruchung als Grundlage einer technischen Zerkleinerungswissenschaft. Chem Ing Tec. 1965;37:187–202.
16. Schilde C, Beinert S, Kwade A. Comparison of the micromechanical aggregate properties of nanostructured aggregates with the stress conditions during stirred media milling. Chem Eng Sci. 2011;66:4943–52.
17. Kwade A. Physical model to describe and select comminution and dispersion processes. Chem Ing Tec. 2001;73(6):703.
18. Kwade A. A stressing model for the description and optimization of grinding processes. Chem Eng Technol. 2003;26(2):199–205.
19. Rumpf H, Raasch J. Desagglomeration in Strömungen. In: 1. Europ. Symp. Zerkleinern 10.-13.04.1962. 1962. Frankfurt a. M: Verlag Chemie, Weinheim und VDI-Verlag, Düsseldorf.
20. Kolmogorov AN. Die lokale Struktur der Turbulenz in einer inkompressiblen zähen Flüssigkeit bei sehr großen Reynoldsschen Zahlen. Sammelband zur statischen Theorie der turbulenz 1958, Berlin: Akademie Verlag. p. 71–6.
21. Rumpf H. Grundlagen und Methoden des Granulierens. Chem Ing Tec. 1958;30(3):144–58.
22. Zellmer S, et al. Influence of surface modification on the micromechanical propertics of spray-dried silica aggregates. J Colloid Interface Sci. 2015;464:183–90.
23. Knieke C, et al. Nanoparticle production with stirred-media mills: opportunities and limits. Chem Eng Technol. 2010;33(9):1401–11.
24. Schilde C, Breitung-Faes S, Kwade A. Grinding kinetics of nano-sized particles for different electrostatical stabilizing acids in a stirred media mill. Powder Technol. 2013;235:1008 16.
25. Sauter C, et al. Influence of hydrostatic pressure and sound amplitude on the ultrasound induced dispersion and de-agglomeration of nanoparticles. Ultrason Sonochem. 2008;15(4):517–23.
26. Kim YG, et al. Efficient light harvesting polymers for nanocrystalline TiO$_2$ photovoltaic cells. Nano Lett. 2003;3(4):523–5.
27. Biedermann A, Henzler H-J. Beanspruchung von Partikeln in Rührreaktoren. Chem Ing Tec. 1994;66(2):209–11.
28. Winkler J. Nanopigmente dispergieren. Farbe und Lack. 2006;2:35–9.
29. Bittmann B, Haupert F, Schlarp AK. Ultrasonic dispersion of inorganic nanoparticles in epoxy resin. Ultrason Sonochem. 2009;16:622–8.
30. Mende S. Mechanische Erzeugung von Nanopartikeln in Rührwerkskugelmühlen. Braunscheig: TU Braunschweig; 2004.
31. Sommer MM. Mechanical production of nanoparticles in stirred media mills. Technische Fakultät der Universität Nürnberg-Erlangen; 2007.
32. Derjaguin BV, Landau LD. Theory of the stability of strongly charged lyophobic sols and the adhesion of strongly charged particles in solutions of electrolytes. Acta Physicochimica U.R.S.S., 1941;14:633–62.
33. Verwey EJW, Overbeek JTG. Theory of stability of loyophobic colloids: the interaction of sol particles having an electrical double layer. New York: Elsevier; 1948.
34. Segets D, et al. Experimental and theoretical studies of the colloidal stability of nanoparticles— a general interpretation based on stability maps. ACS Nano. 2011;5(6):4658–69.
35. Reindl A, Peukert W. Intrinsically stable dispersions of silicon nanoparticles. J Colloid Interface Sci. 2008;325:173–8.
36. Stenger F, et al. Nanomilling in stirred media mills. Chem Eng Sci. 2005;60(16):4557–65.

37. Mende S, et al. Production of sub-micron particles by wet comminution in stirred media mills. J Mater Sci. 2004;39:5223–6.
38. Mende S, et al. Mechanical production and stabilization of submicron particles in stirred media mills. Powder Technol. 2003;132:64–73.
39. Barth N, Schilde C, Kwade A. Influence of electrostatic particle interactions on the properties of particulate coatings of titanium dioxide. J Colloid Interface Sci. 2014;420:80–7.
40. Evans R, Napper DH. Steric stabilization I—comparison of theories with experiment. Kolloid-Zeitschrift und Zeitschrift für Polymere. 1972;251(6):409–14.
41. Evans R, Napper DH. Steric stabilization II—a generalization of Fischer's solvency theory. Kolloid-Zeitschrift und Zeitschrift für Polymere. 1972;251(5):329–36.
42. Ali A, et al. Nanoemulsion: an advanced vehicle for efficient drug delivery. Drug Res (Stuttg). 2017;67(11):617–31.
43. Jintapattanakit A. Preparation of nanoemulsions by phase inversion temperature (PIT). Pharm Sci Asia. 2018;42(1):1–12.
44. Yukuyama MN, et al. Nanoemulsion: process selection and application in cosmetics–a review. Int J Cosmet Sci. 2016;38(1):13–24.
45. Villalobos-Hernandez JR, Muller-Goymann CC. Novel nanoparticulate carrier system based on carnauba wax and decyl oleate for the dispersion of inorganic sunscreens in aqueous media. Eur J Pharm Biopharm. 2005;60(1):113–22.
46. Villalobos-Hernandez JR, Muller-Goymann CC. In vitro erythemal UV-A protection factors of inorganic sunscreens distributed in aqueous media using carnauba wax-decyl oleate nanoparticles. Eur J Pharm Biopharm. 2007;65(1):122–5.
47. Richter C, et al. Innovative process chain for the development of wear resistant 3D metal microsystems. Microelectron Eng. 2013;110:392–7.
48. Finke JH, et al. Modular overall microsystem for the integrated production and loading of solid lipid nanoparticles, in mikroPART—Microsystems for Particulate Life Science Products (Concluding results of the DFG Research Group FOR 856) In: Kwade A, Kampen I, Finke JH, editors. Braunschweig; 2014. p. CDCP1-6.
49. Finke JH, et al. The influence of customized geometries and process parameters on nanoemulsion and solid lipid nanoparticle production in microsystems. Chem Eng J. 2012;209:126–37.
50. Gothsch T, et al. Effect of microchannel geometry on high-pressure dispersion and emulsification. Chem Eng Technol. 2011;34(3):335–43.
51. Beinert S, Gothsch T, Kwade A. Numerical evaluation of flow fields and stresses acting on agglomerates dispersed in high-pressure microsystems. Chem Eng Technol. 2012;35 (11):1922–30.
52. Beinert S, Gothsch T, Kwade A. Numerical evaluation of stresses acting on particles in high-pressure microsystems using a Reynolds stress model. Chem Eng Sci. 2015;123:197–206.
53. Gothsch T, et al. High-pressure microfluidic systems (HPMS): flow and cavitation measurements in supported silicon microsystems. Microfluid Nanofluid. 2014;18(1):121–30.
54. Gothsch T, et al. Effect of cavitation on dispersion and emulsification process in high-pressure microsystems (HPMS). Chem Eng Sci. 2016;144:239–48.
55. Finke JH, et al. Multiple orifices in customized microsystem high-pressure emulsification: the impact of design and counter pressure on homogenization efficiency. Chem Eng J. 2014;248:107–21.
56. Schubert MA, Muller-Goymann CC. Characterisation of surface-modified solid lipid nanoparticles (SLN): influence of lecithin and nonionic emulsifier. Eur J Pharm Biopharm. 2005;61(1–2):77–86.
57. Ullmanns Encyclopedia of Industrial Chemistry, Dextran. 6 ed. Weinheim: Wiley-VCH; 2002.
58. Breitung-Faes S, Kwade A. Prediction of energy effective grinding conditions. Miner Eng. 2013;43–44:36–43.

59. Breitung-Faes S, Kwade A. Production of transparent suspensions by real grinding of fused corundum. Powder Technol. 2011;212(3):383–9.
60. Breitung-Faes S, Kwade A. Use of an enhanced stress model for the optimization of wet stirred media milling processes. Chem Eng Technol. 2014;37(5):1–9.
61. Kwade A. A stressing model for the description and optimization of grinding processes. Chem Eng Technol. 2003;26(2):199–205.
62. Kwade A, Blecher L, Schwedes J. Motion and stress intensity of grinding beads in a stirred media mill part II: stress intensity and its effect on comminution. Powder Technol. 1996;86:69–76.
63. Kwade A, Schwedes J. Breaking charakteristics of different materials and their effect on stress intensity and stress number in stirred media mills. Powder Technol. 2002;122:109–21.
64. Flach F, et al. Impact of formulation and operating parameters on particle size and grinding media wear in wet media milling of organic compounds—a case study for pyrene. Adv Powder Technol. 2016;27(6):2507–19.
65. Bitterlich A, et al. Challenges in nanogrinding of active pharmaceutical ingredients. Chem Eng Technol. 2014;37(5):840–6.
66. Steiner D, et al. Breakage, temperature dependency and contamination of lactose during ball milling in ethanol. Adv Powder Technol. 2016;27(4):1700–9.
67. Kumar S, Burgess DJ. Wet milling induced physical and chemical instabilities of naproxen nano-crystalline suspensions. Int J Pharm. 2014;466(1):223–32.
68. Flach F, Breitung-Faes S, Kwade A. Grinding media wear induced agglomeration of electrosteric stabilized particles. Colloids Surf, A: Physicochem Eng Asp. 2017;522:140–51.
69. Knieke C, et al. Nanoparticle production with stirred-media mills: opportunities and limits. Chem Eng Technol. 2010;33(9):1401–11.

Part IV
Governance and Potentials of Nanocosmetics

Safety and Toxicity Counts of Nanocosmetics

14

Gunjan Jeswani, Swarnali Das Paul, Lipika Chablani and Ajazuddin

Abstract

The advent of nanotechnology has led to advances in the cosmetic industry and is expected to grow further in the near future. Nanotechnology-driven products cater to the expectations of both consumers and manufactures in terms of better quality and effectiveness along with improved stability and easy scale-up. Several organic and inorganic materials are being utilized for the preparation of nanocosmetics having improved characteristics. At the same time, the safety aspects of nanocosmetics are also being pondered. Physicochemical properties play a significant role in controlling the toxicity of nanomaterials. Several mechanisms have been studied for nanomaterial generated toxicity; out of all, reactive oxygen species, generation is the most important mechanism. This chapter discusses all the relevant aspects which are required for safety and toxicity assessments of nano-ingredients for cosmetic use. Regulatory issues are also discussed because of their relevance in preventing the unforeseen toxicity of nanocosmetics.

Keywords

Nanocosmetics · Reactive oxygen species · Genotoxicity · Cytotoxicity · Regulatory

G. Jeswani (✉) · S. Das Paul
Department of Pharmaceutics, Faculty of Pharmaceutical Sciences, Shri Shankaracharya Group of Institutions, SSTC, Bhilai, Chhattisgarh, India
e-mail: gunjanjeswani@gmail.com

L. Chablani
St. John Fisher College, Rochester, NY 14618, USA

Ajazuddin
Rungta College of Pharmaceutical Sciences and Research, Kohaka, Bhilai, India

© Springer Nature Switzerland AG 2019
J. Cornier et al. (eds.), *Nanocosmetics*,
https://doi.org/10.1007/978-3-030-16573-4_14

14.1   Introduction

The increased complexity of today's environment poses several challenges to the cosmetic industry in a highly exigent market. Latest trends in makeup and skincare demand increased knowledge, specialization, innovation, and adaptation of advanced technology, which play an important role in the product formulation. At the same time, the consumers are more concerned with the quality, efficiency, safety, sustainability, and instant results of the product. In view of this, nanotechnology has greatly contributed to advances in the cosmetic industry. The major areas where nanotechnology has the potential for use in cosmetic products are titanium dioxide (TiO_2) in sunscreens, gold nanoparticles in creams for revitalizing skin, and antibacterial properties of silver nanoparticles in soaps/shampoos, wet wipes, makeup, lotions, and shampoos. The second use of nanotechnology is in the delivery of actives. Vesicular structures such as liposomes, ethosomes, transfersomes, and niosomes have been commercially used for the delivery of cosmetic ingredients and other technological impressions since more than 25 years [120]. Moreover, it is expected that this novel technology will be employed in much more areas as the years go by. By statistical analysis, it has been observed that the contribution of nanoproducts to the global cosmetic economy will be approximately more than $1 trillion in the next three years.

In particular, nanotechnology-driven products are designated as "Nanocosmetics." The researchers and formulators are constantly striving to produce "out of the box" nanoproducts that maintain the highest quality standards and effectiveness to ensure consumer satisfaction. Moreover, creations of nanotechnology are turning in both predictable and unpredicted ways. Major players of cosmetic industries and research organizations are utilizing nanotechnology approaches as reflected by the numerous publications and patents being granted in the last five years. Besides, all this hue and cry, the most important feature lies in the safety of the products and the ingredients.

As compared to traditional cosmetics, nanocosmetics exhibit improved characteristics owing to their unique phenomena such as surface properties, small size, quantum, and its tunneling effect. The nanocosmetics have the capability to exist in very desirable particulate forms like aerosols (deodorant and sunscreen spray) and suspension. The activity of the nanocosmetics products is enhanced mainly because of these forms. The properties of nano-ingredients can be altered as per the specific requirements like targeted delivery, step-down stability problems, and enhanced aesthetics [94]. This helps in reconstructing nanocosmetics for many other functions including improved skin hydration, better skin incursion, and enhanced bioavailability owing to their extraordinary or unique physicochemical properties. Different nanomaterials, for example, particles, rods, tubes, composites, and film, are used in the formation of nano-objects according to their characteristic accounts.

14.2 State of the Art in Nanotoxicity

Many investigations regarding the safety of nanomaterials have been carried out in the field of cosmetics and pharmaceuticals during the last few years [2, 82, 116]. Since many years, nanomaterials have been used to produce common consumer products and their scope is expanding day by day. Therefore, it becomes necessary to ponder if at all they present any toxicity. As a result of this, an upcoming field of "nanotoxicity" has emerged. It deals with the study of toxic implications of nanomaterials. The present section provides the latest state-of-the-art information as obtained from the articles and reports provided by experts on nanotechnology and the various advisory committees. The safety aspects of nanomaterials as used in cosmetics and other regulated products have been dealt with in this section.

The first published report on nanotoxicity was about the in vivo study of iron oxide magnetic nanoparticles in 1989 by Weissleder et al. Since then, there have been numerous scientific reports regarding safety assessments, such as skin irritations, in vitro and in vivo tested cytotoxicity and genotoxicity, and epigenicity.

The process of nanotoxicity is multifactorial and influenced by the fundamental interaction between product and cells. Factors involved in this process include tissue compatibility, cytotoxic behavior, and physicochemical properties such as particle size, degree of crystallinity, morphological changes, hydrophobicity, and zeta potential [100]. The study about these factors reveals contradictory status about nanoparticle generated toxicity. Many have reported that nanoparticles are safe in the cases when they are applied over the intact skin. Generally, nanoparticles are restricted to the superficial layers of epidermis; however, results from some in vitro and in vivo studies have shown that nanoparticles administered for a long period of time on hairless mouse skin were able to permeate stratum corneum and reached deeper epidermal layers [61]. Additionally, short-term, systemic administration of nanomaterials shows little toxicity with very large doses exceptionally [143]. Moreover, specific size-dependent effects and relative permeability through membranes also play a major role in their permeation behavior.

Several inorganic nanoparticles have been used in cosmetics and scientific researches due to their anticipated rich functionality and good biocompatibility, other than skin restoring effects. However, when presented in molecular form, they work as substrate or catalyst and accelerate toxic reactions. In an attempt to probe the unusual size-dependent properties, gold particles were subjected to toxicity studies. The results were conflicting as many studies suggested nontoxicity of gold nanoparticles and others reported dose-dependent toxicity of gold particle [121, 136]. During a study conducted for eight days, it was found that the dosage of under <400 µg/kg showed no toxic effects when given by intraperitoneal injection route [79]. Thus, the nanotoxicity can be controlled by ascertaining the following factors, e.g., size, shape, surface charge, and other testing characteristics.

Villiers et al. described this perplex reaction and the complicated state of affairs in their work. The study aimed to assess in detail the effect of gold nanoparticle on the immune system. The gold nanoparticles of zeta potential −13.0 mV and size

10 nm were co-cultured with bone marrow, dendrite cells of C57BL/6 mice. The result of this study showed that the cell viability remains unchanged even at high dose. However, there was an increase in particle accumulation in endocytic compartments which could modify the secretion of cytokines [139]. Similar endocytic accumulation was reported by Nadie Pernodet et al. using citrate/gold nanoparticles and human dermal fibroblast cells. The study showed that very small particle (4 nm) can penetrate the membrane and result into cellular death [109]. It is generally observed that the particles of less than 30 nm easily follow endocytosis [2], and the particles of diameter 12 nm or less can even cross the blood–brain barrier [103]. Moreover, gold nanoparticles can induce oxidative stress via the generation of the cellular reactive oxygen species. This results into the activation of macrophagic process. These findings suggest that the particles larger than 30 nm can be considered as safe but there is a need to comprehensively study the effects of gold nanoparticles on the basis of their size distribution for their safe application in cosmetics.

So far, several explanations have been worked out for describing the cellular mechanism of nanotoxicity. However, it can be substantiated that no single pathway is completely responsible for nanotoxicity generation. Different mechanisms involved in nanotoxicity include biochemical interaction with cellular proteins, direct or indirect influence on inflammatory processes, and generation of free radical formation by reactive oxygen species (ROS) [122]. Out of all, the formation of free radical is the most commonly observed cause as depicted in Fig. 14.1. When cosmetics containing nanoparticles are applied on the dermis layer, they get absorbed and free radicals are generated due to the formation of ROS, which reacts with cellular machinery to cause several dysfunctions. ROS cause direct damage to

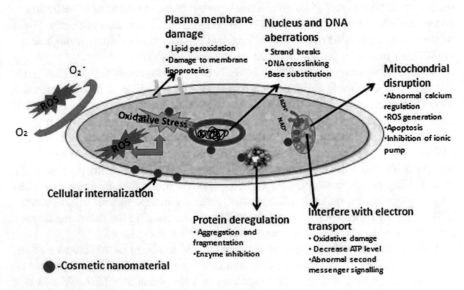

Fig. 14.1 Cellular targets of cosmetic nanomaterials

proteins, lipids, and DNA. Genotoxic and other degenerative effects also arise as a result of oxidative stress induced by ROS [115]. Some other mechanisms are also involved like induced apoptosis, transcriptional changes, and enzyme down-regulation [70, 89].

It has been observed that during in vitro studies, overproduction of ROS results into breakage of DNA helices, unregulated gene expression, cell motility, and mitochondrial disturbance through an oxidative stress-related mechanism, whereas during in vivo studies, they hasten inflammation and induce or prohibit the immune system activity.

Several studies have been done for determining the cellular mechanism of oxidative stress-related toxicity like determination of decreased superoxide dismutase catalase activity, glutathione levels, ROS levels, and mitochondrial transmembrane potential [70]. All evaluations are done under control conditions. Recently, Dubey A et al. studied the effect of oxidative stress produced by two metallic nanoparticles, ZnO and TiO_2, on newly developed cell line, derived from aquatic organism, Wallago attu [34]. The results showed that the levels of ROS, protein carbonyl content, and lipid peroxidation increased with the higher doses. Lipid peroxidation indicates cellular membrane damage, whereas increase in protein carbonyl content reflects damage to cellular proteins.

Measurement of a potent free radical scavenger like glutathione and malondialdehyde is usually performed to assess oxidative stress conditions. Khan H et al. measured glutathione levels in lung, liver, and heart of rats and found depletion of total glutathione to be dose and size dependent [71]. Oxidative stress leads to change in normal tGSH level in cells [18]. It gets converted to oxidized form glutathione disulfide. NADPH-dependent flavoenzyme, catalase, glutathione peroxidase, and superoxide dismutase are other prominent ROS-metabolizing, antioxidant enzymes which are measured for toxicity assessment [44]. In this context, Shvedova A. A et al. demonstrated that the carbon nanotubes generate ROS through NADPH pathway. The OH generated during mitochondrial dysfunction leads to the activation of signaling cascades associated with cell proliferation and tumor progression in vitro [128]. OH is a potent radical that reacts with DNA and causes single strand breakage. It forms 8-hydroxyl-2'-deoxyguanosine (8-OHdG) DNA adduct that acts as a biomarker for oxidative stress. G. Lenaz et al. reported that the metal oxide nanoparticles induce apoptosis through mitochondrial membrane depolarization in a dose-dependent manner, mediated by the damage to membrane phospholipids [88]. However, over-expression of antioxidant enzymes indicated a stage before toxic oxidative stress [147]. ZnO and TiO_2 nanoparticles produce elevated inflammatory response and activation of cellular mechanisms that are closely associated with the transcription of genes, leading to cytotoxicity and genotoxicity [22].

14.3 Properties of Nanoparticles for Risk Estimation

Nanoparticles can have an intense effect on the human body system. This is because of their distinctive properties owing to their dimensional characteristic. Most nanomaterials have the ability to reach deep situated tissues or organs, where their bulk counterparts fail to reach. At the same time, it becomes obligatory to check extrinsic and intrinsic properties like (i) zeta potential, (ii) size, (iii) coating materials, (iv) physicochemical aspects, (v) shape, and (vi) microenvironment, which can manipulate the toxicity of nanomaterials. Changes in these properties influence their interactions with biomolecules, proteins, cells lines, and tissues. All together they influence cellular uptake and target binding. It is also noteworthy that these characteristics form the basis of guidance which prevents the use of toxic materials, for example, guidance documents issued by FDA (available at https://www.fda.gov/ScienceResearch/SpecialTopics/Nanotechnology/ucm301093. htm). The below-mentioned properties provide details about differences in interactions with biological systems and safety prospects. Most accordant methods supported by International Cooperation on Cosmetic Regulation for the characterization include spectroscopy, chromatography, microscopy, and physical techniques [4]. However, for a comprehensive understanding, results from more than one method should be investigated.

Zeta potential: Zeta potential is a computation of the extent of the electrostatic charge between particles. It is the basic factors identified to affect stability. Surface charge on nanoparticles helps them to remain isolated, as the appealing properties of nanoparticles seize on agglomeration. Moreover, surface charge status is critical to cellular internalization and might also be responsible for cytotoxicity. Goodman et al. used 3-[4,5-dimethylthiazol-2-yl]-2,5-diphenyl tetrazolium bromide (MTT), hemolysis, and bacterial viability assays to check toxicity arising of their surface charges [48]. The study results revealed that positively charged particles are moderately toxic, and it helps the conjugation of nanoparticles with negative surfaces on cells during the toxicity assay, whereas negatively charged particles are less toxic. A vesicular delivery system called as "Niosomes" are nonionic surfactant vesicles prepared by using nonionic surfactant as bilayer component [69]. Niosomes are found to be efficient in topical delivery of cosmetic ingredients as they possess uncharged single-chain molecules and relatively nontoxic characteristics which preclude systemic toxicity. Currently, there is a great concern on nanopolysiloxane polymers which are used at different concentrations and for different intentions in cosmetic products meant for leave-on and rinse-off, like hair, skin, lip, face, and nail products [98]. Depending on the size of the molecule and the charge of the system, dermal penetration of nano-silica, especially through damaged skin, is possible. Therefore, the European Commission demanded the Scientific Committee on Consumer Safety to review the safety of silica in nanoform (plain silica, silica silylate, silica dimethyl silylate, and hydrated silica). The opinion of the committee is discussed in the respective section of silica.

Size and surface area: Nanoparticle size serves as a key feature that facilitates transport through the cell junctions and biological membranes. On the other hand, they can disrupt mitochondrial machinery or get deposited in nucleus and cause severe DNA damage leading to apoptosis [143]. Other formulation characteristics affected by size include sedimentation, diffusion, optical, mechanical, magnetic, electrical, and chemical. In general, as reported by Schaefer et al., particles follow mainly two routes to gain entry into the cell: intercellular pathway and follicular pathway. Particles of 3 μm diameter or less size find access through intercellular pathway, whereas particles greater than 3 μm but less than 10 μm collect in sebaceous or hair follicles. Large particles are, however, restricted over the stratum corneum and sometimes form a film [123]. This assertion does not always hold true; it is contradicted by several experimental observations where size invariably affects the absorption, pharmacokinetics, clearance, cellular uptake, and cytotoxicity [42].

Size has a close relation with surface area. As known, as the size of a particle decreases, its surface area increases. Consequently, it allows a greater ratio of its molecules to be displayed onto the face rather than the core of the material contributing to higher chances of complete reactivity [150]. Surface chemistry of nanoparticles involves many catalytic and oxidative reactions resulting in cytotoxicity [20]. Thus, the resultant toxicity could be greater than its counterparts. Depending on the nanoparticle surface characteristics, their affinity toward skin components varies. Thus, engineered nanoparticles are helpful in passive targeting to specific parts like hair follicles. There are some evidences to prove that the particle of size between 320 and 750 nm can easily penetrate into the hair follicles. Moreover, the particles of 320 nm can be retained in the hair follicles for up to 10 days [76]. Moreover, lipid nanoparticles used in cosmetics, like liposomes, ethosomes, and transfersomes are comparatively flexible. When applied topically, they easily penetrate even with greater size due to their reversible nature [59].

Considering the cutaneous route of administration for the improvement of the cosmetic esthetics recently, Yin et al. tried to assess the influence of size on the toxicity of typical sunscreen pigments. They studied the effect of UVA irradiation on nano-TiO_2 by MTS assay ((3-(4,5-dimethylthiazol-2-yl)-5-(3-carboxymethoxy-phenyl)-2-(4-sulfophenyl)-2H-tetrazolium)). They evaluated two polymorphic forms, anatase and rutile, and four different sizes of nanoparticles in HaCaT keratinocytes (human skin model). Upon photoirradiation, cytotoxicity and cell membrane damaging effects were produced by all nano-TiO_2 particles. However, nano-TiO_2 particles with the smallest particle size (less than 25 nm) induced greater cell damage. Additionally, the rutile crystal shape proved to be safer than anatase form. From what is discussed above, it is clear that the reactivity of nanoparticles is largely dependent on their particle size, shape, and crystal structure [30]. On the other hand, it is worth noting that the size specifications do not exclusively indicate toxicity, as observed from large published data [19, 58]. Therefore, no bright line limitation can be created for size.

Shape and charge: Nanoparticles can occur in different geometric shapes like simple solid sphere, simple rod, closed shell, cage, nanocluster, and complex truncated icosahedrons. Some researchers have reported gold nanospheres to be less cytotoxic than gold nanorods. Yinan Zhang et al. compared gold nanorods encapsulated with cetyltrimethylammonium bromide and polymer-capped and citrate-stabilized gold nanospheres [151]. The study findings supported that the small molecular citrate produces discreetly monodisperse spherical metal nanoparticles and polymers, such as poly (acrylic acid), poly (diallyldimethylammonium chloride)-poly (4-styrenesulfonic acid), and poly (ethylene glycol), which prevent the direct interaction of particles with biomembranes. Moreover, citrate acts as reducing agent and also determines the growth of the particles without causing instability and cytotoxicity with low dosage levels. In contrast, some researchers concluded that a set of gold nanoparticles with varying shapes had similar cytotoxicity [142]. However, it is still worthwhile to consider the aspect ratio as some respective studies have described that the higher aspect ratio (i.e., proportion of width to height) can encourage more toxicity like immune response, oxidative stress, DNA damage, and hypoxia response [38].

In another study, ZnO with different morphologies was compared and found that snowflake particles seem to perform better among others forms [129]. Similarly, nanocrystalline ZnO having hexagonal plate shape was also accounted to exhibit significant activity as compared to rod-shaped crystal [92]. Interestingly, the order of uptake into cells depending on morphology may be accounted as rods/spheres > cylinders > cubes [29, 114]. In addition, the aspect ratio also affects cellular internalization of nanorods during toxicity studies. However, another player of oxidative stress includes surface electric charge. Electric charges on surface also activate scavengers like macrophages [54].

To explore the relation between surface charge and toxicity, Shuguang Wang et al. used absorption spectroscopy, TEM, and MTT test assays [142]. The study results revealed that cationic particles are fairly toxic, whereas negatively charged particles are to a certain extent nontoxic. This phenomenon can be explained by two ways. Firstly, positive charge facilitates the transport of nanoparticles into cells, and secondly, they have a good chance of being directly absorbed through the diffusion pathway, while anionic particles follow endocytosis. These findings highlight the influence of specific shapes and charge on toxicological responses.

Surface porosity of nanoparticles: Porosity is defined as a ratio of pore volume to the total volume of the particle. It is used to describe the surface morphology or the pore architecture of the nanoparticles. Specific surface area and pore volume are the two components of porosity measurement. Degree of nanotoxicity depends on the pore architecture of nanomaterials. This is due to their unusual efficiency of cellular internalization and immune response. However, pores of at least 1–2 nm are essential for entry of biological fluids.

In the field of cosmetic applications, non-porous silica and mesoporous silica have recently gained increased interest due to their tremendous potential for providing soft slip, easy spreadability, and precise dryness. These characteristics are basic requirements for various formulations including creams, foundations, mascara, lotions, lipsticks, emulsions [83]. Mesoporous silica and colloidal silica are differentiated on the basis of pore structure and pore size (Slowing et al., n.d.). Soyoung Lee et al. used MTT and flow cytometry and fluorescence-activated cell sorting analysis to explore different degrees of toxicity depending on their pore conditions. The study results revealed that mesoporous silica nanoparticles induced lesser generation of pro-inflammatory cytokines—interleukin-1β, tumor necrosis factor-α, and interleukin-6 in macrophages, whereas colloidal silica nanoparticles initiate immunogenic activities and contact hypersensitivity. A classical example of non-nanometer size porous material is zeolite. They are widely used as absorbents and catalyst due to their large internal surfaces.

Solubility: Solubility is expressed as a chemical property that enables a given substance called as solute to dissolve in a specific solvent at specified temperature. Optimization of penetration of a substance across a membrane depends on the knowledge of solubility. Solubility in respect to cosmetic application encompasses the ability of the particle to get solubilized in the lipids of stratum corneum, corneocytes, etc. Inorganic solids (TiO_2, fullerenes, and quantum dots) used in cosmetics generally have low solubility. Moreover, when present in bulk, they do not allow the dissolution of any material, whereas soft carriers such as lipid nanocarriers, liposomes, ethosomes, and nanoemulsions fuse with the cell membrane and allow the encapsulated material to permeate the membrane barrier. Hence, the speed of permeation is largely dependent on the differential solubility of a substance in membrane lipids.

Solubility is also a key physical determinant in cytotoxicity as it provides a point of contact with the cell materials and promotes intracellular dissolution. Shen et al. reported that the dissolved nano-ZnO present in biological system greatly induced cytotoxicity in human immune cells as compared to freely present in intracellular zinc ions [124]. Similarly, Franklin et al. determined that there is a high-quality correlation between the dissolution profile and disruption of cellular homeostasis. ZnO_2 ions dispersed in aqueous media are highly toxic to Pseudokirchneriella subcapitata, freshwater microalgae [41]. Similar studies were done by Xia et al. using TiO_2, ZnO, and CeO_2 nanoparticles [147].

Most studies carried out with sunscreen agents like ZnO or TiO_2 nanoparticles showed limited skin permeation. For a detailed report, please visit TGA (Therapeutic Goods Administration, Department of Health, Australian Government) Web site, https://www.tga.gov.au/literature-review-safety-titanium-dioxide-and-zinc-oxide-nanoparticles-sunscreens. However, several studies have reported that solubilized zinc can penetrate the skin. Thus, cosmetic companies are recommended to prevent the addition of solubilized zinc in the formulations to avoid the cellular toxicity.

14.4 Potential Toxic Nanomaterials Used in Cosmetics

Nanomaterials used in cosmetics are typically based on organic or inorganic compounds. Organic-based nanoparticles include polymeric systems and lipid systems, whereas nanoparticles based on inorganic compounds include metal nanoparticles and carbon-based nanoparticles. The toxicity of nanoparticles depends on the nature, physicochemical properties, and bio-interaction of the nanoparticles. When cosmetics are applied, the skin gets exposed to various formulation materials including nanomaterials if present, which exhibit different toxic potency. Therefore, associated risks of nanotoxicity need to be kept in mind while exploring different cosmetics containing nanoparticles.

Nano-Titanium dioxide: Titanium dioxide (TiO_2) is addressed as titanium (IV) oxide or titania. It occurs naturally as oxide of titanium. It has been used extensively because of its multifaceted actions. It is used as physical UV screening agent in sunscreens, pressed powders, and loose powders and as pigments in lotions and creams. When used as physical or chemical suns blocking agent, it reflects and disperses sun rays. UVB (290–320 nm) and UVA (320–400 nm) radiations are the primary cause of malignant growth in skin tumor [36]. As pigment, it provides optical protection by scattering sun rays. For this, a thin layer of formulation is applied on the parts to be protected like skin or lips.

TiO_2 naturally occurs in three different crystalline forms, namely rutile, ilmenite, and anatase [96]. Rutile form of TiO_2 nanoparticle can decrease the cell viability of human amnion cell, epidermoid carcinoma cells, and human skin fibroblasts, but cytotoxicity is not observed in Chinese hamster fibroblasts [23]. Similarly, anatase form of TiO_2 can initiate cellular dysfunctions like cell death, oxidative stress, and impair mitochondrial activity in in vitro human obtained glial and lung cells [7]. It can also lead to nucleus interactions and break DNA strands along with chromosome aberrations in human intestinal, amniotic, and epidermal cells. Release of hydroxyl species which act as photocatalysts is considered as responsible factor [127]. However in few studies, anatase did not show cellular toxicity in three different cells: human intestinal cells [25], nasal mucosa, and lymphocytes [56].

Often, to reduce the photocatalytic properties of TiO_2, they are completely coated with nanoscale substances like alumina or silica for water or stain-repellent properties. This is done before addition to sunscreens. Fatty acids (lauric and oleic) or silicones are also used to make them dispersible in organic solvent.

Nano-Gold: Gold (Au) in cosmetics has been used in many forms such as creams, packs, or directly applicable leaves or foils. Other than the traditional uses gold is used for revitalizing the fiber tissues in skin, accelerating blood circulation, antiaging, antiseptic, and improving skin tone. As per the currently ongoing cancer studies, gold nanoparticles help to wipe out/to kill malignant cells in the body, and therefore, it has become exciting to investigate the applications of gold nanoparticles in skin products.

Generally, nanosized gold colloids are perceived as relatively inert toward biological systems because of its higher LD50 or LD80 values [111]. However on the bigger side, biodegradability is a big question in terms of their accumulation in cells or long-term circulation in plasma, whether as individual particles or agglomerates. Based on the evidence that gold nanoparticle can easily accumulate in the lung cells, the A549 lung cells were used for in vitro test. In fact, chirally active poly(acryloyl-l-valine)-capped gold nanoparticles in cube shape were preferentially internalized in addition to D and octahedra form of valine-capped gold nanoparticles. This showed chirality-associated and shape-dependent internalization process of gold nanoparticles [27].

Gold nanoparticles can activate necrosis, induce mitochondrial injury, and stimulate an oxidative stress with size of 1.4 nm [107], but no evidence of cellular damage is observed for 15-nm gold nanospheres. Similarly, nanoalloy of gold with copper produces some changes in tumor-inducting genes which promote micronuclei formation and generation of 8-OHdG DNA adduct as well as impair glutathione peroxidase activity [46]. Thus, a possible dose, size, and external nature-dependent toxicity of gold nanoparticles can be inferred from the above reports.

On the contrary, gold nanoparticles maintained the viability of C57BL/6 mice bone marrow extracted dendritic cells. Similarly, insufficient cytotoxicity at higher doses was reported by Villiers et al. However, at intracellular level, gold nanoparticles can accumulate in endocytic compartments and alter the secretion of cytokines [139]. From the discussion, it can be inferred that the activity of gold nanoparticle is largely dependent on its physicochemical properties. Thus, careful optimization of physicochemical properties can help in avoiding the toxic effects of gold nanoparticles.

Nano-Silica: Silica is a ubiquitous and most primitive mineral discovered by the human race. It is chemically expressed as SiO_2 (silicon dioxide), an oxide of silicon. All silica forms are identical in chemical composition, but have different atomic arrangements. Silica can be of crystalline or amorphous nature. Crystalline silica is not functionally useful as it induces pulmonary diseases, silicosis, and lung cancer [132]. Amorphous form is found in nature as components of granite, sandstone, and clay and in certain plants and animals. Certainly, it is a multifaceted material in the cosmetic formulations because of its distinctive properties like surface-to-volume ratio, higher surface reactivity, and hardness. It therefore bears capacity to serve as an anticaking agent, bulking agent, suspending agent, abrasive, opacifying agent, and as thickener. It occurs in many forms like hydrated silica, silica silylate, and silica dimethyl silylate (modified with alkyl groups).

Scientific Committee on Consumer Safety (of the Directorate-General for Health and Consumer Protection of the European Commission) in its opinion draft stated the maximum non-lethal doses of two sample of hydrophobic precipitated synthetic amorphous silica. Following oral gavage to Sprague-Dawley rats, the doses were above 7900 and 5000 mg/kg, respectively. However, no results are drawn for acute toxicity by intraperitoneal, inhalation, and dermal exposure route. No toxicity conclusions are reported for irritation and corrosivity test, and dermal and percutaneous absorption. Thus, silica has no definite potential to produce toxicity when used under specified application. For further information, please visit https://ec.europa.eu/health/scientific_committees/consumer_safety/docs/sccs_o_175.pdf.

Silica is also used as an ingredient in powder perfume because the porous spheres can deliver fragrance over a longer period of time. Recently, the use of silica for coating metal oxide particle has been discovered. Silica coating provides better feel factors and absence of photocatalytic effect which would otherwise be a serious drawback of metal nanoparticles in generating reactive oxygen species. Other than the cosmetic benefits, it has immunosuppressant, hypocholesterolemic, antiallergenic, antiapoptotic, and antihistamine effects [86].

Some in vitro studies have reported that silica nanoparticles encourage ROS generation after endocytosis leading to DNA destruction. The formation and accumulation of protein aggregates, nuclear inclusions, and membrane disruption promote apoptosis and cell cycle arrest [99]. Large particles (60 nm) produce more cytotoxicity than smaller particles at higher doses (200–500 µg/ml) and most ROS production, whereas the 20 nm size particles produce greater damage to intracellular structures at different concentrations. But studies on different cell lines have produced disagreements with the present findings [73]. These data thus indicate that the interaction with different cells is not similar. Nevertheless, there are dose, size, and cell type dependence, whereas negative charges on nanoparticles also seem to suppress the immune response and play a key role in cytotoxicity.

Nano-Zirconium oxide—Zirconium (Zr) is a silver-gray metal. It is exceptionally malleable, strong, ductile, and lustrous and has properties similar to titanium, extremely resistant to heat and corrosion. Complexes containing zirconium have been used in cosmetics as deodorants or aerosol cosmetic antiperspirants. Evidences from human data indicate that such compounds can cause human skin granulomas [130]. Aerosolized form enables zirconium to reach the inner parts of lungs and is even able to cross the blood brain and the placental barrier. Genotoxicity studies suggest that zirconium oxychloride is clastogenic in vitro and mutagenic in vivo [45]. Clastogenic substances are agents that give rise to disturbances or changes in integrity or arrangement of chromosomes.

According to Federal Food, Drug, and Cosmetic Act, Section 601(a), any zirconium containing cosmetic aerosol is deemed to be adulterated and is subjected to regulatory action for interstate business. It is used in many forms like amino acid complex of aluminum zirconium trichlorohydrex, aluminum zirconium tetrachlorohydrex, aluminum zirconium pentachlorohydrex, and aluminum zirconium octachlorohydrex. Nanosize zirconium dioxide is used as opacifier but as demonstrated particle size greater than 530 nm is cytotoxic to fibroblasts and macrophages [125]. Zirconium coating renders TiO_2 nanoparticles less photoreactive. Hence, the use of coated oxides in commercial products eliminates the distress over the generation of ROS.

Nano-Silver: Silver (Ag) has been used extensively throughout the immortalized times for a variety of cosmetic products like soaps, nail polish, toothpastes, facial moisturizer, antiseptic creams, wet wipes, deosprays, lip care products, around-eye creams as well as shower foams. Recently, silver nanoparticle form has achieved great interest in cosmetics, and medicinals for their antibacterial properties and their unique electrical, optical, and thermal properties. Extensively utilized silver nanomaterials are the field of exceptional concern, with several reports indicating that they could make their way from the outer to inner organs of the body and accumulate there.

Silver nanoparticles are usually designated as colloidal silver solutions, which contain silver nanoparticles and are allegedly used for their antioxidant and antibacterial properties in toothpastes and lotions as preservatives. However, argyria—a condition caused by excessive exposure to silver particles, which causes blue colored skin—is developed on prolonged exposure. At the same time, it is also reported that silver nanoparticles penetrate the stratum corneum but reach not farther than superficial part of epidermis.

Moreover, on coming in contact with cell machinery, silver nanoparticles produce dose-dependent elevation in the generation of ROS and induce mitochondrial disruption by reducing the ATP levels [104]. Since it has been accounted that silver nanoparticles at even lower concentration can reduce cell viability in HeLa cells as compared to silver ions, their usage concentration has been intensely monitored. The cytotoxicity and genotoxicity studies of silver nanoparticles indicate that they are cytotoxic and DNA destructors but most likely not mutagenic [67]. However, no such specific toxicity study could be identified for proving the toxicity or other serious complications with silver nanoparticles.

Nano-Aluminum: Aluminum oxide is an inorganic compound of aluminum and oxygen, chemically identified as Al_2O_3. Recent applications of aluminum oxide nanoparticles in cosmetics include fillers and additives. Its applications are steadily increasing in recent years due to the high adsorption ability of nanoparticles, nontoxic profile, antibacterial properties, and stability. Higher transparency and soft focus properties are also present.

The cytotoxic effects of aluminum oxide nanoparticles on various mammalian cell lines, L929 and BJ, were studied by E Radziun et al. by EZ4U assay (activity of a mitochondrial enzyme, succinate dehydrogenase was measured), and the oxide nanoparticle was observed to produce minor, less than ten percent cytotoxic effects [74]. Aluminum oxide nanoparticles are commonly used in face masks, nail makeup, and sun protection products. It is interesting to note that all are routes of dermal exposure but toothpastes provide oral route of exposure for hydrated aluminum oxides, where they are present as abrasives and dispersing agent.

In another study, toxicity of inhaled alumina nanoparticles was studied on the basis of three parameters (lactate dehydrogenase release, tumor necrosis factor alpha production, and ROS production) using RAW 264.7 macrophage cell and these were observed to have no notable cytotoxicity as compared to positive control (quartz)

Nano-Copper: Copper (Cu) is an easily available mineral and is known to have many beneficial biological applications. It is cosmetically used in chemical form of copper carbonate, copper silicate, copper sulfate, copper oxide, and copper chloride. Pure metallic copper oxide (CuO) nanoparticles are spherical in shape and have been successfully used in antiaging creams; around-eye cream; facial and body cleanser for men; facial treatment for both men and women. They are good at stimulating collagen and elastin regeneration while ameliorating the skin firmness and elasticity. They also improve skin condition by increasing microcirculation in the skin microarteries and also act as cicatrizing agent [14]. Interestingly, cosmetic market flaunts many textile products, embedded with microscopic copper oxide particles like pillows, socks, linens, undergarments, baby diapers, containing copper oxide for skin rejuvenation and brightness [14, 15]. In the biological system, copper is responsible for the activity of enzymes, whereas animal studies suggest that if present in large quantity and very small size, they may cause damage to the kidney, liver, spleen, and brain tissues. This is because as the size is reduced, they easily intrude the body via the skin and respiratory organs.

Cytotoxicity of copper nanoparticles of size 40–80 nm has been observed in dorsal root ganglion neurons by lactate dehydrogenase assay method, especially for evaluating effects on sensory neurons. Overall, results have shown copper nanoparticles to be cytotoxic in concentration- and size-dependent manner. Particularly, outcomes of MTS assays suggested that vulnerability to copper nanoparticles in higher concentrations (40 µM) contributed to a considerable reduction in cell viability via mitochondrial activity and metabolic disturbances [113].

Carbon nanotubes and fullerenes—Carbon nanotubes and fullerenes are two valuable allotropic nanoforms of carbon molecule. Carbon nanotubes are available in different forms depending on the number of layered sheets present. Multiwalled carbon nanotubes contain more than one cylindrical sheet of graphene, and single-walled cylindrical tubes contain only one rolled sheet of grapheme [3]. Fullerenes are spherical molecules with superior radical scavenging properties, having mean diameters of <100 nm [64]. Currently, the most popular use for carbon nanotubes is in hair colorants owing to their enhanced affinity for hair fibers. High ratio of surface area to volume results in long-lasting effect of hair colorants.

So far, the penetration of nanotubes from cosmetic formulations into the skin has not been accounted in the literature. However, the results of some experimental studies have disclosed that carbon nanotubes can damage abdominal walls and produce cytotoxicity. Poland et al. demonstrated that the inflammation of the abdominal wall occurs on long-term exposure to lengthy carbon nanotubes [112]. Moreover, a dose-dependent increase in cytotoxicity and genomic DNA damage with multiwalled carbon nanotubes has also been reported, when tested on normal human dermal fibroblast cells using three different concentrations. The results suggested that carbon nanotubes are toxic only at high concentrations (dose 40 µg/ml) and that careful monitoring of toxicity studies can preclude the risk potential [108]. Additionally, it has been reported that the length of multiwalled carbon nanotube is a key characteristic for its lung toxicity [78].

Spherical fullerenes, or buckyballs as they are also known due to resemblance with football, are carbon allotropes that are only about one carbon atom thick, discovered in 1985 and named after Buckminster Fuller. Their structure includes 20 hexagonal and 12 pentagonal rings, resulting into the formation of closed cage structure with icosahedral symmetry. They are allegedly used for antioxidant and smoothing properties in moisturizers and a number of antiaging creams. Fullerenes are able to scavenge free radicals and inhibit the related malfunctions. For this, they can be used as inhibitors of hair follicle apoptosis and aging. Uses in cosmetics involve fullerenes wrapped by polyvinylpyrrolidone (radical sponge) for skin whitening purpose. "CosIng" the European Commission database furnishes information on cosmetic ingredients. "Cosmetics Directive" (76/768/EEC) lists fullerenes and hydroxyl fullerenes as a special ingredient and labels it as antimicrobials and for skin conditioning.

Despite being an impressive material, there are some concerns about its penetration into dermis. Studies have been carried out to determine the influence of various factors on penetration using flexed and unflexed skin. Results showed that the fullerenes are incapable to penetrate the skin when used alone, but when used with specific vehicles, they penetrate easily. Hence, their penetration seems to be dependent on the vehicle characteristics like polarity, log P value, density, and viscosity. The solvent flux associated with different solvent systems like toluene, cyclohexane, or chloroform and particle aggregation state are also contributing factors in penetration efficacy [148].

14.5 Toxicological Concerns for Dermal Exposure and Percutaneous Penetration

The most important uptake route of nanomaterials contained in cosmetics is through dermal exposure. The upper dermis is the inner layer of the skin having dense connective tissue, whereas lower dermis also called as reticular dermis has blood vessel, mast cells, lymphatics, dendritic cells, and numerous sensory nerve endings. Dermis integrity is breached by age, vulnerability to solvents, poor skin care products, health situations, and environmental components. Further increase in delivery of compounds across the cell membrane can be made through modification or adjustment of epidermal lipid biosynthesis or use of chemical enhancers. Moreover, broken or bruised skin further allows unrestricted entry of micro- and macromolecules [47]. The propensity for nanomaterials to pass through the skin barrier is a primary indicating factor of its dermatologic perspective.

The uptake of nanoparticles is mediated by a reversible deformation in the bilayer structure of the skin [123]. Wound dressings and cosmetics are the primary sources of dermal exposure of nanomaterials. For example, according to a study, TiO_2 nanoparticles in sunscreens perpetually passed through the stratum corneum and accumulated in the deeper parts of hair follicles [28].

Among the other routes of exposure, inhalation, accessed via aerosolized products, is most hazardous, whereas exposure through dermis is generally considered secure; at the same time, it is generally accepted that nanoparticles greater than 20 nm do not penetrate through intact sound skin. Accordingly, dermal route has acquired most attention for penetration studies using easily available skin care products, while the case studies on inhalation route have been comparatively limited. However, on longer duration exposure (more than two months), it has been demonstrated that nanoparticles can intrude the skin of hairless mice. Liver and skin are the most common targets for pathological changes [62]. Other than skin, nanoparticles have been documented to interact with the blood–brain and blood–testis barrier, but there has been no evidence of causing any pathological changes there [16]. Similarly, studies on sunscreens containing metal oxide nanoparticle have demonstrated the absence of nanoparticle penetration into deeper dermal layers after small duration exposure. However, comprehensive studies on interaction with body cells help in excluding any potential damage.

There is some evidence which suggests the nanoparticle can penetrate into the epidermis or dermis depending on their flexibility. Recently, the in vitro and in vivo cutaneous permeation investigation has indicated nanoparticle penetration into the deeper layer of skin. Nanostructured lipid carrier is more flexible and easily passes through photodamaged skin and reaches dermis level, whereas polymeric nanoparticles are restrained at epidermal level, suggestive of the safety of the polymeric nanoparticles during the studied period. The vivo study on skin absorption of nanoparticles was done on nude mice and tracked by fluorescence and confocal microscopic technique. Solvatochromism of Nile red was used for fluorescence microscopy. Further, the study compared the effects of UVA and UVB induced damage on skin with control. While the skin segments for lipidic nanoparticle containing Nile red showed dark red color in dermis, polymeric nanoparticles did not show any color segments beyond epidermis and hair follicles. Results also point out that the stain was not able to reach the deeper layers in intact skin due to the membrane barrier exerted by stratum corneum, thus protected the internal tissues. Furthermore, polymeric nanoparticles were better than lipid particles, as they did not affect the human keratinocytes cells, whereas lipid nanoparticles showed dose-dependent decrease in cell viability on the same cell line [61]

In a study, topical exposure of ZnO nanoparticles showed limited penetration into the inner layer of epidermis (stratum granulosum). This leads to the understanding that if nanoparticles gain entry into epidermis, it can reach blood circulation. In an earlier study, penetration of oil-in-water emulsion (as in sunscreen lotions) containing TiO_2 nanoparticle was calculated using the tape stripping method [10]. In this study, penetration from emulsion was comparatively more than from aqueous suspension due to the lipophilic nature of carrier system and that penetration was even better on hairy skin, which also indicates that TiO_2 nanoparticles penetrate via hair follicles or skin pores.

At the end, it is worthwhile to note that dermal absorption and percutaneous penetration of organic and inorganic nanoparticles take place in a different manner. Inorganic molecules generally penetrate easily because of the absence of surface charge, functional groups, and small size.

14.6 Nanosafety and Toxicity Testing

Past two decades have experienced a remarkable growth in guidance documents contributing to nanoproducts' safety protocols. Nanosafety is a term describing the safety of nanotechnology-enabled products. It involves the multidisciplinary safety approaches and testing protocols.

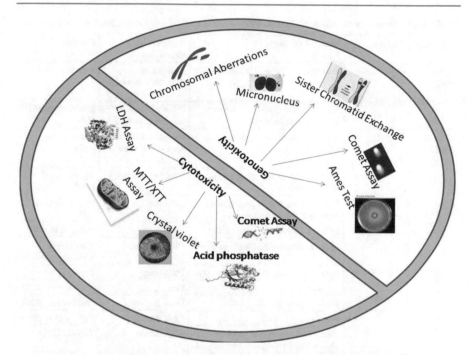

Fig. 14.2 Various in vitro and in vivo tests for genotoxicity and cytotoxicity

Toxicity testing is usually based on toxicological profile and route of exposure. Duration of exposure along with physicochemical properties is participating factor which influences the toxicokinetics of nanomaterials. Adverse effects are mainly observed in continuously exposed models, while no significant signs of toxicity are noted in infrequently exposed animals. As per the guidelines issued by the Cosmetic, Toiletry and Fragrance Association (CTFA) and the Organization for Economic Cooperation and Development (OECD), safety studies for nanocosmetics are done by testing for skin irritation, acute toxicity, ocular irritation, dermal photoirritation, skin sensitization, mutagenicity/genotoxicity, repeated dose (21–28 days) toxicity, and subchronic (90 days) toxicity. Out of these, few are discussed here.

Genotoxicity: Genotoxicity is expressed as a consequence of the damaging effect on a cell's genetic material. The integrity of DNA and RNA is compromised on the exposure of genotoxic agents [13]. Genotoxins or genotoxic agents are substances which lead to mutations and are called mutagens. Genotoxins involve radioactive waste materials, radiations, and chemical substances. Several in vitro and in vivo studies for genotoxicity and cytotoxicity have been designed (Fig. 14.2) and performed (Table 14.1). Their endpoint recognition involves detecting DNA damage or other associated consequences in diverse biological systems like bacterial (e.g., of prokaryotic) or mammalian, avian or yeast (e.g., of eukaryotic) cells or different organisms like rodents, humans, and aquatic species, such as zebrafish. In one study, TiO_2 nanoparticles displayed genotoxicity to prokaryotic cells. For

Table 14.1 Summary of cytotoxic and genotoxic tests carried out on cosmetic nanomaterials

Ingredient	Activity	Nanotoxicity study	Proposed toxicity mechanism	References
Carbon black	Colorant	MTT assay—cytotoxic to THP-1 cells[a]	Afflicted the phagocytic capacity of monocytes	Sahu et al. [119]
		MTT reduction assay—cytotoxic to human dermal fibroblasts	Breached plasma membrane integrity	Grudzinski et al. [49]
Titanium dioxide	UV filter	In vivo comet assay—genotoxic to bone marrow and liver	Oxidative stress and agglomeration inside the cell	Sycheva et al. [133]
		Micronucleus assay—genotoxic to B6C3F1 mice	Disturbed cell cycle	Trouiller et al. [137]
		In vitro comet assay—genotoxic to bottlenose dolphin leukocytes	DNA damage	Bernardeschi et al. [11]
		HEK293 (embryonic kidney cells)	DNA damage	Demir et al. [26]
		Chinese hamster lung fibroblast cells	DNA damage	Hamzeh and Sunahara [57]
		Human peripheral blood lymphocytes	DNA damage	Turkez [138]
		Micronucleus assay—HepG2 cells	DNA damage	Shukla et al. [126]
		Human epidermal cells (A431)	DNA damage	Shukla et al. [127]
		Human lymphocytes	DNA damage	Tavares et al. [134]
		Sister chromatid exchange assay Human peripheral blood lymphocytes	DNA damage	Turkez [138]
Zinc oxide	UV filter	Comet assay—genotoxic to L02 and HEK293 cells	DNA damage	Guan et al. [50]
		MTT assay—cytotoxic to human negroid cervix carcinoma (HEp-2 cell)		Osman et al. [106]
		Comet and micronucleus assays—genotoxic to human negroid cervix carcinoma (HEp-2 cell)	Increase in tyrosine phosphorylation	Osman et al. [105]
		MTT assay—cytotoxic to human dermal fibroblast cells	Upregulation of p53 and phospho-p38 proteins	Meyer et al. [93]
Gold	Revitalizing	Comet assay—genotoxic to yeast *Saccharomyces cerevisiae*	DNA damage	de Alteriis et al. [24]
		Human bronchial epithelial cells HBEC3-KT	Increased DNA strand breaks	Lebedová et al. [80]
		Human hepatoma, HepG2 cells	Dose-dependent DNA damage	Jia et al. [65]
		MTT assay—human dermal fibroblasts	ROS production	Mateo et al., n.d.
Silver	Antibacterial	Cytotoxic to Caco-2 cells	ROS production	Kaiser et al. [68]
		Cytotoxic and genotoxic to human lung fibroblast cells (IMR-90) and human glioblastoma cells (U251)	DNA damage and chromosomal aberrations	Asharani et al. [6]
		Micronucleus assay—genotoxic to Wistar rat bone marrow cells		Dobrzynska et al. [31]
		Comet assay—genotoxic to mouse blood leukocytes	ROS-induced	Plotnikov et al. [110]

(continued)

Table 14.1 (continued)

Ingredient	Activity	Nanotoxicity study	Proposed toxicity mechanism	References
Silica	Opacifying agent	Comet assay—genotoxic to mouse embryo fibroblast cell	Oxidative stress	Yang et al. [149]
		Genotoxic to mouse fibroblast cells	Induction of chromosomal damage	Kwon et al. [75]
		Micronuclei assay—genotoxic to V79 cells	Oxidative DNA damage	Liu et al. [87]
Zirconium oxide	Antiperspirants	Male Wistar rats—increase in malondialdehyde concentration level and decrease in catalase, glutathione peroxidase, and superoxide dismutase activities	ROS generation	Arefian et al. [5]
Aluminum	Filler	Comet assay and micronucleus test—genotoxic to female Wistar rat peripheral blood cells	Discharge of DNase from the lysosomes	Balasubramanyam et al. [9]
		MTT assay—cytotoxic to human brain endothelial cell line	Alterations in blood–brain barrier tight junction protein expression	Chen et al. [21]
Copper	Colorant	Inhibition in the activity of catalase, superoxide dismutase, and glutathione peroxidase of liver tissues of freshwater fish Nile tilapia	Oxidative stress	Abdel-Khalek et al. [1]
		Cytotoxic to A549 human lung cell line	ROS generation	Wongrakpanich et al. [145]
		Cytotoxic to airway epithelial cells	CuO-induced oxidative stress	Fahmy and Cormier [39]

*THP-1—human monocytic cell line

nanomaterial assay, mammalian cell line is preferred over a bacterial system due to lack of understanding that whether nanoparticles can penetrate through bacterial cell wall or not. These assays are performed to evaluate the safety of cosmetic ingredients and beauty products. They also help in elucidating the mechanism of action of genotoxins. Mechanism of action of genotoxins involves their pro-metabolic activation. The reactive products formed bind covalently to DNA, and form DNA adducts, which are detected in laboratory tissues through sensitive analytical techniques.

The nephrotoxicity and genotoxicity of zinc oxide nanoparticles have been investigated using the MTT, Trypan blue (TB) and Neutral red uptake (NRU) assays and comet assay using NRK-52E cells (rat kidney epithelial). The mean of inhibition concentration or half maximal inhibitory concentration (IC50) values of zinc oxide nanoparticles were determined at 25–100 g/mL exposure concentrations, and the results (73.05; 75.39;80.46 for MTT; NRU; TB, respectively) of this study indicated that these particles caused invariably significant cell death and DNA damage which were likely to be oxidative stress mediated [63].

Table 14.2 Summary of negative genotoxic (in vitro and in vivo) tests carried out on cosmetic nanomaterials

Ingredient	Activity	Characteristic feature	Size (nm)	Nanotoxicity study	References
TiO$_2$ nanoparticles	Sun protection	84% anatase and 16% rutile	23	Genotoxicity—Ames test	Jomini et al. [66]
		Anatase	<100	Genotoxicity—comet assay—human lung fibroblasts and human bronchial Fibroblasts	Bhattacharya et al. [12]
		Rutile	62	Genotoxicity—comet assay—Syrian hamster embryo cells	Guichard et al. [52]
		Polyacrylate-coated	1–10	Genotoxicity—comet assay—Chinese hamster lung fibroblast cells	Hamzeh and Sunahara [57]
		Rutile in T-Lite	10–50	Genotoxicity—comet assay V79 cells	Landsiedel et al. [77]
		Anatase	10	Genotoxicity—comet assay—Chinese hamster ovary cells	Wang et al. [141]
			10	Genotoxicity—Ames test—TK6 human lymphocytes	Woodruff et al. [146]
		Rutile	25	Genotoxicity—micronucleus assay—Syrian hamster embryo cells	Guichard et al. [52]
		85% anatase and 15% rutile	20	Genotoxicity—micronucleus assay—human lymphocytes	Tavares et al. [135]
		Anatase	10	Genotoxicity—HPRT mutation assay—Chinese hamster ovary cells	Wang et al. [141]
			43–213	Genotoxicity—micronucleus assay—female male Wistar rats—peripheral blood reticulocytes	Donner et al. [32]
		74% anatase and 26% brookite	84.5–89.2	Genotoxicity—micronucleus assay and comet assay—male C57BL/6J mice	Lindberg et al. [85]
		Anatase	5	Genotoxicity—comet assay—(lung tissues—male Crl: CD (SD) rats	Naya et al. [102]
		Anatase	12.1	Genotoxicity—micronucleus assay—male B6C3F1 mice—peripheral blood reticulocytes	Sadiq et al. [118]
Silicon dioxide nanoparticle	Opacifying agent	–	17.7–24.7	Genotoxicity—comet assay—male Sprague-Dawley rats	Guichard et al. [51]
			15–55	Genotoxicity—Comet assay—lung tissues—Wistar Hannover rats	Maser et al. [90]

(continued)

Table 14.2 (continued)

Ingredient	Activity	Characteristic feature	Size (nm)	Nanotoxicity study	References
Gold nanoparticle	Antiaging		2–200	Comet assay—male Wistar rat blood, liver, and lung tissues	Downs et al. [33]
Carbon nanotubes	Hair colorants and antiaging		44	Comet assay—lung cells	Ema et al. [37]

Similarly, in chromosome aberrations and micronucleus assays, ZnO nanomaterials dispersed in (heparinized whole blood sample with chromosome medium B along with phyto-hemagglutinin L) cultured human peripheral lymphocytes showed four different cases of chromosome aberrations namely chromatid and chromosome breaks, fragment and dicentric chromosomes. A decrease in the mitotic index indicates cellular apoptosis [55]. These outcomes reveal that ZnO is clastogenic in human lymphocytes in vitro. A clastogen is basically a mutagenic agent that induces chromosome distortions. However, often there are discrepancies or contradiction in such studies. For example, in a SOS chromotest study, no induction of DNA-repairing enzymes was detected for silver, gold, and titanium dioxide nanoparticles and ions [101]. Interestingly, it was found in many studies that there was no genotoxicity after dermal exposure and the effects were seen only on exposure through oral route. Table 14.2 presents a brief description of negative results of nanomaterial toxicity tests.

An increase in micronucleus (MN) assay frequency showing decrease in GSH level, SOD, and catalase enzyme activity has been reported in human epidermal cells exposed with ZnO nanoparticles [55]. Moreover, the observations of increase in ROS and lactate dehydrogenase are indicative that nanoparticles of ZnO cause cytotoxicity and induce oxidative stress through Zinc ions.

Over the years, there have been serious discussions about the genotoxicity of silver nanoparticles as silver ions have not shown any harm to genetic material in spite of showing cytotoxic effects in many assays. To investigate the extent of apoptosis and ROS generation on exposure to ZnO nanoparticles, human lymphoblastoid thymidine kinase heterozygote (TK6) cell lines have been used in customized PCR array and in vitro micronucleus assay using flow cytometric analysis. The gene expression study was used to determine activities of 89 genes which were related to cytotoxic and genotoxic effects like oxidative stress, metal ion binding, disturbed cell cycle, DNA damage and consequent repair, mitochondrial and endoplasmic reticulum stress, deformed protein responses, apoptosis, and proliferation. All the results revealed that silver nanoparticles induced deregulation of genes involved, indicating that silver ions induced oxidative stress act as the major player in genotoxicity mechanism [84]. Moreover, the kind of toxic effects, dose, and cell lines used for studying nanoparticles must also be considered.

Skin sensitization: As already discussed, by and large, the usable data indicate that most nanoparticles are unable to penetrate intact skin deep enough to get absorbed systemically. Therefore, for the dermal route of exposure, qualitative toxicity endpoints for localized applications are probably more relevant than quantitative toxicity endpoints. Therefore, the endpoint irritation and sensitization assays are of better relevance. The action of nanosized chemical substances is not just drive a change in the barrier properties of the skin but may also lead to irritation or sensitization of skin. According to OECD guidelines, all products containing skin-irritating chemicals (also can be nanotechnology-driven products) can be tested for skin irritation through in vitro skin irritation test procedures (http://www. oecd.org/chemicalsafety/news-nanomaterial-safety.htm). One method of irritation testing is by using reconstructed human epidermis (RHE) models (OECD 439). Other models are as per USFDA safety guidelines on Nanomaterials in Cosmetic Products are Epiderm™ and Episkin™. These models are intended to symbolize a genuine epidermis with specific layers of basal, spinous, and granulosum, enclosed by an uppermost stratum corneum. Furthermore, they are incorporated human keratinocytes cells.

Sensitization test involved three important tests: Guinea Pig Maximization Test, Buehler test, and the Local Lymph Node Assay. As all the above tests involve experimental animals alternative in vitro assays have been developed. In vitro and in vivo test procedure already used for assessing toxic chemicals can be suitable for nanomaterials also. For example, the KeratinoSens™ assay (OECD 442D) and Direct Peptide Reactivity Assay (OECD 442C), can have great applicability as in vitro tests for sensitization testing of nanocosmetic ingredients. Optional, in vitro tests include human cell line activation test (h-CLAT), and gene expression in the VITOSENS® model.

Positive results of at least two tests out of three performed are considered for designating a molecule as sensitizer and represent the outcome as yes or no for marketing. Data from patch testing and available human experiences are also used for identification. These assays compare the cytotoxic behavior of test molecules on exposure to UV light with control (without exposure) in fibroblasts and other tissues [117]. Cytotoxicity is measured as a function of dose with dependent reduction in the viability of cells measured by the uptake concentration of the vital dye, neutral red after 22 h of the treatment. The results of 3T3 NRU phototoxicity assay indicate the absence of phototoxic potential. The negative results obviate the requirement of phototoxicity testing in vivo. Moreover, a chemical is considered to be irritant if the human skin equivalent model (HSEM) preferably derived from keratinocytes viability is less than 50% [95].

Recently, skin irritation study on 3D epidermal model, Episkin was conducted to evaluate the irritation potential of 500 μg/mL of different nanoparticle and non-nanoparticle of ZnO after 24 h of application time. No significant irritation events were found on the treated models; however, reduction in the cell viability of HaCaT cells was observed in dose- and time-dependent manner [140]. Additionally, histological studies on human skin models showed no difference from the control state. These results corroborate with other authors who found significant cytotoxicity in HaCaT cells treated with ZnO after 24 h exposure time [81].

Previously, KeraSkin$_{TM}$ was used to evaluate the safety of mixture of zinc oxide and titanium oxide nanoparticles. Skin irritation was assessed by using 3-(4, 5-dimethylthiazol-2-yl)-2.5-diphenyl tetrazolium bromide assay. The viability was generally found to be higher than 50% on 3 min exposure time and 15% after 60 min exposure time. Cytokine release behavior and histopathological remarks supplemented the results of cell viability test [97]. These findings suggested that mixtures of zinc oxide nanoparticles and titanium oxide nanoparticles are safe (non-corrosive and non-irritant) for the human skin.

Phototoxicity: Phototoxicity or photoirritation is a kind of skin inflammatory response, inducted by initial exposure of drugs or cosmetic substances through dermal or systemic route, followed by UVA radiations [72]. The test for phototoxicity is intended to compare the reduction in cell viability on exposure to phototoxic agents in the presence of light and control conditions (without light). Substances that show photocytotoxicity are called as "photocytotoxic agents" Mechanism of assay depends on the amount of uptake of neutral red dye which is inversely proportionate to the concentration of phototoxin present.

Effect is measured after 24 h of treatment. Neutral red dye is a weak cationic dye that easily penetrates cell membranes by binding to anionic cell membrane structure and accumulates intracellularly in lysosomes. On exposure to phototoxicins, structure of cell surface of the lysosomes gets altered that lead to lysosomal fragility. Such changes are readily detected by modified uptake and binding of dye. It hence differentiates between viable and nonviable cells.

Resultant harmful effects depend on the dose, time, and the type of photoexcited molecule. Some adverse reactions of phototoxicity include erythema/edema, pigmentary changes and visual defects, ocular tissue damage, photoallergy, and photocarcinogenicity [43]. It is essential to check the phototoxic potential of the ingredients that absorb light in the range of 290–700 nm with the help of appropriate assays as prescribed by OECD guidelines. A validated in vitro test for phototoxicity testing is 3T3 fibroblasts neutral red uptake phototoxicity testing (3T3 NRPT). It is applicable to ultraviolet absorbing substances and should be carried out as recommended by USFDA: Guidance for Industry: Safety of Nanomaterials in Cosmetic Products. To facilitate assessment of the data, a photoirritation factor (PIF) is calculated.

Titanium dioxide and zinc oxide nanoparticles can generate ROS by UV-induced photocatalysis. ROS are cytotoxic if produced in excess. To obstruct the production of ROS, inorganic nanoparticles are first often coated with antioxidant compounds and then used in sunscreens.

While oxidative stress-induced cytotoxicity has been documented in a variety of cell types for inorganic nanoparticles, their potential in phototoxicity is still unclear. Some studies have demonstrated nanoparticle-mediated photocytotoxicity in skin or skin-derived cells. M. Horie and co-workers have studied the phototoxicity induced by photocatalytic activity of anatase and rutile form of titanium dioxide nanoparticles. When applied to human keratinocyte cells followed by UVA irradiation, rutile titanium dioxide nanoparticles did not show any cellular effects, whereas anatase form reduced mitochondrial activity along with cell membrane damage and oxidative stress, suggesting that these results can help us in anticipating phototoxic potential of titanium dioxide. But at the same time, no significant toxicity was observed when the particles were exposed to EpiDerm™ [60]. Therefore, even with phototoxic activity in cell cultures, the safety of titanium dioxide nanoparticle in terms of phototoxicity and skin irritation can be guaranteed with the help of more advanced procedures.

Anne J.Wyrwoll et al. established that particle size and primary ionic strength of observed medium are critical factors for the generation of free radical-mediated phototoxicity. The effect was determined on crustaceans and daphnia magna. The phototoxic effect increases as the size of nanoparticles decreases [8].

14.7   Legal Regulations and Guidance for Nanocosmetics

Given the impact that regulation and guidance can have on developing nanocosmetic market and consequently on revenue generation, we undertake a section that focuses on safety standards and legal frameworks for cosmetics that contain nanomaterials

Thus, the objective of this section is to understand the regulatory guidelines that are applied to cater to the various uncertainties that could arise from the use of nanocosmetics. There has been an outgrowth of regulatory guidelines in recent years. Due to reported cases of undesired adverse affects, it has become obligatory to collect data for assessing the risks of nanoparticles. There are two core areas to focus: regulations on nanocosmetics and legal obligations.

USA: In USA, cosmetics in general do not require FDA premarket approval with the exception of color improvers. Cosmetic governance is done through two prime laws: the Federal Food, Drug, and Cosmetic Act (FD&C Act) and the Fair Packaging and Labeling Act (FPLA) [35]. Manufacturer solely bears the responsibility of ensuring the safety and efficacy of the product. Regarding the regulation of nanomaterials used in cosmetics following time frame with key developments has been highlighted:

2006: FDA established Nanotechnology Task Force.
2007: Report submitted by Task Force authorized FDA to encourage producers to gather scientific data required to assess the safety and adequacy of nanomaterials containing products, including data on long-term and chronic toxicity.
2007: The Working Party on Manufactured Nanomaterials of the Organization for Economic Cooperation and Development launched the Sponsorship Program for the Testing of Manufactured Nanomaterials (Testing Programme).
2011: Issued draft guidance documents on issues related to Nanotechnology Applications in Regulated Products, Including Cosmetics and Food Substances and invited comments on it.

2014: US FDA on the basis of comments received on aforementioned draft guidance issued three final intensive guidance documents associated with safety issues of nanotechnology-based products, including cosmetics and food substances.

The first guidance document (in Docket Number FDA-2011-D-0489, Federal Register notice: 79 FR 36534 (June 27, 2014) explains FDA's current thought process on determining the basis of nanotechnology principles in cosmetics. This document gives recommendations for technical assessments of different nanoproducts. It urges that assessment should be based on the specification of the product, emphasizing on the biological and mechanical effect exhibited. Furthermore, it gives relevance to two key points: particle size parameters and size-dependent properties or phenomena ("Guidance Documents > Guidance for Industry: Safety of Nanomaterials in Cosmetic Products," n.d.).

The second guidance document (Federal register notice: 79 FR 36532 (June 27, 2014) deals with the safety assessment of nanomaterials used in cosmetic products. It comprehends toxicological data after evaluating the physicochemical characteristics ("Federal Register : Guidance for Industry: Safety of Nanomaterials in Cosmetic Products; Availability," n.d.). Route of exposure, absorption, and toxicity testing are discussed in detail with some alternatives suggested for critical and unconventional cases like use of pigskin in a diffusion cell for dermal absorption, reconstructed human skin such as Episkin™ and Epiderm™ for skin irritation and corrosion testing, Bovine Corneal Opacity and Permeability (BCOP) and the Isolated Chicken Eye (ICE) for ocular irritation, and genotoxicity testing. The third guidance document deals with evaluating the property of important manufacturing procedure modifications, together with promising technologies. It does not deal with cosmetics but with food.

FDA does not necessitate manufacturers to disclose safety data to get approval before a cosmetic product is marketed and also name the nanomaterials on label. The reason behind is that since there are no substantial proofs that particle size does correlate with toxicity so it is quite possible that labeling may bother the consumers about its safety. Also, FDA governs productions based on their legal classification instead of the technology they utilize. Moreover, Section 301(a) forbids the marketing of adulterated or misbranded cosmetics according to FFDCA [144].

FDA has certain regulations and procedures for cosmetics with which manufacturers voluntarily may choose to comply with. FDA in conjugation with Personal Care Products Council has regulations on voluntary facility registration, Voluntary Cosmetic Registration Program and voluntarily reporting for ingredients and associated adverse reactions. FDA through this program updates manufacturers about risk possessing materials which helps them to remove such substances from their product. Moreover, in general, FDA restrains or provide limits for certain chemicals which may possess the risk of toxicity; e.g., zinc oxide is permissible in concentrations up to 25% only ("CFR—Code of Federal Regulations Title 21," n.d.).

European Union: To promote safe operation of cosmetic products in the European market, regulations and directives have been made in agreement with the European regulations in the last forty years. The key partners of the European Union (EU) are the European Directives and Regulations. Regulations are useful to all twenty-eight countries, member states. The standard suggestions for the screening of cosmetic materials are given in the SCCP's "Notes of Guidance for the Testing of Cosmetic Ingredients and their Safety Evaluation."

In the year 2009, the European "Cosmetic Regulation" (1223/2009) was adopted. The Regulation replaced Directive 76/768/EC, which was adopted in 1976. It defined nanomaterials as "an insoluble or biopersistant and intentionally manufactured material with one or more external dimensions, or an internal structure, on the scale from 1 to 100 nm." Substances included are those that are man-made, fullerene, single- or multiwalled carbon nanotubes and flakes of grapheme. Although there are differences in precise definition of nanomaterial among the various bodies, there are certain similarities. The agreements include 1–100 nm size, aggregate form, and intentional manufacturing criteria.

Article 16 of the European regulation dictates all issues associated with nanomaterials in cosmetics except as colorants, UV filters, or preservatives used in nanosize range. These are explicitly regulated under Article 14. Article 16 is due for the first revision in 2018.

According to articles 3, 4 and 13, a "responsible person" holds the sole responsibility to notify the commission by electronic means at cosmetic products notification portal. A responsible person could be any person appointed, i.e., manufacturers, importers, or third persons appointed. This should be done six months prior to the product being placed on the market. The required specifications supplied include the IUPAC chemical name of the nanomaterial, size, and physiochemical properties along with toxicological profile, yearly market placed expected quantity of nanomaterials contained in cosmetic products, accumulated safety data (by skin and eye irritation, and skin sensitization test), and anticipated exposure routes. The commission at any point of time if desired can take the assistance of SCCP for evaluating the safety of nanomaterials and make the sufficed report public. So far, titanium dioxide, zinc oxide, and tris-biphenyl triazine have been endorsed by the Commission as nano-UV filters. Carbon black in nanoform has also been authorized as a colorant in cosmetic products. Essentially, nanomaterials ought to be labeled with the word "nano" in brackets following the name of the substance in order to allow consumers to make a vigilant choice.

According to Regulation (EC) No. 1223/2009 Article 16 (10a), the Commission has recently on June 15, 2017, published a catalog of nanomaterials used in cosmetic products. The catalog contains information on name of the nanomaterials, applied category of cosmetic, route of exposure, rinse-off or leave-on product. The list is divided into three main sections, i.e., colorants, UV filters, and preservatives. The delay in publishing could be due to the paucity of the information received from the industry. According to catalog, most nanomaterials are exposed through dermal route other than few which are exposed to oral and inhalation route like nail varnish, nail makeup, lip care, soap, and eye contour products. Moreover, the

European Commission does not declare them as authorized nanomaterials but only a source of information.

Nanomaterials are also covered under REACH (Registration, Evaluation, Authorization and Restriction of Chemical substances) Regulation No. 1907/2006, Biocidal Products Regulation No. 528/2012, Novel Food Regulation No. 2015/2283, Food Additives Regulation No. 1333/2008, Plastic Food Contact Materials Regulation No. 10/2011, Active and Intelligent Food Contact Materials Regulation No. 450/2009, Provision of Food Information to Consumers Regulation No. 1169/2011, Medical Devices Regulation Proposal COM(2012) 542, and Scientific Committee on Consumer Safety (SCCS). The SCCS furnishes judgments on health and safety risks of cosmetic products and their ingredients. SCCP also underscores the importance of validated in vitro methods for case-by-case assessments and risk evaluations, thus avoiding undue animal testing. According to SCCP/1005/06, an uncertainty factor is applicable during safety evaluations. It is called as Margin of Safety (MOS). For the application of nanomaterials in cosmetics, even this parameter is inadequate in providing risk scales due to the unlikely exposure of nanomaterial to the abnormal skin and in cases of early clearance from epidermis to vital tissues and systemic circulation.

14.8 Conclusion

Over the last decade, nanotechnology has improved the performance of cosmetics to a large extent. It has enabled the molecules of organic and inorganic materials to produce desired effects due to their unique properties and small size. For example, the most used materials are TiO_2, silver, and gold nanoparticles. Nanotechnology-driven products have been used in various segments of cosmetics like skin care, hair coloring, sun protection, and novel applications like foot sprays and wet wipes. The interaction of nanomaterial with the biomolecules or the biochemical process plays a major role in determining the effect of the material on the human body. Therefore, proper exploitation of physicochemical properties like size, shape, surface charge, solubility in a desirable way helps in producing more benefit. Different types of experimental methods have been carried out to judge the toxic potential of nanomaterials. However, most studies have been done in general (not using cosmetic formulation but pure nanomaterials in most cases) and are not purely toxicological but more mechanics oriented in nature. Therefore, the results obtained through these experimental sets are mystifying and cannot be directly correlated with the application of nanomaterials in cosmetics. It can be considered that nanomaterials being used in cosmetics are less likely to produce any untoward condition because of the given reasons:

1. Most cosmetics are applied on the surface of the skin and skin act as a strong barrier to penetration.
2. Generally, skin cosmetics are used in very small quantity and others are rinse off (where the contact time is very less).

3. Toxic potential of nanomaterial is being taken care of by adjusting the physicochemical characteristics like size, surface charge, functional group, and polarity.
4. The results of toxicity assay largely depend on dose metrics and exposure route. The dose of the nanomaterials used in cosmetic products is not the same as used for the toxicity studies. Therefore, the obtained results cannot be directly extrapolated to cosmetic applications.
5. Moreover, the characterization of only in-product nanomaterials without actual simulation of realistic exposure cannot insure certainty of the toxicity potential of nanomaterial in cosmetics.

The regulatory bodies are judiciously providing guidance to the manufacturers for necessary precautionary steps that could help in eliminating any potential toxic condition. Steps are being taken to create standardization and worldwide consensus on experimental procedures and requirements for use of specific nanomaterial for cosmetic application, e.g., The International Cooperation on Cosmetic Regulation having USA, Japan, the European Union, and Canada as the core members.

Moreover, it can be suggested that more emphasis should be given to cosmetic dosage forms which are capable of avoiding high penetration, e.g., aqueous suspensions, solutions, non-aerosol sprays, and which can provide better action with less concentration of active ingredient. Furthermore, the preparation of nanoparticles through green synthesis can also help in avoiding the presence of toxic chemicals within the nanomaterials.

Lastly, the nanomaterials discussed in this chapter have been safely used in cosmetics since many decades with no critical adverse effects reported by the consumers. It therefore can be safely concluded that nanoparticles of various inorganic and organic materials can still be recommended for cosmetic purpose.

References

1. Abdel-Khalek AA, Badran SR, Marie M-AS. Toxicity evaluation of copper oxide bulk and nanoparticles in Nile tilapia, Oreochromis niloticus, using hematological, bioaccumulation and histological biomarkers. Fish Physiol Biochem. 2016;42(4):1225–36. https://doi.org/10.1007/s10695-016-0212-8.
2. Alkilany AM, Murphy CJ. Toxicity and cellular uptake of gold nanoparticles: what we have learned so far? J Nanoparticle Res An Interdisc Forum Nanoscale Sci Technol. 2010;12 (7):2313–33. https://doi.org/10.1007/s11051-010-9911-8.
3. Ando Y, Zhao X, Shimoyama H, Sakai G, Kaneto K. Physical properties of multiwalled carbon nanotubes. Int J Inorg Mater. 1999;1(1):77–82. https://doi.org/10.1016/S1463-0176 (99)00012-5.
4. Ansell J, Care P, Council P, Rauscher H. Report of the Joint Regulator—Industry Ad Hoc Working Group: Currently Available Methods for Characterization of Nanomaterials, 35 (2011). Retrieved from http://ec.europa.eu/consumers/sectors/cosmetics/files/pdf/iccr5_char_nano_en.pdf.
5. Arefian Z, Pishbin F, Negahdary M, Ajdary M. Potential toxic effects of Zirconia Oxide nanoparticles on liver and kidney factors. Biomed Res. 2015;26(1):89–97.

6. Asharani PV, Low G, Mun K, Hande MP, Valiyaveettil S. Cytotoxicity and Genotoxicity of Silver. 2009;3(2):279–90.
7. Aueviriyavit S, Phummiratch D, Kulthong K, Maniratanachote R. Titanium Dioxide nanoparticles-mediated in vitro cytotoxicity does not induce Hsp70 and Grp78 expression in human bronchial epithelial A549 cells. Biol Trace Element Res. 2012;149(1):123–32. https://doi.org/10.1007/s12011-012-9403-z.
8. Bach A, Hellack B, Wyrwoll AJ, Lautenschl P, Sch A. Size matters e the phototoxicity of TiO$_2$ nanomaterials. 2016;208:859–67. https://doi.org/10.1016/j.envpol.2015.10.035.
9. Balasubramanyam A, Sailaja N, Mahboob M, Rahman F, Hussain SM, Grover P. In vivo genotoxicity assessment of aluminium oxide nanomaterials in rat peripheral blood cells using the comet assay and micronucleus test. Mutagenesis. 2009;24(3):245–51. https://doi.org/10.1093/mutage/gep003.
10. Bennat C, Müller-Goymann CC. Skin penetration and stabilization of formulations containing microfine titanium dioxide as physical UV filter. Int J Cosmet Sci. 2000;22 (4):271–83. https://doi.org/10.1046/j.1467-2494.2000.00009.x.
11. Bernardeschi M, Guidi P, Scarcelli V, Frenzilli G, Nigro M. Genotoxic potential of TiO$_2$ on bottlenose dolphin leukocytes. Anal Bioanal Chem. 2010;396(2):619–23. https://doi.org/10.1007/s00216-009-3261-3.
12. Bhattacharya K, Davoren M, Boertz J, Schins RP, Hoffmann E, Dopp E. Titanium dioxide nanoparticles induce oxidative stress and DNA-adduct formation but not DNA-breakage in human lung cells. Particle Fibre Toxicol. 2009;6(1):17. https://doi.org/10.1186/1743-8977-6-17.
13. Bolt HM. Genotoxicity–threshold or not? Introduction of cases of industrial chemicals. Toxicol Lett. 2003;140–141:43–51. Retrieved from http://www.ncbi.nlm.nih.gov/pubmed/12676450.
14. Borkow G. Using Copper to improve the well-being of the skin. Curr Chem Biol. 2014;47:89–102.
15. Borkow G, Del A, Elías C. Facial skin lifting and brightening following sleep on Copper Oxide containing pillowcases. 2016. https://doi.org/10.3390/cosmetics3030024.
16. Brohi RD, Wang L, Talpur HS, Wu D, Khan FA, Bhattarai D, Huo L-J, et al. Toxicity of nanoparticles on the reproductive system in animal models: a review. Front Pharmacol. 2017;8:606. https://doi.org/10.3389/fphar.2017.00606.
17. CFR—Code of Federal Regulations Title 21. (n.d.). Retrieved from https://www.accessdata.fda.gov/scripts/cdrh/cfdocs/cfcfr/CFRSearch.cfm?fr=352.50.
18. Calabrese V, Cornelius C, Leso V, Trovato-Salinaro A, Ventimiglia B, Cavallaro M, Castellino P, et al. Oxidative stress, glutathione status, sirtuin and cellular stress response in type 2 diabetes. Biochim Biophys Acta (BBA)—Mol Basis Disease. 2012;1822(5):729–36. https://doi.org/10.1016/j.bbadis.2011.12.003.
19. Chambers C, Degen G, Dubakiene R, Grimalt R, Jazwiec-Kanyion B, Kapoulas V, White I, et al. Scientific committee on consumer products SCCP preliminary opinion on safety of nanomaterials in cosmetic products. 2007.
20. Chandran P, Riviere JE, Monteiro-Riviere NA. Surface chemistry of gold nanoparticles determines the biocorona composition impacting cellular uptake, toxicity and gene expression profiles in human endothelial cells. Nanotoxicology. 2017;11(4):507–19. https://doi.org/10.1080/17435390.2017.1314036.
21. Chen L, Yokel RA, Hennig B, Toborek M. Manufactured aluminum oxide nanoparticles decrease expression of tight junction proteins in brain vasculature. J Neuroimmune Pharmacol Official J Soc NeuroImmune Pharmacol. 2008;3(4):286–95. https://doi.org/10.1007/s11481-008-9131-5.
22. Chen T, Yan J, Li Y. Genotoxicity of titanium dioxide nanoparticles. J Food Drug Anal. 2014. https://doi.org/10.1016/j.jfda.2014.01.008.

23. Crosera M, Prodi A, Mauro M, Pelin M, Florio C, Bellomo F, Filon FL, et al. Titanium Dioxide nanoparticle penetration into the skin and effects on HaCaT cells. Int J Environ Res Public Health. 2015;12(8):9282–97. https://doi.org/10.3390/ijerph120809282.

24. De Alteriis E, Falanga A, Galdiero S, Guida M, Maselli C, Galdiero E. Genotoxicity of gold nanoparticles functionalized with indolicidin towards Saccharomyces cerevisiae. J Environ Sci. 2017. https://doi.org/10.1016/J.JES.2017.04.034.

25. De Angelis I, Barone F, Zijno A, Bizzarri L, Russo MT, Pozzi R, De Berardis B, et al. Comparative study of ZnO and TiO$_2$ nanoparticles: physicochemical characterisation and toxicological effects on human colon carcinoma cells. Nanotoxicology. 2013;7(8):1361–72. https://doi.org/10.3109/17435390.2012.741724.

26. Demir E, Akça H, Turna F, Aksakal S, Burgucu D, Kaya B, Marcos R, et al. Genotoxic and cell-transforming effects of titanium dioxide nanoparticles. Environ Res. 2015;136:300–8. https://doi.org/10.1016/j.envres.2014.10.032.

27. Deng J, Yao M, Gao C. Cytotoxicity of gold nanoparticles with different structures and surface-anchored chiral polymers. Acta Biomater. 2017;53:610–8. https://doi.org/10.1016/j.actbio.2017.01.082.

28. Desai P, Patlolla RR, Singh M. Interaction of nanoparticles and cell-penetrating peptides with skin for transdermal drug delivery. Mol Membr Biol. 2010;27(7):247–59. https://doi.org/10.3109/09687688.2010.522203.

29. Devika Chithrani B, Ghazani AA, Chan WCW. Determining the size and shape dependence of Gold nanoparticle uptake into mammalian cells. 2006. https://doi.org/10.1021/NL052396O.

30. Djuris AB, Leung YH, Ng AMC, Xu XY, Lee PKH, Degger N, Wu RSS. Toxicity of Metal Oxide nanoparticles : mechanisms, characterization, and avoiding experimental artefacts. 2014:1–19. https://doi.org/10.1002/smll.201303947.

31. Dobrzyńska MM, Gajowik A, Radzikowska J, Lankoff A, Dušinská M, Kruszewski M. Genotoxicity of Silver and Titanium Dioxide nanoparticles in bone marrow cells of rats in vivo. Toxicology. 2014;315:86–91. https://doi.org/10.1016/j.tox.2013.11.012.

32. Donner EM, Myhre A, Brown SC, Boatman R, Warheit DB. In vivo micronucleus studies with 6 Titanium Dioxide materials (3 pigment-grade & 3 nanoscale) in orally-exposed rats. Regul Toxicol Pharmacol. 2016;74:64–74. https://doi.org/10.1016/j.yrtph.2015.11.003.

33. Downs TR, Crosby ME, Hu T, Kumar S, Sullivan A, Sarlo K, Pfuhler S, et al. Silica nanoparticles administered at the maximum tolerated dose induce genotoxic effects through an inflammatory reaction while gold nanoparticles do not. Mutat Res/Genetic Toxicol Environ Mutagen. 2012;745(1–2):38–50. https://doi.org/10.1016/j.mrgentox.2012.03.012.

34. Dubey A, Goswami M, Yadav K, Chaudhary D. Oxidative stress and nano-toxicity induced by TiO$_2$ and ZnO on WAG cell line. 2015:1–26. https://doi.org/10.1371/journal.pone.0127493.

35. Duvall MN, Knight K. FDA regulation of nanotechnology. Food and Drug Agency (FDA) USA, (December). 2011. Retrieved from http://www.fda.gov/nanotechnology/regulation.html.

36. Egerton TA, Tooley IR. UV absorption and scattering properties of inorganic-based sunscreens. Int J Cosmet Sci. 2012;34(2):117–22. https://doi.org/10.1111/J.1468-2494.2011.00689.X.

37. Ema M, Masumori S, Kobayashi N, Naya M, Endoh S, Maru J, Nakanishi J, et al. *In vivo* comet assay of multi-walled carbon nanotubes using lung cells of rats intratracheally instilled. J Appl Toxicol. 2013;33(10):1053–60. https://doi.org/10.1002/jat.2810.

38. Eom H, Jeong J, Choi J. Effect of aspect ratio on the uptake and toxicity of hydroxylated-multi walled carbon nanotubes in the nematode. Caenorhabditis Elegans. 2015:1–8.

39. Fahmy B, Cormier SA. Copper oxide nanoparticles induce oxidative stress and cytotoxicity in airway epithelial cells. Toxicol In Vitro. 2009;23(7):1365–71. https://doi.org/10.1016/j.tiv.2009.08.005.

40. Federal Register : Guidance for Industry: Safety of Nanomaterials in Cosmetic Products; Availability. (n.d.). Retrieved November 17, 2017, from https://www.federalregister.gov/documents/2014/06/27/2014-15032/guidance-for-industry-safety-of-nanomaterials-in-cosmetic-products-availability.

41. Franklin NM, Rogers NJ, Apte SC, Batley GE, Gadd GE, Casey PS. Comparative toxicity of nanoparticulate ZnO, Bulk ZnO, and $ZnCl_2$ to a freshwater microalga (Pseudokirchneriella subcapitata): the importance of particle solubility. Environ Sci Technol. 2007;41(24):8484–90. https://doi.org/10.1021/es071445r.

42. Fu PP, Xia Q, Hwang H, Ray PC. Sciencedirect mechanisms of nanotoxicity: generation of reactive oxygen species 5. J Food Drug Anal. 2014;22(1):64–75. https://doi.org/10.1016/j.jfda.2014.01.005.

43. Gaspar LR, Tharmann J, Maia Campos PMBG, Liebsch M. Skin phototoxicity of cosmetic formulations containing photounstable and photostable UV-filters and vitamin A palmitate. Toxicol In Vitro. 2013;27(1):418–25. https://doi.org/10.1016/j.tiv.2012.08.006.

44. Gałecka E, Jacewicz R, Mrowicka M, Florkowski A, Gałecki P. [Antioxidative enzymes–structure, properties, functions]. Polski Merkuriusz Lekarski: Organ Polskiego Towarzystwa Lekarskiego. 2008;25(147):266–8. Retrieved from http://www.ncbi.nlm.nih.gov/pubmed/19112846.

45. Ghosh S, Sharma A, Talukder G. Relationship of clastogenic effects of zirconium oxychloride to dose and duration of exposure in bone marrow cells of mice in vivo. Toxicol Lett. 1991;55(2):195–201. Retrieved from http://www.ncbi.nlm.nih.gov/pubmed/1998207.

46. Girgis E, Khalil WKB, Emam AN, Mohamed MB, Rao KV. Nanotoxicity of Gold and Gold-Cobalt nanoalloy. Chem Res Toxicol. 2012;25(5):1086–98. https://doi.org/10.1021/tx300053h.

47. Gontier E, Ynsa M-D, Bíró T, Hunyadi J, Kiss B, Gáspár K, Surlève-Bazeille J-E, et al. Is there penetration of titania nanoparticles in sunscreens through skin? A comparative electron and ion microscopy study. Nanotoxicology. 2008;2(4):218–31. https://doi.org/10.1080/17435390802538508.

48. Goodman CM, McCusker CD, Yilmaz T, Rotello VM. Toxicity of Gold nanoparticles functionalized with cationic and anionic side chains. Bioconjug Chem. 2004;15(4):897–900. https://doi.org/10.1021/bc049951i.

49. Grudzinski IP, Bystrzejewski M, Cywinska MA, Kosmider A, Poplawska M, Cieszanowski A, Ostrowska A. Cytotoxicity evaluation of carbon-encapsulated iron nanoparticles in melanoma cells and dermal fibroblasts. 2013. https://doi.org/10.1007/s11051-013-1835-7.

50. Guan R, Kang T, Lu F, Zhang Z, Shen H, Liu M. Cytotoxicity, oxidative stress, and genotoxicity in human hepatocyte and embryonic kidney cells exposed to ZnO nanoparticles. Nanoscale Res Lett. 2012;7(1):602. https://doi.org/10.1186/1556-276X-7-602.

51. Guichard Y, Maire M-A, Sébillaud S, Fontana C, Langlais C, Micillino J-C, Gaté L, et al. Genotoxicity of synthetic amorphous silica nanoparticles in rats following short-term exposure, part 2: intratracheal instillation and intravenous injection. Environ Mol Mutagen. 2015;56(2):228–44. https://doi.org/10.1002/em.21928.

52. Guichard Y, Schmit J, Darne C, Gaté L, Goutet M, Rousset D, Binet S, et al. Cytotoxicity and genotoxicity of nanosized and microsized Titanium Dioxide and Iron Oxide particles in Syrian Hamster embryo cells. Ann Occup Hyg. 2012;56(5):631–44. https://doi.org/10.1093/annhyg/mes006.

53. Guidance Documents > Guidance for Industry: Safety of Nanomaterials in Cosmetic Products. (n.d.). Retrieved November 17, 2017, from https://www.fda.gov/Cosmetics/GuidanceRegulation/GuidanceDocuments/ucm300886.htm.

54. Gustafson HH, Holt-Casper D, Grainger DW, Ghandehari H. Nanoparticle uptake: the phagocyte problem. Nano Today. 2015;10(4):487–510. https://www.sciencedirect.com/science/article/pii/S1748013215000766.
55. Gümüş D, Berber AA, Ada K, Aksoy H. In vitro genotoxic effects of ZnO nanomaterials in human peripheral lymphocytes. Cytotechnology. 2014;66(2):317–25. https://doi.org/10.1007/s10616-013-9575-1.
56. Hackenberg S, Friehs G, Kessler M, Froelich K, Ginzkey C, Koehler C, Kleinsasser N, et al. Nanosized titanium dioxide particles do not induce DNA damage in human peripheral blood lymphocytes. Environ Mol Mutagen. 2011;52(4):264–8. https://doi.org/10.1002/em.20615.
57. Hamzeh M, Sunahara GI. In vitro cytotoxicity and genotoxicity studies of titanium dioxide (TiO$_2$) nanoparticles in Chinese hamster lung fibroblast cells. Toxicol In Vitro. 2013;27 (2):864–73. https://doi.org/10.1016/j.tiv.2012.12.018.
58. He X, Hwang HM. Nanotechnology in food science: functionality, applicability, and safety assessment. J Food Drug Anal. 2016. https://doi.org/10.1016/j.jfda.2016.06.001.
59. Honeywell-Nguyen PL, Groenink HWW, De Graaff AM, Bouwstra JA. The in vivo transport of elastic vesicles into human skin: effects of occlusion, volume and duration of application. J Controlled Release. 2003;90(2):243–55. https://doi.org/10.1016/S0168-3659 (03)00202-5.
60. Horie M, Sugino S, Kato H, Tabei Y, Nakamura A, Yoshida Y. Does photocatalytic activity of TiO$_2$ nanoparticles correspond to photo-cytotoxicity? Cellular uptake of TiO$_2$ nanoparticles is important in their photo-cytotoxicity. Toxicol Mech Methods. 2016. https://doi.org/10.1080/15376516.2016.1175530.
61. Hung C, Chen W, Hsu C, Aljuffali IA, Shih H, Fang J. European Journal of Pharmaceutics and Biopharmaceutics Cutaneous penetration of soft nanoparticles via photo damaged skin : Lipid-based and polymer-based nanocarriers for drug delivery. Eur J Pharm Biopharm (May). 2015. https://doi.org/10.1016/j.ejpb.2015.05.005.
62. Iavicoli I, Leso V, Bergamaschi A. Toxicological effects of Titanium Dioxide nanoparticles: a review of *In Vivo* studies. J Nanomaterials. 2012;2012:1–36. https://doi.org/10.1155/2012/964381.
63. Ivask A, Voelcker NH, Seabrook SA, Hor M, Kirby JK, Fenech M, Ke PC. DNA melting and genotoxicity induced by Silver nanoparticles and graphene. 2015. https://doi.org/10.1021/acs.chemrestox.5b00052.
64. Jaishree V, Gupta PD. Nanotechnology: a revolution in cancer diagnosis. Indian J Clin Biochem. 2012;27(3):214–20. https://doi.org/10.1007/s12291-012-0221-z.
65. Jia Y-P, Ma B-Y, Wei X-W, Qian Z-Y. The in vitro and in vivo toxicity of gold nanoparticles. Chin Chem Letters. 2017. https://doi.org/10.1016/j.cclet.2017.01.021.
66. Jomini S, Labille J, Bauda P, Pagnout C. Modifications of the bacterial reverse mutation test reveals mutagenicity of TiO$_2$ nanoparticles and byproducts from a sunscreen TiO$_2$-based nanocomposite. Toxicol Lett. 2012;215(1):54–61. https://doi.org/10.1016/J.TOXLET.2012.09.012.
67. Kaba SI, Egorova EM. In vitro studies of the toxic effects of silver nanoparticles on HeLa and U937 cells. Nanotechnol Sci Appl. 2015;8:19–29. https://doi.org/10.2147/NSA.S78134.
68. Kaiser JP, Roesslein M, Diener L, Wichser A, Nowack B, Wick P. Cytotoxic effects of nanosilver are highly dependent on the chloride concentration and the presence of organic compounds in the cell culture media. J Nanobiotechnol. 1–11. 2017. https://doi.org/10.1186/s12951-016-0244-3.
69. Kazi KM, Mandal AS, Biswas N, Guha A, Chatterjee S, Behera M, Kuotsu K. Niosome: a future of targeted drug delivery systems. J Adv Pharm Technol Res. 2010;1(4):374–80. https://doi.org/10.4103/0110-5558.76435.
70. Khalili Fard J, Jafari S, Eghbal MA. A review of molecular mechanisms involved in toxicity of nanoparticles. Adv Pharm Bull. 2015;5(4):447–54. https://doi.org/10.15171/apb.2015.061.

71. Khan HA, Abdelhalim MAK, Al-Ayed MS, Alhomida AS. Effect of gold nanoparticles on glutathione and malondialdehyde levels in liver, lung and heart of rats. Saudi J Biol Sci. 2012;19(4):461–4. https://doi.org/10.1016/j.sjbs.2012.06.005.

72. Kim K, Park H, Lim K-M. Phototoxicity: its mechanism and animal alternative test methods. Toxicol Res. 2015;31(2):97–104. https://doi.org/10.5487/TR.2015.31.2.097.

73. Kim IY, Joachim E, Choi H, Kim K. Toxicity of silica nanoparticles depends on size, dose, and cell type. Nanomed Nanotechnol Biol Med. 2015. https://doi.org/10.1016/j.nano.2015. 03.004.

74. Ksia I, Radziun E, Wilczyn JD. Toxicology in Vitro assessment of the cytotoxicity of aluminium oxide nanoparticles on selected mammalian cells. 2011;25:1694–700. https://doi. org/10.1016/j.tiv.2011.07.010.

75. Kwon JY, Koedrith P, Seo YR. Current investigations into the genotoxicity of zinc oxide and silica nanoparticles in mammalian models in vitro and in vivo: carcinogenic/genotoxic potential, relevant mechanisms and biomarkers, artifacts, and limitations. Int J Nanomed. 2014. https://doi.org/10.2147/IJN.S57918.

76. Lademann J, Richter H, Schanzer S, Knorr F, Meinke M, Sterry W, Patzelt A. Penetration and storage of particles in human skin: perspectives and safety aspects. Eur J Pharm Biopharm. 2011;77(3):465–8. https://doi.org/10.1016/j.ejpb.2010.10.015.

77. Landsiedel R, Ma-Hock L, Van Ravenzwaay B, Schulz M, Wiench K, Champ S, Oesch F, et al. Gene toxicity studies on titanium dioxide and zinc oxide nanomaterials used for UV-protection in cosmetic formulations. Nanotoxicology. 2010;4(4):364–81. https://doi.org/ 10.3109/17435390.2010.506694.

78. Landsiedel R, Sauer UG, Ma-Hock L, Schnekenburger J, Wiemann M. Pulmonary toxicity of nanomaterials: a critical comparison of published in vitro assays and in vivo inhalation or instillation studies. Nanomedicine. 2014;9(16):2557–85. https://www.futuremedicine.com/ doi/abs/10.2217/nnm.14.149.

79. Lasagna-Reeves C, Gonzalez-Romero D, Barria MA, Olmedo I, Clos A, Ramanujam VS, Soto C. Bioaccumulation and toxicity of gold nanoparticles after repeated administration in mice. Biochem Biophys Res Communications. 2010;393(4):649–55. https://www. sciencedirect.com/science/article/pii/S0006291X10002573.

80. Lebedová J, Hedberg YS, Wallinder IO, Karlsson HL. Size-dependent genotoxicity of silver, gold and platinum nanoparticles studied using the mini-gel comet assay and micronucleus scoring with flow cytometry. 2017;(2):1–9. https://doi.org/10.1093/mutage/gcx027.

81. Lee SH, Lee HR, Kim Y-R, Kim M-K. Toxic response of Zinc Oxide nanoparticles in human epidermal keratinocyte HaCaT cells. Toxicol Environ Health Sci. 2012;4(1):14–8. https://doi.org/10.1007/s13530-012-0112-y.

82. Lee J, Mahendra S, Alvarez PJJ. Nanomaterials in the construction industry: a review of their applications and environmental health and safety considerations. ACS Nano. 2010;4 (7):3580–90. https://doi.org/10.1021/nn100866w.

83. Lee S, Yun H-S, Kim S-H. The comparative effects of mesoporous silica nanoparticles and colloidal silica on inflammation and apoptosis. Biomaterials. 2011;32(35):9434–43. https:// doi.org/10.1016/j.biomaterials.2011.08.042.

84. Li Y, Qin T, Ingle T, Yan J, He W, Yin J-J, Chen T. Differential genotoxicity mechanisms of silver nanoparticles and silver ions. Arch Toxicol. 2017;91(1):509–19. https://doi.org/10. 1007/s00204-016-1730-y.

85. Lindberg HK, Falck GC-M, Catalán J, Koivisto AJ, Suhonen S, Järventaus H, Norppa H, et al. Genotoxicity of inhaled nanosized TiO$_2$ in mice. Mutat Res/Genetic Toxicol Environ Mutagen. 2012;745(1–2):58–64. https://doi.org/10.1016/J.MRGENTOX.2011.10.011.

86. Linskaya AN, Dobrovolskaia MA. Immunosuppressive and anti-inflammatory properties of engineered nanomaterials. Br J Pharmacol. 2014;171(17):3988–4000. http://onlinelibrary. wiley.com/doi/10.1111/bph.12722/full.

87. Liu X, Keane MJ, Zhong BZ, Ong TM, Wallace WE. Micronucleus formation in V79 cells treated with respirable silica dispersed in medium and in simulated pulmonary surfactant.

Mutat Res. 1996;361(2–3):89–94. Retrieved from http://www.ncbi.nlm.nih.gov/pubmed/8980693.

88. Ma D-D, Yang W-X. Engineered nanoparticles induce cell apoptosis: potential for cancer therapy. Oncotarget. 2016;7(26):40882–903. https://doi.org/10.18632/oncotarget.8553.

89. Manke A, Wang L, Rojanasakul Y. Mechanisms of nanoparticle-induced oxidative stress and toxicity. Biomed Res Int. 2013;2013:942916. https://doi.org/10.1155/2013/942916.

90. Maser E, Schulz M, Sauer UG, Wiemann M, Ma-Hock L, Wohlleben W, Landsiedel R, et al. In vitro and in vivo genotoxicity investigations of differently sized amorphous SiO_2 nanomaterials. Mutat Res/Genetic Toxicol Environ Mutagen. 2015;794:57–74. https://doi.org/10.1016/j.mrgentox.2015.10.005.

91. Mateo D, Morales P, Avalos A, Haza AI. (n.d.). Comparative cytotoxicity evaluation of different size gold nanoparticles in human dermal fibroblasts. https://doi.org/10.1080/17458080.2015.1014934.

92. Mclaren A, Valdes-Solis T, Li G, Tsang SC. Shape and size effects of ZnO nanocrystals on photocatalytic activity. J Am Chem Soc. 2009;131(35):12540–1. https://doi.org/10.1021/ja9052703.

93. Meyer K, Rajanahalli P, Ahamed M, Rowe JJ, Hong Y. ZnO nanoparticles induce apoptosis in human dermal fibroblasts via p53 and p38 pathways. Toxicol In Vitro. 2011;25(8):1721–6. https://doi.org/10.1016/j.tiv.2011.08.011.

94. Mihranyan A, Ferraz N, Strømme M. Current status and future prospects of nanotechnology in cosmetics. Prog Mater Sci. 2012;57(5):875–910. https://doi.org/10.1016/j.pmatsci.2011.10.001.

95. Miyani VA, Hughes MF. Assessment of the in vitro dermal irritation potential of cerium, silver, and titanium nanoparticles in a human skin equivalent model. Cutan Ocular Toxicol. 2017;36(2):145–51. https://doi.org/10.1080/15569527.2016.1211671.

96. Mo S-D, Ching WY. Electronic and optical properties of three phases of titanium dioxide: rutile, anatase, and brookite. Phys Rev B. 1995;51(19):13023–32. https://doi.org/10.1103/PhysRevB.51.13023.

97. Moia C, Zhu H. In vitro toxicological assessment of amorphous silica particles in relation to their characteristics and mode of action in human skin cells. 2012. Retrieved from https://dspace.lib.cranfield.ac.uk/bitstream/1826/9760/1/Moia_C_2015.pdf.

98. Morganti P. Use and potential of nanotechnology in cosmetic dermatology. Clin Cosmet Investig Dermatol. 2010;3:5–13. Retrieved from http://www.ncbi.nlm.nih.gov/pubmed/21437055.

99. Murugadoss S, Lison D, Godderis L, Van Den Brule S, Mast J, Brassinne F, Sebaihi N, Hoet PH. Toxicology of silica nanoparticles: an update. Archives Toxicol. 2017;91(9):2967–3010.

100. Nafisi S, Maibach HI. Nanotechnology in cosmetics. 2017. https://doi.org/10.1016/B978-0-12-802005-0.00022-7.

101. Nam S-H, Kim SW, An Y-J. No evidence of the genotoxic potential of gold, silver, zinc oxide and titanium dioxide nanoparticles in the SOS chromotest. J Appl Toxicol. 2013;33(10):1061–1069.

102. Naya M, Kobayashi N, Ema M, Kasamoto S, Fukumuro M, Takami S, Nakanishi J, et al. In vivo genotoxicity study of titanium dioxide nanoparticles using comet assay following intratracheal instillation in rats. Regul Toxicol Pharmacol. 2012;62(1):1–6. https://doi.org/10.1016/j.yrtph.2011.12.002.

103. Oberdörster G, Oberdörster E, Oberdörster J. Nanotoxicology: an emerging discipline evolving from studies of ultrafine particles. Environ Health Perspect. 2005;113(7):823–39. https://doi.org/10.1289/EHP.7339.

104. Onodera A, Nishiumi F, Kakiguchi K, Tanaka A, Tanabe N, Honma A, Yanagihara I, et al. Short-term changes in intracellular ROS localisation after the silver nanoparticles exposure depending on particle size. Toxicol Rep. 2015;2:574–9. https://www.sciencedirect.com/science/article/pii/S2214750015000360.

105. Osman IF, Baumgartner A, Cemeli E, Fletcher JN, Anderson D. Genotoxicity and cytotoxicity of zinc oxide and titanium dioxide in HEp-2 cells. Nanomedicine. 2010;5 (8):1193–203. https://doi.org/10.2217/nnm.10.52.
106. Osman IF, Baumgartner A, Cemeli E, Fletcher JN, Anderson D. Genotoxicity and cytotoxicity of zinc oxide and titanium dioxide in HEp-2 cells. Nanomedicine. 2010;5 (8):1193–203. https://doi.org/10.2217/nnm.10.52.
107. Pan Y, Leifert A, Ruau D, Neuss S, Bornemann J, Schmid G, Jahnen-Dechent W, et al. Gold nanoparticles of diameter 1.4 nm trigger necrosis by oxidative stress and mitochondrial damage. Small. 2009;5(18):2067–76. https://doi.org/10.1002/smll.200900466.
108. Patlolla A, Knighten B, Tchounwou P. Multi-walled carbon nanotubes induce cytotoxicity, genotoxicity and apoptosis in normal human dermal fibroblast cells. Ethn Dis. 2010;20(1 Suppl 1):S1-65–72. Retrieved from http://www.ncbi.nlm.nih.gov/pubmed/20521388.
109. Pernodet N, Fang X, Sun Y, Bakhtina A, Ramakrishnan A, Sokolov J, Rafailovich M, et al. Adverse effects of Citrate/Gold nanoparticles on human dermal fibroblasts. Small. 2006;2 (6):766–73. https://doi.org/10.1002/smll.200500492.
110. Plotnikov E, Zhuravkov S, Gapeyev A, Plotnikov V, Martemianova I, Martemianov D. Comparative study of genotoxicity of Silver and Gold nanoparticles prepared by the electric spark dispersion method. J Appl Pharm Sci. 2017;7(7):35–9. https://doi.org/10.7324/JAPS. 2017.70705.
111. Pokharkar V, Dhar S, Bhumkar D, Mali V, Bodhankar S, Prasad BLV. Acute and subacute toxicity studies of chitosan reduced gold nanoparticles: a novel carrier for therapeutic agents. J Biomed Nanotechnol. 2009;5(3):233–9. Retrieved from http://www.ncbi.nlm.nih.gov/ pubmed/20055004.
112. Poland CA, Duffin R, Kinloch I, Maynard A, Wallace WAH, Seaton A, Donaldson K, et al. Carbon nanotubes introduced into the abdominal cavity of mice show asbestos like pathogenicity in a pilot study. Nat Nanotechnol. 2008;3(7):423–8. https://doi.org/10.1038/ nnano.2008.111.
113. Prabhu BM, Ali SF, Murdock RC, Hussain SM, Srivatsan M. Copper nanoparticles exert size and concentration dependent toxicity on somatosensory neurons of rat. Nanotoxicology. 2010. https://doi.org/10.3109/17435390903337693.
114. Qiu Y, Liu Y, Wang L, Xu L, Bai R, Ji Y, Chen C, et al. Surface chemistry and aspect ratio mediated cellular uptake of Au nanorods. Biomaterials. 2010;31(30):7606–19. https://doi. org/10.1016/j.biomaterials.2010.06.051.
115. Rahman T, Hosen I, Islam MMT, Shekhar HU. Oxidative stress and human health. Adv Biosci Biotechnol. 2012;3:997–1019. https://doi.org/10.4236/abb.2012.327123.
116. Ray PC, Yu H, Fu PP. Toxicity and environmental risks of nanomaterials: challenges and future needs. J Environ Sci Health Part C Environ Carcinog Ecotoxicol Rev. 2009;27(1):1– 35. https://doi.org/10.1080/10590500802708267.
117. Reisinger K, Hoffmann S, Alépée N, Ashikaga T, Barroso J, Elcombe C, Maxwell G, et al. Systematic evaluation of non-animal test methods for skin sensitisation safety assessment. Toxicol In Vitro. 2015;29(1):259–70. https://doi.org/10.1016/j.tiv.2014.10.018.
118. Sadiq R, Bhalli JA, Yan J, Woodruff RS, Pearce MG, Li Y, Chen T, et al. Genotoxicity of TiO(2) anatase nanoparticles in B6C3F1 male mice evaluated using Pig-a and flow cytometric micronucleus assays. Mutat Res. 2012;745(1–2):65–72. Retrieved from http:// www.ncbi.nlm.nih.gov/pubmed/22712079.
119. Sahu D, Kannan GM, Vijayaraghavan R. Carbon black particle exhibits size dependent toxicity in human monocytes. Int J Inflammation. 2014;2014:827019. https://doi.org/10. 1155/2014/827019.
120. Saraf S, Jeswani G, Kaur C. Development of novel herbal cosmetic cream with curcuma longa extract loaded transfersomes for antiwrinkle effect. Afr J Pharm Pharmacol. 2011:1054–1062, 5(August). Retrieved from http://scholar.google.com/scholar?hl= en&btnG=Search&q=intitle:Development+of+novel+herbal+cosmetic+cream+with +Curcuma+longa+extract+loaded+transfersomes+for+antiwrinkle+effect#0.

121. Sardar R, Funston AM, Mulvaney P, Murray RW. Gold nanoparticles: past, present, and future. Langmuir. 2009;25(24):13840–51. https://doi.org/10.1021/la9019475.
122. Schneider M, Stracke F, Hansen S, Schaefer UF. Nanoparticles and their interactions with the dermal barrier. Dermato-Endocrinology. 2009a;1(4):197–206. Retrieved from http://www.ncbi.nlm.nih.gov/pubmed/20592791.
123. Schneider M, Stracke F, Hansen S, Schaefer UF. Nanoparticles and their interactions with the dermal barrier. Dermato-Endocrinology. 2009b;1(4):197–206. Retrieved from http://www.ncbi.nlm.nih.gov/pubmed/20592791.
124. Shen C, James SA, de Jonge MD, Turney TW, Wright PFA, Feltis BN. Relating cytotoxicity, Zinc ions, and reactive oxygen in ZnO nanoparticle-exposed human immune cells. Toxicol Sci. 2013;136(1):120–30. https://doi.org/10.1093/toxsci/kft187.
125. Shin H, Ko H, Kim M. Cytotoxicity and biocompatibility of Zirconia (Y-TZP) posts with various dental cements. Restorative Dent Endod. 2016;41(3):167–75. https://doi.org/10.5395/rde.2016.41.3.167.
126. Shukla RK, Kumar A, Gurbani D, Pandey AK, Singh S, Dhawan A. TiO$_2$ nanoparticles induce oxidative DNA damage and apoptosis in human liver cells. Nanotoxicology. 2013;7 (1):48–60. https://doi.org/10.3109/17435390.2011.629747.
127. Shukla RK, Sharma V, Pandey AK, Singh S, Sultana S, Dhawan A. ROS-mediated genotoxicity induced by titanium dioxide nanoparticles in human epidermal cells. Toxicol In Vitro. 2011;25(1):231–41. https://doi.org/10.1016/j.tiv.2010.11.008.
128. Shvedova A, Castranova V, Kisin E, Schwegler-Berry D, Murray A, Gandelsman V, Baron P, et al. Exposure to carbon nanotube material: assessment of nanotube cytotoxicity using human keratinocyte cells. J Toxicol Environ Health Part A. 2003;66(20):1909–26. https://doi.org/10.1080/713853956.
129. Sirelkhatim A, Mahmud S, Seeni A, Kaus NHM, Ann LC, Bakhori SKM, Mohamad D, et al. Review on Zinc Oxide nanoparticles: antibacterial activity and toxicity mechanism. Nano-Micro Lett. 2015;7(3):219–42. https://doi.org/10.1007/s40820-015-0040-x.
130. Skelton HG, Smith KJ, Johnson FB, Cooper CR, Tyler WF, Lupton GP. Zirconium granuloma resulting from an aluminum zirconium complex: a previously unrecognized agent in the development of hypersensitivity granulomas. J Am Acad Dermatol. 1993;28(5 Pt 2):874–6. Retrieved from http://www.ncbi.nlm.nih.gov/pubmed/8491884.
131. Slowing II, Vivero-Escoto JL, Trewyn BG, Lin VSY, Vivero-Escoto JL, Lin VS-Y. (n.d.). Mesoporous silica nanoparticles: structural design and applications Mesoporous silica nanoparticles: structural design and applications Mesoporous silica nanoparticles: structural design and applications. J Mater Chem. Retrieved from http://lib.dr.iastate.edu/chem_pubs.
132. Steenland K, Sanderson W. Lung cancer among industrial sand workers exposed to crystalline silica. Am J Epidemiol. 2001;153(7):695–703. https://doi.org/10.1093/aje/153.7.695.
133. Sycheva LP, Zhurkov VS, Iurchenko VV, Daugel-Dauge NO, Kovalenko MA, Krivtsova EK, Durnev AD. Investigation of genotoxic and cytotoxic effects of micro- and nanosized titanium dioxide in six organs of mice in vivo. Mutat Res/Genetic Toxicol Environ Mutagen. 2011;726(1):8–14. https://doi.org/10.1016/j.mrgentox.2011.07.010.
134. Tavares AM, Louro H, Antunes S, Quarré S, Simar S, De Temmerman P-J, Silva MJ, et al. Genotoxicity evaluation of nanosized titanium dioxide, synthetic amorphous silica and multi-walled carbon nanotubes in human lymphocytes. Toxicol In Vitro. 2014;28(1):60–9. https://doi.org/10.1016/j.tiv.2013.06.009.
135. Tavares AM, Louro H, Antunes S, Quarré S, Simar S, De Temmerman P-J, Silva MJ, et al. Genotoxicity evaluation of nanosized titanium dioxide, synthetic amorphous silica and multi-walled carbon nanotubes in human lymphocytes. Toxicol In Vitro. 2014;28(1):60–9. https://doi.org/10.1016/j.tiv.2013.06.009.
136. Taylor MR. Nanotechnology: does FDA have the tools it needs? Nanotechnology. 2006 (October).

137. Trouiller B, Reliene R, Westbrook A, Solaimani P, Schiestl RH. Titanium dioxide nanoparticles induce DNA damage and genetic instability in vivo in mice. Cancer Res. 2009;69(22):8784–9. https://doi.org/10.1158/0008-5472.CAN-09-2496.

138. Turkez H. The role of ascorbic acid on titanium dioxide-induced genetic damage assessed by the comet assay and cytogenetic tests. Exp Toxicol Pathol. 2011;63(5):453–7. https://doi.org/10.1016/j.etp.2010.03.004.

139. Villiers C, Freitas H, Couderc R, Villiers M-B, Marche P. Analysis of the toxicity of gold nano particles on the immune system: effect on dendritic cell functions. J Nanoparticle Res Interdisc Forum Nanoscale Sci Technol. 2010;12(1):55–60. https://doi.org/10.1007/s11051-009-9692-0.

140. Vinardell M, Llanas H, Marics L, Mitjans M. In Vitro comparative skin irritation induced by nano and non-nano Zinc Oxide. Nanomaterials. 2017;7(3):56. https://doi.org/10.3390/nano7030056.

141. Wang S, Hunter LA, Arslan Z, Wilkerson MG, Wickliffe JK. Chronic exposure to nanosized, anatase titanium dioxide is not cyto- or genotoxic to Chinese hamster ovary cells. Environ Mol Mutagen. 2011;52(8):614–22. https://doi.org/10.1002/em.20660.

142. Wang S, Lu W, Tovmachenko O, Rai US, Yu H, Ray PC. Challenge in understanding size and shape dependent toxicity of Gold nanomaterials in human skin keratinocytes. Chem Phys Lett. 2008;463(1–3):145–9. https://doi.org/10.1016/j.cplett.2008.08.039.

143. Wani MY, Hashim MA, Nabi F, Malik MA. Nanotoxicity: dimensional and morphological concerns. 2011. https://doi.org/10.1155/2011/450912.

144. Williams E. Food and Drug Administration (FDA): overview and issues. Congressional Research Service. 2009. Retrieved from http://www.healthpolicyfellows.org/pdfs/FoodandDrugAdministrationFDA-OverviewandIssues.pdf.

145. Wongrakpanich A, Mudunkotuwa IA, Geary SM, Morris AS, Mapuskar KA, Spitz DR, Salem AK. Size-dependent cytotoxicity of copper oxide nanoparticles in lung epithelial cells. Environ Sci Nano. 2016;3(2):365–74. https://doi.org/10.1039/C5EN00271K.

146. Woodruff RS, Li Y, Yan J, Bishop M, Jones MY, Watanabe F, Chen T, et al. Genotoxicity evaluation of titanium dioxide nanoparticles using the Ames test and Comet assay. J Appl Toxicol. 2012;32(11):934–43. https://doi.org/10.1002/jat.2781.

147. Xia T, Kovochich M, Liong M, Mädler L, Gilbert B, Shi H, Nel AE, et al. Comparison of the mechanism of toxicity of Zinc Oxide and Cerium Oxide nanoparticles based on dissolution and oxidative stress properties. ACS Nano. 2008;2(10):2121–34. https://doi.org/10.1021/nn800511k.

148. Xia XR, Monteiro-Riviere NA, Riviere JE. Skin penetration and kinetics of pristine fullerenes (C60) topically exposed in industrial organic solvents. Toxicol Appl Pharmacol. 2010;242(1):29–37. https://doi.org/10.1016/j.taap.2009.09.011.

149. Yang H, Liu C, Yang D, Zhang H, Xi Z. Comparative study of cytotoxicity, oxidative stress and genotoxicity induced by four typical nanomaterials: the role of particle size, shape and composition. J Appl Toxicol. 2009;29(1):69–78. https://doi.org/10.1002/jat.1385.

150. Zhang Y, Xu D, Li W, Yu J, Chen Y. Effect of size, shape, and surface modification on cytotoxicity of gold nanoparticles to human HEp-2 and Canine MDCK Cells. J Nanomater. 2012a. https://doi.org/10.1155/2012/375496.

151. Zhang Y, Xu D, Li W, Yu J, Chen Y. Effect of size, shape, and surface modification on cytotoxicity of gold nanoparticles to human HEp-2 and canine MDCK cells. 2012b. https://doi.org/10.1155/2012/375496.

Regulation of Nanomaterials in Cosmetic Products on the EU Market

15

Florian Schellauf

Abstract

The Cosmetics Directive 76/768/EC underwent a revision, and the use of nanomaterials in cosmetic products was high on the agenda of the discussions. The outcome was that regulatory measures specifically focused on nanomaterials used in cosmetic products were included in the resulting legal text (Regulation 1223/2009/EC): (1) Definition of a nanomaterial (2) Product Notification: Indication of presence of nanomaterials (3) Notification of certain products containing nanomaterials (4) Reporting requirements for the Commission (5) Regular review of nanorelated requirements (5) Nanolabelling. This chapter will provide an overview over these requirements and some pointers for their practical implementation.

Keywords

Catalogue of nanomaterials · Cosmetic nanomaterial · Cosmetic product regulation · CPNP (Nano) notification · Nanolabelling · Nanodefinition · Regulation 1223/2009/EC

15.1 Introduction

Nanomaterials have been used for a long time in cosmetic products, much longer than the EU Cosmetics Regulation exists, even longer than Europe exists as a political or historical entity.

F. Schellauf (✉)
Cosmetics Europe, Avenue Hermann Debroux 40, 1160 Brussels, Belgium
e-mail: fschellauf@cosmeticseurope.eu

© Springer Nature Switzerland AG 2019
J. Cornier et al. (eds.), *Nanocosmetics*,
https://doi.org/10.1007/978-3-030-16573-4_15

Table 15.1 Overview of the main compliance requirements for nanomaterials used in cosmetics

Reference	Requirement	Applicable to:
Article 13	Declaration of nanomaterial use under the general product notification procedure (CPNP)	**All** products containing nanomaterials
Article 16	Nanonotification of products containing nanomaterials six months prior to placing on the market	Products containing nanomaterials not listed in Annex IV, V or VI and not listed in the nanoform in Annex III
Article 19	Nanolabelling	**All** products containing nanomaterials

The use of cosmetic products was already widespread in the Antique world, and discoveries from graves in Ancient Egypt show that nanomaterials were used as ingredients in some of the cosmetics found there.

The current focus of societal and regulatory concerns on nanomaterials started to emerge before the last change of the century. Intense political debate led to the publication of a first Commission communication "Towards a European strategy for Nanotechnology", published on 12 May 2004 [1]. It sought to bring the discussion on nanosciences and nanotechnologies to an institutional level and proposed an integrated and responsible approach for Europe.

One year later, on 7 June 2005, the EU Commission published its Action Plan for Europe 2005–2009 [2]. While this plan proposed many actions to further Research and Development in this promising field, it also responded to increasing concerns. One of the actions proposed was to "examine and, where appropriate, propose adaptations of EU regulations in relevant sectors…".

At that time, the Cosmetics Directive 76/768/EC was undergoing a revision and use of nanomaterials in cosmetic products was high on the agenda of the discussions. The outcome was that regulatory measures specifically focused on nanomaterials used in cosmetic products should be included.

The following requirements were introduced:

- Definition of a nanomaterial Article 2(1)k
- Product Notification: Indication of presence of nanomaterials Article 13(1)f
- Notification of certain products containing nanomaterials Article 16, 1-9
- Reporting requirements for the Commission Article 16(10)
- Regular review of nanorelated requirements Article 16(11)
- Nanolabelling Article 19(1)g (Table 15.1).

This chapter will provide an overview over these requirements and some pointers for their practical application.

15.2 Definition of a Nanomaterial Under the EU Cosmetics Regulation

The definition brought forward in the Cosmetic Products Regulation 1223/2009/EC ("CPR") [3] has a central role in the legislation of nanomaterials in cosmetics. It focuses the specific regulatory regime for nanomaterials under this legislation on those materials, which, according to regulators, *may* pose a risk for consumers and therefore warrant special attention.

A nanomaterial is defined according to the CPR as follows: "*An insoluble or biopersistent and intentionally manufactured material with one or more external dimensions, or an internal structure, on the scale from 1 to 100 nm*".

This definition seems in some aspects rather superficial, but it needs to be noted that it is of a regulatory nature and is the basis for a regulatory decision whether a certain ingredient is within the scope of application of the nanospecific requirements or not. It does not constitute a scientific description or characterisation of these materials.

When examining this definition, it will become clear quickly that it does not capture all possible nanomaterials. This is intentional, as scientific advice obtained by the Commission at this time indicates that not all possible nanomaterials may pose such a risk that they need to be subject to these requirements.

For example, cosmetic formulations often contain micelles, i.e. lipid structures that temporarily encapsulate certain ingredients to better protect them. These micelles can have sizes in the nanomaterial range (1–100 nm). However, the aim of the formulator is that these micelles disintegrate into their molecular components upon contact with the skin and release the encapsulated molecules. Given this behaviour, it would not make sense to subject these micelles to the increased safety requirements of the legislation, if the consumer is actually never exposed to them.

It has to be noted that in a horizontal activity, the Commission has issued a Recommendation for a broad, multi-sectorial definition, which includes a very large variety of nanomaterials [4]. This definition is meant to describe and capture as many types of nanomaterials as possible in one all-encompassing definition. However, this Recommendation, in line with scientific advice, allows the possibility of maintaining sector-specific definitions of the term "nanomaterial".

Consequently, as far as cosmetic products and ingredients are concerned, the two regulatory definitions of nanomaterials can perfectly coexist in the EU: the multi-sectorial recommendation and the specific definition included in the CPR.[1]

A major challenge of working with the definition is the need to ensure correct interpretation and implementation and to ensure that easily accessible and standardised methods for the characterisation of the materials are available to economic operators. It should be noted that cosmetics are regulated in the EU under a post-market surveillance system and therefore authorities performing in-market

---

[1]At the time of writing of this text, the Recommendation is in the process of being reviewed and revised. However, as far as it can be predicted no major changes to the Recommendation can be expected.

controls in the EU are facing the same challenges with characterising nanomaterials as manufacturers.

In the absence of official guidance from authorities, industry needs to rely on best practice and pragmatism. Various industry associations and bodies have issued implementation guidelines, which are intended to help cosmetic companies justifying their decision whether or not they consider that an ingredient falls under the regulatory definition of a "nanomaterial" in the CPR.

One example for a practical implementation challenge is the fact that particulate materials in general are always present in a distribution of particle sizes where a small number of particles can extend into small size ranges that have little relation with the median or average size of the distribution. Consequently, materials may contain particles in the nano range, although most particles are larger than 100 nm.

Applying the size range criterion in the definition (1–100 nm) to these materials would include them in the definition, although they do not have much in common with nanomaterials. It is therefore generally recognised that a cut-off level needs to be applied. This cut-off level defines the fraction of nanomaterials in a particulate material above which the material is considered a nanomaterial in a regulatory sense and the nano-related requirements of the legislation need to be applied.

The values such cut-off levels should have are subject to intense regulatory and scientific debate with sometimes very conservative expectations by regulators on the one side and limitations in routine analytical and characterisation methods on the other side.

The EU Joint Research Centre (JRC) has recently prepared an excellent overview over this complex issue [5–7].

A second characteristic of nanomaterials is that they can be found in several physical forms, from isolated nanoparticles (primary particles, nano-objects) to aggregates and agglomerates (nanostructured materials). One possible definition of these structures was proposed by the SCENHIR in 2007 as aggregates being "*groups of particles held together by strong forces such as those associated with covalent or metallic bonds*" and agglomerates as "*(…) by weak forces such as Van der Waals forces, some electrostatic forces and surface tension*" [8]. Additionally, some composites and nanoporous materials may be considered to have internal structures in the nanorange.

However, the regulator's intention for cosmetic products is to focus on materials that are present in the form of free elements between 1-100 nm, irrespective of the exact physical state of such elements. Thus, it is usually proposed that stable elements with constitutive elements having a dimension in the nanorange (e.g. aggregates, agglomerates, composites) but that are themselves greater than 100 nm in size should not be considered as nanomaterials unless they release particles or aggregates of less than 100 nm in size in cosmetic products under normal use conditions.

Macromolecules with a three-dimensional structure or in the form of molecular associations larger than 1 nm used as cosmetic ingredients are normally not considered as nanomaterials (e.g. globulin, myoglobin, casein).

Similarly, fullerenes that have a diameter below 1 nm are usually considered to be nanomaterials by default, although the use of these materials in cosmetic products is to date practically non-existing in the EU.

15.3   Product Notification: Indication of Presence of Nanomaterials

Product Notification is an essential element of the CPR, and Recital 23 describes the purpose: "In order to allow for rapid and appropriate medical treatment in the event of difficulties, the necessary information about the product formulation should be submitted to poison control centres and assimilated entities, where such centres have been established by Member States to that end".

The CPR is therefore a repository of information on all cosmetic products on the EU market that is relevant to poison control centres and to which these centres can access quickly in case of an emergency.

Article 13 of the CPR then goes on to describe the requirements and the information that needs to be notified. To minimise the administrative burden, the EU Commission established an electronic centralised notification portal, called Cosmetic Product Notification Portal (CPNP).

Article 13 requires the manufacturer to submit this information, i.e. "notify the product" to the European Commission via CPNP before placing each cosmetic product on the EU market.

Amongst any other elements the presence of any substance in the form of nanomaterials has to be notified, including information on its identification and the reasonably foreseeable exposure conditions. This requirement applies to all products containing nanomaterials without exception.

15.4   Article 16 (Nano)notification

Article 16 of the CPR introduces a specific notification mechanism for products containing certain nanomaterials. This mechanism is separate from the Article 13 notification described above.

The purpose of this Article 16 notification can be better understood, if one looks first at what products and ingredients are exempt from this notification.

Nanomaterials that are used as colorants, preservatives or UV filters have to undergo a true pre-market authorisation to be allowed for use. Products containing them are therefore never subject to the Article 16 notification requirements, irrespective of the size of the ingredients. In these cases, the positive listing and pre-market authorisation supersede the need for the nano notification.

Products containing ingredients listed on Annex III in the form of a nanomaterial have also undergone already a review by the Scientific Committee on Consumer Safety (SCCS), an independent advisory committee to the European Commission, and need not be notified under Article 16.

It follows therefore that the Article 16 notification is a mechanism that brings those products to the attention of the European Commission, which contain nanomaterials whose safety has not been specifically assessed by the SCCS.

Due to the basic premise of the CPR that cosmetic products do not need pre-market authorisation, the Article 16 notification does not constitute such a system, but is rather a pre-notification.

The nanonotification must be submitted six months in advance of placing the product on the market, which is challenging for the cosmetic sector due to the fast innovation cycles and product launches. The submission has to be done by electronic means, usually via a dedicated separate section of the CPNP and has to contain the following information:

(a) the identification of the nanomaterial including its chemical name (IUPAC) and other descriptors as specified in point 2 of the Preamble to Annexes II to VI;
(b) the specification of the nanomaterial including size of particles, physical and chemical properties;
(c) an estimate of the quantity of nanomaterial contained in cosmetic products intended to be placed on the market per year;
(d) the toxicological profile of the nanomaterial;
     Comment: This point is generally assumed to refer to the hazard profile of the ingredient.
(e) the safety data of the nanomaterial relating to the category of cosmetic product, as used in such products;
     Comment: This point is generally assumed to refer to the safety assessment taking into account the hazard profile (point d) and the exposure assessment (point f)
(f) the reasonably foreseeable exposure conditions.

The Commission will examine the submissions and can decide (but is not obliged) to submit the notification dossier to the SCCS for a safety assessment. Based on the outcome, the Commission may or may not decide to propose regulatory measures for the respective ingredient.

After the six months of pre-notification period is over, the company is allowed to market the notified product, irrespective of whether the Commission has decided to submit the notification dossier to the SCCS or not. It is important to note that the Commission can take this decision during the six months pre-notification period, but also at any time after that period. In other words, no feedback from the Commission after six months does not mean that the product is "approved".

More information on the practical workings of the CPNP for Article 16 notifications can be found in the Commission user manual, which updated regularly and can be accessed online [9].

One design element of this system turned out to be problematic in its practical application: the legislation requires notifications of *products* containing the respective nanomaterial, but the review that may be requested is related to the *ingredient*. The Commission/SCCS is therefore faced with many Product Notification dossiers that potentially contain very similar or even redundant information. Nevertheless, all dossiers need to be examined in order to be able to puzzle together the complete picture for the ingredient under review. It might be worth considering converting the Article 16 notification into an ingredient notification to make life easier for the Commission and the SCCS.

15.5 Catalogue of Nanomaterials

The European Commission is required by the CPR to publish a catalogue of all nanomaterials used in cosmetic products and to indicate the reasonably foreseeable exposure conditions. This catalogue has to be updated regularly.

A first version of this catalogue was published in June 2017 and is based on information notified electronically to the European Commission through the CPNP [10].

This catalogue contains to date twelve colourants, six UV filters and 25 other ingredients. However, based on current knowledge on nanomaterials, there may be some uncertainties regarding the status of some notified substances under the CPR nanodefinition.

The European Commission states in its preamble to the catalogue that "the catalogue remains a work in progress subject to modifications and will be updated regularly. This catalogue has an informative value only and is not in any case a list of authorised nanomaterials".

The European Commission announced a first update of the catalogue to be published before the summer break 2018.

The uncertainties with regard to the notified information on nanomaterials in the CPNP system reflect the fact that the area of nanomaterials, and especially their characterisation and the decision whether a material falls under the regulatory definition, is highly complex even for experienced persons. In addition, some information displayed in the CPNP may have been misleading, but has been improved in the meantime.

The Commission is equally required to submit a status report and annual updates to the European Parliament and the Council, which provide information on developments in the use of nanomaterials in cosmetic products. The updates shall then summarise the new nanomaterials in new categories of cosmetic products, the number of notifications, the progress made in developing nanospecific assessment methods and safety assessment guides, and information on international cooperation programmes.

15.6 Nanolabelling

Before speaking about specific aspects of the nano labelling, it is worth to recall why certain information is required by law to appear on the packaging.

Two categories can be distinguished:

(1) Warnings for the consumer: this is information that provides advice to the consumer on specific ways to use products or that informs about the content of certain ingredients. These measures are usually based on the product safety assessment by the manufacturer and are an integral part of product safety and risk management by the manufacturer or legislators.

(2) Consumer information: These are elements that are unrelated to safety considerations as, for example, the nominal content of the container.

Cosmetic products containing nanomaterials have to undergo a safety assessment by the manufacturer and, if found safe, can be placed on the market. This is a normal process that each product has to go through. But the fact that some products contain a nanomaterial or not is per se not a risk and does not require a warning labelling.

The requirement for nanolabelling falls firmly into the second category of consumer information. The cosmetic industry was requested to provide this information to consumers to enable them to make a choice whether to buy products containing nanomaterials. It is not a warning that the consumer should be especially cautious in using these products.

However, it is also true that the considerations that lead the consumer to a decision not to use a safe product containing a nanomaterial is often based on the perception of a lesser level of product safety. These perceptions are sometimes fuelled by unbalanced reporting in media and campaigns by NGOs, which leads to an effective gap between the intention of nanolabelling and its perception by the consumer.

The official text on nanolabelling in Article 19(1)g reads: "All ingredients present in the form of nanomaterials shall be clearly indicated in the list of ingredients. The names of such ingredients shall be followed by the word 'nano' in brackets".

As always, the devil is hidden in the detail: most language versions of the Regulation do not specify the type of brackets to be used. However, some do.

Examples of nanolabelling with different types of brackets:

TITANIUM DIOXIDE (NANO)
TITANIUM DIOXIDE [NANO]
TITANIUM DIOXIDE {NANO}
TITANIUM DIOXIDE < NANO>

In the author's view, all types of brackets specified in the various language versions should be allowed on the basis that all language versions of the CPR are equally valid, and that Member States shall not refuse, prohibit or restrict the making available on the market of cosmetic products which comply with the requirements of this Regulation (Article 9 CPR).

With regard to the position of nanomaterials in the ingredient list, they are treated in the same way as any other ingredient. The relevant extract from article 19.1(g) requiring ingredient labelling is:

- "The list of ingredients shall be established in descending order of weight of the ingredients at the time they are added to the cosmetic product. Ingredients in concentrations of less than 1% may be listed in any order after those in concentrations of more than 1%."

For substances present at exactly 1% the text is usually interpreted as meaning "1% or more".

The list of ingredients may be printed only on the secondary packaging, i.e. the packaging designed to contain one or more containers, including protective materials, if any.

15.7 Review of Provisions

The Commission is required to review the provisions of the CPR concerning nanomaterials regularly in the light of scientific progress. Where necessary, the Commission may propose suitable amendments to those provisions.

The CPR indicates that the first review should be completed by 11 July 2018. At the time of writing this text, no information on details or intentions by the Commission with regard to this review were available, yet.

References

1. Communication from the European Commission—Towards a European strategy for Nanotechnology, 12-05-2004. https://ec.europa.eu/research/industrial_technologies/pdf/policy/nano_com_en_new.pdf.
2. Communication from the Commission to the Council, the European Parliament and the economic and social committee—Nanosciences and nanotechnologies: An action plan for Europe 2005–2009, COM(2005) 243 final, 7-06-2005.
3. Regulation (EC) No 1223/2009 of the European Parliament and the European Council of 30 November 2009 on cosmetic products. http://eur-lex.europa.eu/legal-content/EN/TXT/?uri=CELEX:02009R1223-20160812&from=EN.
4. Commission Recommendation of 18 October 2011 on the definition of nanomaterial. https://eur-lex.europa.eu/legal-content/EN/TXT/?uri=CELEX:32011H0696.
5. Towards a review of the EC Recommendation for a definition of the term "nanomaterial" Part 1: Compilation of information concerning the experience with the definition. JRC EUR 26567 EN. http://publications.jrc.ec.europa.eu/repository/bitstream/JRC89369/lbna26567enn.pdf.
6. Towards a review of the EC Recommendation for a definition of the term "nanomaterial" Part 2: Assessment of collected information concerning the experience with the definition.

JRC EUR 26744 EN. http://publications.jrc.ec.europa.eu/repository/bitstream/JRC91377/jrc_nm-def_report2_eur26744.pdf.

7. Towards a review of the EC Recommendation for a definition of the term "nanomaterial": Part 3: Scientific-technical evaluation of options to clarify the definition and to facilitate its implementatio. JRC EUR 27240 EN. http://publications.jrc.ec.europa.eu/repository/bitstream/JRC95675/towards%20review%20ec%20rec%20def%20nanomaterial%20-%20part%203_report_online%20id.pdf.

8. SCENIHR Opinion on the scientific aspects of the existing and proposed definitions relating to products of nanoscience and nanotechnologies; Nov 2007. http://ec.europa.eu/health/archive/ph_risk/committees/04_scenihr/docs/scenihr_o_012.pdf.

9. CPNP user manual referring to Article 16. http://ec.europa.eu/DocsRoom/documents/13129/attachments/1/translations.

10. Catalogue of nanomaterials used in cosmetic products. Version 1. http://ec.europa.eu/docsroom/documents/24521.

Why Nanotechnology in Dermal Products?—Advantages, Challenges, and Market Aspects

16

Rainer H. Müller and Sung Min Pyo

16.1 Why Nanotechnology—Is There a Need?

Dermal products in pharma are a relatively small market when compared to, e.g., oral products like tablets and capsules. Considering the small market size (annual sales per dermal product) together with the high costs for introducing a new drug product to the market, pharma companies consider twice if they replace a well selling product with "old" technology by a successor product, having only a limited improvement in performance. Thus, we find dermal products on the market being sold since decades, e.g., Ichtholan® ointment [1]. However, distinct improvements can be realized by using nanotechnology. A nice example is a dermal econazole formulated into liposomes, the product Pevaryl® Lipogel [2] was advertised with "dreimal so schnell pilzfrei" (curing of fungal infections in only one-third of the previous treatment time). Thus, nanotechnology provides the chance to develop products with a distinct improvement in therapy, justifying the higher costs for such products, and therefore being refundable by the national health systems. In addition, such products provide a market differentiation and the chance to take over market segments from less well-performing competitor products. Of high interest is nanotechnology, if it works as an enabling technology. That means, if molecules can be successfully delivered to the skin which were not deliverable before, new classes of products can be generated.

The situation is different for the extremely competitive cosmetic market. Companies need to come up with product innovations to bind their customers, and preferentially to attract new ones. The classical example is again the liposomes, introduced to the market by Dior in 1986 (product Capture). In reply, L´Oreal

R. H. Müller · S. M. Pyo (✉)
Department of Pharmaceutical Technology, Freie Universität Berlin,
Institute of Pharmacy, Kelchstr. 31, 12169 Berlin, Germany
e-mail: pyo.sungmin@fu-berlin.de

© Springer Nature Switzerland AG 2019
J. Cornier et al. (eds.), *Nanocosmetics*,
https://doi.org/10.1007/978-3-030-16573-4_16

presented its niosomes as the alternative development (e.g., Niosôme in 1986 [3]). The market lifetime of many cosmetic products is very short, sometimes not even longer than one year. Especially, the home shopping market on TV, e.g., TV channels like QVC and HSE24 (Germany) are affected. The cosmetic market is innovation-driven, and thus novel nanodelivery systems are a market opportunity. This is reflected by the market share of nanotechnology-based cosmetic products, being estimated to be 460 billion USD in 2014. For 2020, a market share of 675 billion USD is predicted. Of course, they should not only be a "market promise," but provide clear product advantages such as protection of labile cosmetic actives, improved skin penetration, prolonged action, or better skin tolerability. Of very high interest are again nanocarriers, when they can enable efficient delivery of molecules with poor or non-existing penetration (e.g., poorly soluble antioxidants). Using nanotechnology, new classes of molecules can be introduced to the cosmetic market, not being listed on the International Nomenclature of Cosmetic Ingredients (INCI) list of the packaging but being in the product as a skin-active (!) form.

16.2 Nanotechnology: Principle, Technical Terms, and Legal Definitions

The word nanotechnology was created by the Japanese Norio Taniguchi in the year 1974. He defined nanotechnology as: "Nano-technology mainly consists of the processing of separation, consolidation, and deformation of materials by one atom or one molecule" [4]. Nanotechnology deals with nanoparticles (i.e., particles below 1000 nm). Before Taniguchi, these particles were called colloidal particles or colloids, and their science "colloid science". These terms were introduced already 1861 by Scottish scientist Thomas Graham [5]. Colloidal particles were defined as particles in the range up to 1000 nm [6]. Nowadays, these old terms are increasingly displaced by the more fashionable terms nanotechnology and nanoparticles.

The principle of nanotechnology is that the physico-chemical properties of materials change when their size is reduced below approx. 1 μm. Examples of this are color, melting point, and saturation solubility. The border limit of 1 μm is not strict and is more an approximate value also depending on the nature of the respective material. Below 1 μm, we have the "nanodimension," ranging from a few nm to <1000 nm. Technically, particles in this size range are called nanoparticles, based on their size unit nm.

The legal definition is different. The EU regulation for cosmetics says: "Nano-material means an insoluble or biopersistant and intentionally manufactured material with one or more external dimensions, or an internal structure, on the scale from 1 to 100 nm." [7]. Particles are only nanoparticles when the majority of a particle population (in number) is below 100 nm. They only need to be labeled as nano, when they are persistent, meaning neither biodegradable nor dissolvable. Only in this case, the addition of the word nano in brackets behind the INCI name is required, e.g., "titanium dioxide (nano)". The legal wording is: "Nanomaterials must be labelled in the list of ingredients with the word 'nano' in brackets following the name of the substance." [8].

Of course, this creates a Babylonian linguistic confusion. If somebody talks about nanoparticles in the legal sense and somebody else about nanoparticles in the technical sense, both are talking about different size ranges. Especially the consumer cannot differentiate! To clarify this situation, the suggestion was made to use the term nanoparticles in products only for particles below 100 nm, and to classify the technical nanoparticles in the range 100 nm to <1000 nm as "**submicron** particles." However, it needs to be pointed out, that both the legal nanoparticles and the submicron particles possess both altered physico-chemical properties since they are both below 1 µm. That means, it is possible to generate particles with nanoproperties, which are legally no nanoparticles.

In this book chapter, the technical term nanoparticles is used, that means we talk about the full nanodimension range from a few nm to just below 1000 nm.

Following the definition of nanomaterials of the EU regulation, porous silica particles with a particle size of a few µm are legally a nanomaterial since their pores are in the nanosize range (e.g., mesopores, 2–50 nm). However, a nanomaterial is legally no nanomaterial when it is dissolvable. And here is the gap in the regulations, because no specification of the solvent type is given. A material with the presence of a nanodimension structure is no nanomaterial when it dissolves in a solvent even when this solvent is not present in the environment (e.g., hydrofluoric acid or alkali metals to dissolve silica particles). Frankly speaking: it can be doubted if this definition via the presence of a nanodimension is really meaningful, because pores in nanodimension are unlikely to present a toxicological risk. And the background of legal definitions of nano is to identify materials or particles being potentially hazardous due to their small size.

16.3 Toxicological and Tolerability Aspects for Skin Application: "Good" Nanos Versus "Bad" Nanos

In many crime movies, we have the "good policeman" and the "bad policeman," similar is it in nano. The background of the legal definition of nanoparticles is, that very small droplets and particles can enter each cell of the body more easily (e.g., via pinocytosis) and potentially cause toxic effects (bad nanos). Larger particles (good nanos) can only be taken up by phagocytosis, which means only by cells being capable of phagocytosis. These cells are very limited in the body, e.g., the macrophages in the blood, the Kupffer cells in the liver, and macrophages in the spleen and the lungs, etc. In addition, these cells are rather difficult to assess by nanoparticles when they are applied dermally. The chance that particles are getting in contact with these cells is generally rather low. An exception is the lungs, e.g., cigarette dust from smoke and air pollution. Thus, the larger particles are less or not risky at all, especially when they are even biodegradable.

The question is, how was this magic size limit of 100 nm found by the politicians? Pinocytosis is easily possible for materials up to about 100 nm, but the pinocytotic process is also described for larger particles. Based on this, a 100 nm limit appears not sensible. The term "nanoscale" was defined by Mihail C. Roco, describing the size range of 1–100 nm. Roco is also the founding chair of the subcommittee on Nanoscale Science of the National Science and Technology Council (NSF) in the USA [9]. This nanodefinition was then used as the basis for the regulatory definition. In 1999, the definition of nanotechnology was defined in the Nanotechnology Research Directions: "Nanotechnology is the ability to control and restructure the matter at the atomic and molecular levels in the range of approximately 1–100 nm and exploiting the distinct properties and phenomena at that scale as compared to those associated with single atoms or molecules or bulk behaviour" [10–12]. That means a rather technical border limit was simply transferred to a biological tox situation. From this, it is not astonishing, that the 100 nm size limit is again under discussion, and the border size might be increased. But the "political mills" mill slowly.

For fast classification of nanoparticles (technical definition), a nanotoxicological classification system (NCS) was developed by Müller and Keck [13]. Particles were categorised into 4 classes accordingly to their size being below or above 100 nm, and accordingly to their persistence being biodegradable or non-biodegradable. A color code based on traffic light was chosen. No risk or minimal risk have particles >100 nm being biodegradable (class I, green), little risk particles >100 nm non-biodegradable (class II) and <100 nm biodegradable (class III, both yellow), and a higher potential risk have particles <100 nm being not biodegradable, i.e., persistent (class IV, red).

It should be pointed out, that non-biodegradable particles <100 nm are not necessarily toxic. For example, gold nanoparticles with a size of about 10 nm are a registered diagnostic for gamma scintigraphy, and they are even injected intravenously [14].

To sum up: Preferred for dermal application are particles of class I after NCS, but also particles of the other classes bear no risk when applied on intact skin. Due to the barrier of the stratum corneum, even very small particles with sizes of 10–20 nm (e.g., TiO_2 in sunscreen products) cannot penetrate, also emphasized in a statement by the German Society for Dermopharmacy [15]. Relevant tox effects by very small particles in sunscreens were not yet reliably confirmed [15]. However, care should be taken for damaged skin, thus the particles for such applications should be biodegradable (e.g., made from lipids or simply dissolvable as nanocrystals) and preferably >100 nm (class I).

16.4 Nanocarriers: "Academic" Systems Versus "Marketable" Carriers

Many different nanoparticulate systems are in development or under investigation in academic laboratories since decades. Sometimes they even work nicely, but cannot be used in products for many different reasons. For example, excipients used are not regulatorily accepted, loading capacity is too low, physical stability in dermal formulations is not given, excipients are too expensive, and cost-effective large-scale production is not possible. Such systems are called the "academic" systems. Systems are also called academic, when they are only in a limited number of dermal products on the market or a contract manufacturer is not available since there is no commercial availability for companies to buy the particles and to incorporate them into their products. From our perspective rather academic systems are:

Ethosomes	Ethnosomes are soft lipid vesicles that contain phospholipids and a high concentration of alcohol and water. In comparison to phospholipid vesicles without ethanol, they have a more loosely packed bilayer [16].
Transferosomes	Transfersomes are soft malleable vesicles. Their bilayer is complex, consisting of phosphatidylcholine and an edge activator and are designed to exhibit the characteristics of a cell vesicle [17, 18].
Layerosomes	Layerosomes are conventional liposomes coated with one or more multiple layers of biocompatible polyelectrolytes in order to stabilize their structure [19].
Sphingosomes	Sphingosomes are a concentric bilayered vesicle in which aqueous volume is entirely enclosed with phospholipid bilayer membrane [20, 21].
Catansomes	Catansomes are vesicles with bilayers consisting of equimolar mixtures of cationic and anionic surfactants forming an ion-pair amphiphile (IPA) [22].

Classical, established marketable nanosystems are nano-emulsions and liposomes. Nano-emulsions can easily be produced by cosmetic companies on their own via high-pressure homogenization. Many differently loaded liposome concentrates are commercially available. More recent carriers, available as concentrates for admixture, are SmartLipids (solid submicron particles made from lipid blends) and SmartCrystals (submicron crystals, both Berg + Schmidt, Hamburg).

16.5 The Liposome Story—An Unbeatable Success?

The liposomes were described by Alec Douglas Bangham in the middle of the 1960s [23], introduced to the market by Dior in 1986 with the product Capture. The liposomes enabled new product claims, provided a product differentiation in the market, and convinced many ordinary consumers to spend 25 €/USD and more for just 50 g of liposome product, to experience the "new dimension in cosmetics." The liposomes provided new product features, because of their nanosize (c.f. Sect. 16.5.). The success can also be seen when looking at the product Capture, in the fast-turning cosmetic market. Various Capture products are still on the market— more than 30 years after the introduction of the first Capture product. The success of liposomes is also confirmed by the ongoing use of them in many cosmetic products.

However, liposomes have also limitations, e.g., the relatively low loading with lipophilic actives in their bilayer. Thus, the market is waiting for a new successor product with superior performance for delivery demands not covered by the liposomes (e.g., poorly soluble actives and highly chemically labile molecules). By now, the academic systems could not meet these demands.

16.6 Advantages of Nanocarriers

In general, many advantages are based on their small size. In addition, some advantages depend on the physical state and matrix structure of the particles. For example, the state of the particle matrix, e.g., solid versus liquid, determines the release velocity (fast vs. prolonged). Different general advantages of nanocarriers are discussed in this section and displayed in Fig. 16.1.

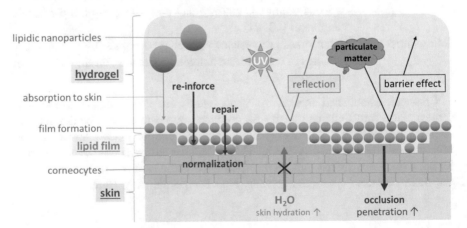

Fig. 16.1 General advantages of lipidic nanoparticles (green dots) are shown, e.g., repair and normalization of the lipid film, UV protection by reflection, and anti-pollution effect by strengthening the barrier function of the skin as well as increased skin hydration and penetration due to occlusion

- adhesion to the skin and prolonged penetration

Reduction in size makes materials more adhesive due to the increased contact area between particles and surface. One classical example from the food industry is the iced sugar sticking much better to bakery than crystalline sugar.

- occlusion, increased hydration, and increased penetration

The adhesion leads to the formation of a film on the skin, in case a sufficient quantity of nanocarriers is in the product [24]. This film formation has occlusive effects and is more pronounced when lipidic particles are used. This leads to increased skin hydration as well as to an increased penetration of actives into the skin.

- repair of the skin barrier, anti-pollution effect

In the case of lipidic nanocarriers, the adhesion to the skin can repair a damaged lipid film on the skin surface [25]. Also, the repaired skin barrier shows an anti-pollution effect.

- controlled release

A prolonged penetration time can be obtained due to both better adhesion and adjusted release profile [26] from the nanocarriers (primary nanocarriers with solid particle matrix, faster release from liquid/fluid systems).

- reduction of side effects, increased tolerability

A slower release can avoid too high local concentrations on the skin, which cause side effects such as irritation, redness, etc. [27].

- stabilization of labile actives

Labile actives can be protected by incorporating them into the particle matrix, whereas the degree of protection depends strongly on the state of the particles. In liquids or semi-liquids like emulsions or liposomes, the protection is limit, where in contrast, the protection of chemically labile active is high for solids, e.g., solid lipid nanoparticles or nanocrystals [28].

- formulation of poorly soluble actives

Poorly water-soluble actives can be incorporated into lipidic carriers, where actives being poorly soluble both in water and simultaneously in lipidic media can be formulated as nanocrystals [29].

16.7   Challenges of Nanocarriers

16.7.1   The Challenge by *D* and *P*

Due to the small size of the nanocarriers, it is more difficult to control/prolong the release. The Einstein equation can be used for considerations how to design or select a suitable nanocarrier for prolonged release:

$$D = \frac{K * T}{6 * \pi r \eta}$$

D   diffusion constant
K   Boltzmann's constant
T   absolute temperature
r   radius of the spherical particle
η   dynamic viscosity

Due to the small size of the carrier, the molecule only needs to diffuse a short way of few 100 nm to leave the particle and release (Fig. 16.2, left). This is easily possible even at low values of *D* (=high molecular weight active). Thus, principle release is fast. To prolong the release, the carrier should be in a solid state with a high viscosity *η*, and a larger size is favorable.

Release is also governed by the partitioning coefficient *P*1 between the particle phase and the surrounding water of the cream, and the partitioning coefficient *P*2 between the particle phase and the skin:

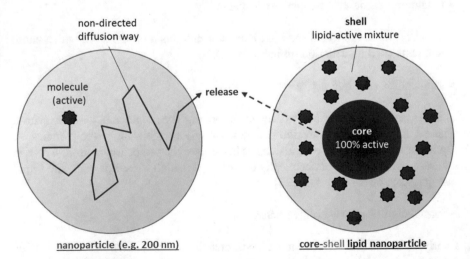

Fig. 16.2 Left: Rapid release of the active molecule (red) from nanoparticles (green) even at low *D* values due to short diffusion path. Right: Controlled/prolonged release of the active molecule from a nanoparticle with an active enriched core

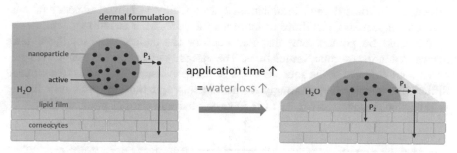

Fig. 16.3 Evaporation of water from the dermal formulation will favor the partitioning of active into the skin due to a second pathway for the active arising

$$P1 = \frac{C(p)}{C(\text{water})} \quad \text{and} \quad P2 = \frac{C(p)}{C(\text{skin})}$$

After skin application, the water phase of dermal formulations evaporates to a large extent, favoring the partitioning of active into the skin (Fig. 16.3), what is the final delivery goal.

A smart solution is the formation of an active core in the center of the nanocarrier surrounded by a dissolution controlling membrane (eutectic mixture lipid and active). However, core-shell models (Fig. 16.2, right), e.g., described for solid lipid nanoparticles by zur Mühlen and Mehnert [26] are technically more complicate to produce.

16.7.2   Physical Stability—the Battle Against Thermodynamics

According to thermodynamics, a system tries to find its lowest energetic state (second law of thermodynamics). The energy of a dispersion increases with increasing surface:

$$E = Y * A$$

E   total energy
Y   interfacial tension
A   surface area

Thus, a suspension tends to minimize the surface by aggregation. This is more pronounced the larger its surface is. The surface of one 10 μm cube is only 600 μm², the surface after splitting this cube into 1,000,000 cubes of 100 nm edge length is 60,000 μm², which means 100 times larger. Thus, nanocarriers need to be efficiently stabilized, preferentially sterically to tolerate potential electrolytes being present in the final dermal formulation. For sterically stabilized nanocrystal suspension concentrates, stabilities of more than 2 years were found (e.g., rutin).

It should be pointed out, that nanocarriers are additionally stabilized after incorporation into dermal formulations. The viscosity increases in vesicles like gels or creams compared to the aqueous phase of the suspension. This leads to a reduced diffusional velocity D accordingly to the Einstein equation and thus to a further stabilization, being reported for SLN incorporated into dermal products [30].

16.8   Prerequisites for Realization of Market Products

There are essential prerequisites for the use of a carrier system by cosmetic companies in dermal products. Ideally, the nanocarrier should be available as a commercial concentrate that can be easily admixed to dermal formulations at the end of their production process. The concentrates should be sufficiently high concentrated, allowing a dilution by a factor 10 or better 20 (i.e., 5 kg concentrate are necessary to produce 100 kg final product).

The excipients used should be regulatorily accepted and skin-friendly (especially the stabilizers used). A sound documentation has to be provided by the product manufacturer (e.g., Certificate of Analysis (CoA), Material Safety Data Sheet (MSDS), etc.).

There should be scientific data supporting the claimed features of the nanocarrier. This does not need to be shown for each single particle composition (e.g., lipid mixture) and not for each cosmetic active. A general proof with other matrices or cosmetic actives is sufficient (e.g., repair of the skin barrier and reduction of side effects of incorporated actives).

The producer should be a reliable company, preferably a company already well established in the cosmetic market. In most cases, producing carriers by oneself is too costly.

16.9   Marketing Aspects

Using a nanocarrier with superior delivery properties is a chance for market differentiation against competitors. Especially in areas where one has a lot of competition (e.g., huge range of dermal Q10 products on the market), it is important to advertise to have the better performing product (e.g., super-Q10 by nanotechnology).

There is an increased trend of cosmeceuticals. One definition of a cosmeceutical is: "a cosmetic product whose active ingredient is meant to have a beneficial physiologic effect resulting from an enhanced pharmacological action" [31]. These products require even better performance of actives than in "normal" cosmetic products to meet the claims made by the companies and to meet the expectations by the consumers. Nanotechnology is one way to achieve this.

For many cosmetic products, the life cycles are relatively short, being 1–5 years. To make an established product further attractive to the consumer, it needs a "face-lift" identical to car industry producing new updated car models in regular time intervals. This face-lift can easily be done by using nanotechnology. One can combine the excellent properties of a consumer accepted, well-established product with the special features of a nanocarrier system, and advertise this accordingly. Another possibility is the creation of a completely new cosmetic product line, tailor-made to the nanocarrier system used.

16.10 Perspectives—The Future

The cosmetic market is getting more and more competitive. To ensure a certain market share, it is essential to have a market differentiation against competitor products. In addition, the products need to be of superior performance and providing the claimed effects. This can be achieved by using nanocarriers, which are science-based, and have a proof for their properties claimed.

The liposomes are still a good choice for hydrophilic actives. For lipophilic actives, suitable lipophilic carriers should be selected. These include the "old" simple o/w nano-emulsions. In case of special skin interaction, release properties, etc., are required; solid lipid particles (e.g., SmartLipids) are the nanocarrier of choice.

Highly attractive appear poorly soluble plant actives. Especially anti-oxidants show no or very little penetration in the traditional formulations, i.e. they are practically not active in the skin. The antioxidants are presently en vogue, e.g., protection against UV/IR radiation and as anti-pollution actives. A solution to make them active in the skin is nanotechnology—using nanocrystals. The new developments in nanotechnology can have a tremendous impact on cosmetic products and might shape the cosmetic market in the next 20 years, especially also with regard to cosmeceuticals.

There is increasing pressure on the cost side in the production of cosmetic products. Investors, especially VC, expect increasing margins. But it needs to be made very clear, that one cannot produce a Ferrari with the money needed to build a simple VW beetle car. Highly effective molecules are expensive, e.g., glabridin costs about 20,000 USD/€ for 1 kg. Producing, e.g., nanocrystals of such compounds is an expensive process. Thus, the latest smart nanotechnology can only be used for medium- to high-priced products and not for a UV protecting nanocarrier for a body sun lotion sold 500 ml for 10 USD/€ in the shop. High-tech nanocarriers are for high-level cosmetic products—and by doing this, good marketing opportunities for cosmetic nanocarrier products are predicted.

References

1. Merck & Co. Ichthyol, its history, properties, and therapeutics. New York Collection: Cornell; 1913.
2. Korting HC, Schäfer-Korting M. Carriers in the topical treatment of skin disease. In: Drug delivery. Berlin: Springer; 2010. p. 435–468.
3. Cerqueira-Coutinho C, dos Santos EP, Mansur CRE. Niosomes as nano-delivery systems in the pharmaceutical field. Crit Rev Ther Drug Carrier Syst. 2016;33(2).
4. Taniguchi N, Arakawa C, Kobayashi T. On the basic concept of nano-technology. In: Proceedings of the International Conference on Production Engineering. 1974;2:18–23.
5. Graham T. X. Liquid diffusion applied to analysis. Philos Trans R Soc Lond. 1861;151:183–224.
6. Levine IN. Physical chemistry. 5th ed. Boston: McGraw-Hill, Boston; 2001.
7. Buzek J, Ask B. Regulation (EC) No 1223/2009 of the European Parliament and of the Council of 30 November 2009 on cosmetic products. Official J Eur Union. 2009;59.
8. http://ec.europa.eu/growth/sectors/cosmetics/products/nanomaterials_en.
9. https://en.wikipedia.org/wiki/Mihail_Roco.
10. Roco MC. The long view of nanotechnology development: the National Nanotechnology Initiative at 10 years. Berlin: Springer; 2011.
11. Roco MC, Williams RS, Alivisatos P, editors. Nanotechnology research directions: IWGN workshop report: vision for nanotechnology in the next decade. Berlin: Springer; 2000.
12. http://www.wtec.org/loyola/nano/IWGN.Research.Directions/.
13. Keck CM, Müller RH. Nanotoxicological classification system (NCS)–a guide for the risk-benefit assessment of nanoparticulate drug delivery systems. Eur J Pharm Biopharm. 2013;84(3):445–8.
14. Cohen Y, Besnard M. Radionuclides. Pharmacokinetics. In: Nuklearmedizin/Nuclear Medicine. Berlin: Springer; 1980. p. 3–76.
15. http://www.gd-online.de/german/veranstalt/images2014/18.GD_JT_GD-Stellungnahme_Nanopartikel_07.04.2014.pdf.
16. Touitou E, Dayan N, Bergelson L, Godin B, Eliaz M. Ethosomes—novel vesicular carriers for enhanced delivery: characterization and skin penetration properties. J Control Release. 2000;65(3):403–18.
17. Cevc G, Blume G, Schätzlein A. Transfersomes-mediated transepidermal delivery improves the regio-specificity and biological activity of corticosteroids in vivo1. J Control Release. 1997;45(3):211–26.
18. Rajan R, Jose S, Mukund VB, Vasudevan DT. Transferosomes-A vesicular transdermal delivery system for enhanced drug permeation. J Adv Pharm Technol Res. 2011;2(3):138.
19. Gupta R, Agrawal A, Anjum MM, Dwivedi H, Kymonil MK, Saraf AS. Lipid nanoformulations for oral delivery of bioactives: an overview. Current Drug Therapy. 2014;9(1):35–46.
20. Webb MS, Bally MB, Mayer LD. Sphingosomes for enhanced drug delivery. US Patent 5,543,152. 1996.
21. Saraf S, Gupta D, Kaur CD, Saraf S, Res IJCS. Sphingosomes a novel approach to vesicular drug delivery. Int J Cur Sci Res. 2011;1(2):63–8.
22. Herrington KL, Kaler EW, Miller DD, Zasadzinski JA, Chiruvolu S. Phase behavior of aqueous mixtures of dodecyltrimethylammonium bromide (DTAB) and sodium dodecyl sulfate (SDS). J Phys Chem. 1993;97(51):13792–802.
23. Bangham AD, Horne RW. Negative staining of phospholipids and their structural modification by surface-active agents as observed in the electron microscope. J Mol Biol. 1964; 8(5):660–IN10.
24. Müller RH, Sinambela P, Keck CM. NLC–the invisible dermal patch for moisturizing & skin protection. EuroCosmetics. 2013;6:20–3.

25. Keck CM, Anantaworasakul P, Patel M, Okonogi S, Singh KK, Roessner D, … , Müller RH. A new concept for the treatment of atopic dermatitis: Silver–nanolipid complex (sNLC). Int J Pharm. 2014;462(1–2):44–51.
26. zur Mühlen A, Schwarz C, Mehnert W. Solid lipid nanoparticles (SLN) for controlled bioactive delivery–bioactive release and release mechanism. Eur J Pharm Biopharm. 1998; 45(2):149–155.
27. Pyo SM, Müller RH. Vitamin A1 smartLipids—improved penetration with reduced side effects. Berlin: Day of Pharmacy; 2018.
28. Olechowski F, Pyo SM, Müller RH. BergaCare SmartLipids—commercial concentrates of solid lipid submicron particles for industrial production of dermal products. PBP World Meeting: Granada; 2018.
29. Petersen RD. Nanocrystals for use in topical formulations and method of production thereof. Germany Patent PCT/EP2007/009943; 2006.
30. Lippacher A, Müller RH, Mäder K. Semisolid SLN™ dispersions for topical application: influence of formulation and production parameters on viscoelastic properties. Eur J Pharm Biopharm. 2002;53(2):155–60.
31. Kligman D. Cosmeceuticals. Dermatol Clin. 2000;18(4):609–15.

Index

© Springer Nature Switzerland AG 2019
J. Cornier et al. (eds.), *Nanocosmetics*,
https://doi.org/10.1007/978-3-030-16573-4

Printed in the United States
By Bookmasters